Science and Soccer

Second edition

Edited by
Thomas Reilly and A. Mark Williams

Routledge
Taylor & Francis Group

LONDON AND NEW YORK

First published 2003
by Routledge
11 New Fetter Lane, London EC4P 4EE

Simultaneously published in the USA and Canada
by Routledge
29 West 35th Street, New York, NY 10001

Routledge is an imprint of the Taylor & Francis Group

© Selection and editorial matter 2003 Thomas Reilly and A. Mark Williams;
individual chapters the contributors

Typeset in Times New Roman by
Newgen Imaging Systems (P) Ltd, Chennai, India
Printed and bound in Great Britain by
TJ International, Padstow, Cornwall

British Library Cataloguing in Publication Data
A catalogue record for this book is available from the British Library

Library of Congress Cataloging in Publication Data
Science and soccer / edited by Thomas Reilly and A. Mark Williams. – 2nd ed.
  p. cm.
  Includes bibliographical references and index.
  1. Soccer. 2. Sports sciences. I. Reilly, Thomas, 1941– II. Williams, A. M. (A. Mark),
1965–

GV943 .S36 2003
796.334–dc21   2002035842

ISBN 0–415–26231–3 (HB)
ISBN 0–415–26232–1 (PB)

Contents

Contributors

Jens Bangsbo
August Krogh Institute
University of Copenhagen
Copenhagen, Denmark

Andy Borrie
Research Institute for Sport and
  Exercise Sciences
Liverpool John Moores University
Henry Cotton Campus, Liverpool, UK

Dominic Doran
Research Institute for Sport and
  Exercise Sciences
Liverpool John Moores University
Henry Cotton Campus, Liverpool, UK

Martin Eubank
Research Institute for Sport and
  Exercise Sciences
Liverpool John Moores University
Henry Cotton Campus, Liverpool, UK

Ian M. Franks
School of Human Kinetics
University of British Columbia
Vancouver, Canada

David Gilbourne
Research Institute for Sport and
  Exercise Sciences
Liverpool John Moores University
Henry Cotton Campus, Liverpool, UK

Nigel Hanchard
School of Health
University of Teesside
Middlesborough, UK

Nicola J. Hodges
Research Institute for Sport and
  Exercise Sciences
Liverpool John Moores University,
Henry Cotton Campus, Liverpool, UK

Robert R. Horn
Research Institute for Sport and
  Exercise Sciences
Liverpool John Moores University,
Henry Cotton Campus, Liverpool, UK

Tracey Howe
School of Health
University of Teesside
Middlesborough, UK

Mike Hughes
School of Sport, PE and Recreation
UWIC
Cardiff, UK

Zoe Knowles
Research Institute for Sport and
  Exercise Sciences
John Moores University
Henry Cotton Campus, Liverpool, UK

Mark Lake
Research Institute for Sport and
  Exercise Sciences
Liverpool John Moores University,
Henry Cotton Campus, Liverpool, UK

Adrian Lees
Research Institute for Sport and
  Exercise Sciences
Liverpool John Moores University
Henry Cotton Campus, Liverpool, UK

Tim McGarry
Faculty of Kinesiology
University of New Brunswick
Fredericton, New Brunswick, Canada

Don MacLaren
Research Institute for Sport and
  Exercise Sciences
Liverpool John Moores University
Henry Cotton Campus, Liverpool, UK

Robert M. Malina
Research Professor,
Tarlton State University,
Texas, USA

John Minten
Head of Sport Science
North East Wales Institute of
  Higher Education
Plascoch Campus
Wrexham, UK

Benny Peiser
Research Institute for Sport and
  Exercise Sciences
Liverpool John Moores University
Henry Cotton Campus, Liverpool, UK

Thomas Reilly
Research Institute for Sport and
  Exercise Sciences
Liverpool John Moores University
Henry Cotton Campus
Liverpool, UK

Dave Richardson
Research Institute for Sport and
  Exercise Sciences
Liverpool John Moores University
Henry Cotton Campus
Liverpool, UK

Frank Sanderson
Business School
Liverpool John Moores University
Mount Pleasant
Liverpool, UK

Tony Shelton
School of Applied Psychology
Liverpool John Moores University
Henry Cotton Campus
Liverpool, UK

A. Mark Williams
Research Institute for Sport and
  Exercise Sciences
Liverpool John Moores University
Henry Cotton Campus
Liverpool, UK

1 Introduction to science and soccer

Thomas Reilly and A. Mark Williams

Introduction

Football is the world's most popular form of sport, being played in every nation without exception. The most widespread code is association football or soccer. The sport has a rich history though it was formalized as we know it today by the establishment of the Football Association in 1863. The game soon spread to continental European countries and later to South America and the other continents. The world's governing body, the Federation of the International Football Association (FIFA), was set up in 1904 and the first Olympic soccer competition was held 4 years later. The United Kingdom (UK) won the final 2–0, defeating Denmark, another nation playing a leading role in the popularization of the game. Uruguay played host to the first World Cup tournament in 1930. This competition is held every 4 years and is arguably the tournament with the most fanatical hold on its spectators and television audiences. So far, only seven nations have won the tournament – Uruguay, Argentina, Brazil, Germany, England, Italy and France. Whilst they may represent the top teams at elite level, the popularity of the game is reflected in the millions who participate in soccer at lower levels of play.

1.1 Development of sports science

In recent years, there has been a remarkable expansion of sports science. The subject area is now recognized both as an academic discipline and a valid area of professional practice. Sports science is well respected within its parent disciplines, for example, biomechanics, biochemistry, physiology, psychology, sociology and so on. A new maturity became apparent as the sports sciences were increasingly applied to address problems in particular sports rather than to sports in general. One of these specific applications has been to soccer.

The applications of science to soccer predated the formal acceptance of sports science as an area of study in university programmes. South American national teams used specialists in psychology, nutrition and physiology in the preparation of squads for the major international tournaments from the early 1970s. The comprehensive systems of scientific support accessible to Eastern European athletes since the 1970s dwarfed the commitments of Western countries to top-level sport. The gulf was notably wide with respect to British soccer, where the sports scientist was more often than not shunned or at best frostily welcomed.

In the 1980s, it became apparent that the football industry and professionals in the game could no longer rely on the traditional methods of previous decades. Coaches and trainers were more open to contemporary scientific approaches to preparing for competition. Methods of management science were applied to organizing the big soccer clubs and the training of players could be formulated on a systematic basis. In general, the clubs that moved with the times were rewarded with success by gaining an advantage over those that did not change.

It has taken some years for the knowledge accumulating within sports science to be translated into a form usable by practitioners. Efforts have been made to compile scientific knowledge and expertise and make them more widely available to the soccer community. In recent years, the majority of professional soccer clubs have employed sports scientists in the quest for a competitive edge. The revision of this textbook is but another step in the direction of updating the relevant knowledge base.

1.2   Science and football

The First World Congress of Science and Football, held at Liverpool in 1987, represented a milestone in the application of science to football. The Congress embraced all the football codes, but a definite attempt was made to establish common threads between them. The broad aim was to bring together those scientists whose research work was directly related to football and practitioners of football interested in obtaining current information about its scientific aspects. Practitioners included players, trainers, coaches, managers and administrators. The list of Congress themes (Table 1.1) demonstrates the scope of topics that were communicated.

The Congress is held every 4 years under the auspices of the World Commission for Science and Sports (WCSS). This meeting at Liverpool was followed by the Second World Congress on Science and Football at Eindhoven in the Netherlands (1991), the Third World Congress at Cardiff (Wales) in 1995; the fourth event was at Sydney, Australia, in 1999 and the fifth event at Lisbon in 2003.

Many national governing bodies of soccer set up their own system of scientific support. Mostly, this was implemented through their sports medicine programmes. An example was the Football Association's National Training and Rehabilitation Centre at Lilleshall in the early 1980s. As the Centre was abandoned when the Football Academies took over the responsibility for nurturing soccer talent within the professional clubs, scientific personnel were appointed to the Academies. The development of the Academies was promoted by the Technical Department of the Football Association and reflected the perceived potential of sports science as a component of systematic support. This trend applied also to the science input to the world's ruling body, FIFA, which historically had been through the medium of its Medical Committee.

A consensus statement concerned with food and nutrition as they applied to soccer was approved at FIFA headquarters in 1994. The event marked another milestone in the progress of scientific information related to the game. A parallel within the Football Association was

Table 1.1 Congress themes at the First World Congress on Science and Football

Clothing and footwear	Structuring football skills and practices
Football surfaces	Physiology of training and match-play
Biomechanics of kicking	Nutritional factors in football
Computer-aided match analysis	Playing in heat or cold
Team management	Football at altitude
Group dynamics in match-play	Coaching the problem player
Decision-making by referees	The injury-prone player
Soccer violence	Post-injury fitness testing
Pre-match stress and performance	Strain in adolescent footballers

the launching of the journal *Insight* in 1997, the official publication of the Football Association's Coaches Association and the more systematic implementation of sport science support services to England international squads.

1.3   Academic programmes in Science and Football

The first academic programmes in sports science were studied in the UK in 1975. The background to and development of these undergraduate courses have been described elsewhere (Reilly, 1992). The disciplines included in the pioneering programmes were biology, biochemistry, physiology, biomechanics, mathematics, psychology and sociology. Contemporary programmes may incorporate economics, recreation, sport development, coaching and computer science but the major thrust of scientific method is maintained.

Whilst the professional preparation of coaches in some countries includes substantial components of sports science, the emphasis is firmly on coaching competence rather than intellectual skills. The first formal academic programme in Science and Football was offered at Diploma level at Liverpool John Moores University in 1991. The syllabus was dedicated to scientific subjects applied to football. The programme was extended in 1997 to a full-blown Bachelor of Science degree in Science and Football. Some of its core modules are shown in Table 1.2.

Formal academic activity is not restricted to undergraduate, or postgraduate courses. The University of Leicester set up its research unit in the 1980s to focus on sociological aspects of soccer. It was funded by the Football Trust and made a major contribution to the study of the 'football hooligan' phenomenon throughout the decade. Later on, in the 1990s, the University of Liverpool set up an MBA course dedicated to the study of football management: each Masters student must conduct a major research project relevant to contemporary soccer management. As the new millennium began, FIFA supported a European-wide Masters course which was directed towards the history of football and the involvement of a number of universities from various countries within the European Union.

The Research Institute for Sport and Exercise Sciences at Liverpool John Moores University supports an active programme of postgraduate research in scientific aspects of football. Most notable achievements elsewhere have included the award of a DSc degree for a thesis on physiological investigations directly related to soccer play (Bangsbo, 1993). In Sweden, as well as in the UK, there were instances of doctoral theses being shaped round physiological investigations of exercise, of the type that occurs in soccer (e.g. Balsom, 1995;

Table 1.2 Core modules for the BSc (Hons) in Science and Football (soccer)

Core programme
Applied sciences and football
Performance analysis in football
Ergonomics of football
Mental training for football
Physiology of football
Skill acquisition in football
Research project

Drust, 1997) and of the unique behavioural characteristics underlying expert performance in the game (e.g. Williams, 1995).

1.4   The field of study

Clearly, there are many aspects of science and soccer and plenty of subject areas which have benefited from scientific knowledge and know-how. These include the natural and physical sciences, the disciplines allied to medicine and the social sciences.

An ergonomics model of the application of science to the game itself is illustrated in Figure 1.1 (Reilly, 1991). It shows how the role of the scientist is to match the characteristics of individuals to the demands of the game. This is a complex problem in team sports where eventual success is determined by how the collection of individuals forms an effective unit. There are implications for fitness testing, training and player selection. The study of the organization of the entire group is also highly relevant.

Similarly, the prediction of performance is more difficult by far in soccer than in individual sports. In competition, success may be determined by choice of tactics of either team. There are also elements of chance that determine the outcome of critical events and tilt the balance of the contest. This makes even the most complex of game theories hard to relate to the outcome of a particular match. Nevertheless, match analysis can be approached from a scientific perspective.

The physical sciences provide insights into the nature and appropriateness of artificial pitches. There have also been applications to the design of shoes and evaluations of the need

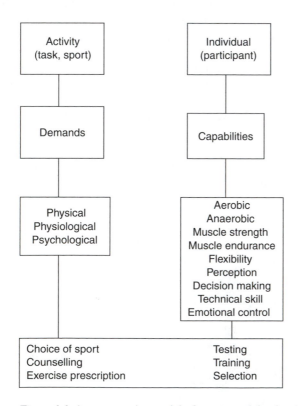

Figure 1.1 An ergonomics model of sports participation (modified from Reilly, 1991).

for protective equipment. Principles of biomechanics are relevant in considering prevention of soccer injuries. The physical sciences also embrace agronomy, the cultivation of grasses and the maintenance of playing conditions in cold and wet weather.

The widest field of application of sports science to soccer is probably apparent in the behavioural disciplines. The many opportunities for investigation include the study of crowds and their control, the management of large groups and the organization of personnel, the stresses on playing officials and on management of the clubs.

1.5   Soccer as an art

Followers of soccer frequently criticize the game as lacking creativity and flair. Some critics may go so far as to blame use of scientific methods by soccer teams for lack of entertainment. Underlying these points is the fact that soccer at top level has an obligation to entertain the viewing public but financial rewards to the players often depend on their securing victory. Consequently, fear of failure to win may motivate players to err on the side of caution and emphasize defence rather than attack. The negative emphasis on preventing the opposition from playing to its strength may leave the 'fans' disenchanted.

The coach and trainer may use scientific information to avoid errors and to maximize the chances of preparing the team well. The style of play and choice of tactics are judgements made by the coach on the basis of the best available information about one's own team, the opposition and the playing conditions. The scientific support may be utilized to guide the right course for the practitioner and so in no sense is science taking over control of the game. What is unarguable is that at elite level, soccer is played at a faster tempo than in previous decades, and players are better prepared all-round for performing their roles.

The professional soccer player is comparable with the actor in that hours of practice or rehearsal underpin the preparation for public performance. The expertise of the player or actor is judged largely on a subjective basis by a critical audience of the public event. Spectators at a soccer match differ from the theatre audience in that their perceptions of the event are mostly partisan and their passion is more overtly expressed.

That soccer itself is an art rather than a science is exemplified by the craft of great players like Zinadine Zidane or Brazil's Rivaldo, the erstwhile guile of Diego Maradona, the precision of David Beckham or the speed of Michael Owen. The game is aleatory and is partly determined by chance or strokes of individual genius. This uncertainty of outcome is part of its appeal.

A scientific approach towards preparation for play can nevertheless enhance the enjoyment of both players and spectators. It can achieve this goal by enabling the team to play to its potential. This realization of possibilities can apply to the recreational player participating for pleasure, or the professional playing for material reward. It can apply to the parents gaining satisfaction from watching talented offspring at play or to the home supporter whose zeal may border on passion and prejudice. It is this microcosm that is subjected to scientific scrutiny in the chapters that follow.

References

Balsom, P.D. (1995). High intensity intermittent exercise: performance and metabolic responses with very high intensity short duration work periods. PhD thesis, Karolinska Institute, Stockholm.

Bangsbo, J. (1993). The physiology of soccer – with special reference to intense intermittent exercise. DSc thesis, August Krogh Institute, University of Copenhagen.

Drust, B. (1997). Metabolic responses to soccer-specific intermittent exercise. PhD thesis, Liverpool John Moores University.

Reilly, T. (1991). Physical fitness – for whom and for what? in *Sport for All* (eds P. Oja and R. Telama), Elsevier Science, Amsterdam, pp. 81–8.

Reilly, T. (1992). *Strategic Directions for Sports Science Research in the United Kingdom*. The Sports Council, London.

Williams, A.M. (1995). Perceptual skill in soccer. PhD thesis, University of Liverpool.

Part 1
Biology and soccer

2 Functional anatomy

Tracey Howe and Nigel Hanchard

Introduction

The part of the musculoskeletal system involved with movement is called the locomotor system. It includes not only most of the bony skeleton but also the joints and the various soft tissues that form or control them. This chapter aims to introduce the major components of the locomotor system, developing an awareness of how their structure underpins their function, in a way which is relevant to those involved in soccer. These include player, coach, trainer and medical staff alike.

For comprehensive coverage of human anatomy, and definitions of the special terminology used in this context, standard texts such as Williams *et al.* (1995) and Moore and Dalley (1999) are recommended.

2.1 Joints

There are several types of joints, but most are *synovial*. On the whole, synovial joints allow more movement than other types, and their structural characteristics include *articular cartilage*, *articular capsule* and *ligaments* (Figure 2.1).

2.1.1 Articular cartilage

Articular cartilage is a white, glistening substance which covers the articulating bone ends, and comprises bundles of *collagen* fibres embedded in a *matrix*. Orientated vertically in the deeper parts of the cartilage, the collagen bundles come to lie more horizontally as they near the surface, and interweave. This serves the dual purpose of presenting a smooth finish to facilitate gliding movements and binding the matrix in place. The matrix itself is largely composed of mutually repellent molecules which resist being pushed together, giving cartilage its characteristic springy resilience.

2.1.2 Articular capsule

The articular capsule comprises an outer *fibrous capsule* lined, on its internal surface, by *synovial membrane*. The fibrous capsule attaches to the margins of the joint – effectively sealing it – and consists of *white fibrous tissue*, a dense, tough, pliant tissue predominantly composed of collagen bundles. The synovial membrane secretes slippery *synovial fluid* into the joint cavity, reducing friction, and facilitating glide between the opposing cartilage surfaces still further.

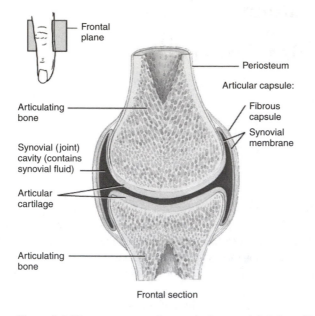

Frontal section

Figure 2.1 The structure of a typical synovial joint. (From *Principles of Anatomy and Physiology*. Tortora, G.T. and Grabowski, S.R. © 1993 Harper Collins. This material is used by permission of John Wiley & Sons Inc.)

2.1.3   Ligaments

Ligaments, like articular capsules, consist of white fibrous tissue, but their constituent collagen bundles are densely packed and aligned parallel to one another, affording them inextensibility and great tensile strength. Some ligaments are, in fact, no more than thickened sections of the fibrous capsule itself (intrinsic ligaments), while others stand clear of the capsule (extrinsic ligaments). In either case they each span a joint, attaching to the bones on either side and, by reason of their position and inextensibility, preventing unwanted movements. The medial and lateral collateral ligaments of the knee, for example, lying on either side of the joint, respectively, connect the femur to the tibia and fibula. This arrangement protects the knee against side-to-side bending (varus and valgus movement) while allowing freedom of forward and backward bending (flexion and extension) and some freedom of rotation.

2.1.4   Mobility versus stability

While it is true to say that, as a group, the synovial joints are freely mobile, it is equally true that considerable variation in mobility is found among them. Those that are more stable will be less mobile, since these attributes are mutually exclusive. Conversely, the more mobile synovial joints will be less stable.

   The elbow is a good example of a synovial joint which does not have a special requirement for mobility. Its function is to lengthen or shorten the arm, for which purpose bending and straightening in one plane (flexion–extension) suffices, and a stable, hinge-like joint is compatible with this modest requirement. Its articular surfaces are very congruent, shaped rather like a spanner gripping a nut, and allow only limited flexion–extension before bony prominences engage against one another, thereby blocking further movement. Movement in other planes is discouraged not only by the shape of the articular surfaces but also by tough collateral ligaments at the inner and outer aspects of the joint.

From a mobility point of view the shoulder joint lies at the other extreme. Being a ball and socket joint it allows movement in all planes, and its mobility is further enhanced by the relative sizes and shapes of its articular surfaces (whereby a large 'ball' articulates with a small, shallow socket), its lax fibrous capsule and its lack of any substantial ligaments. The shoulder joint's extreme mobility is necessary to everyday function but is such, in fact, that the joint would be at constant risk of dislocation were it to rely solely on passive stabilizing mechanisms such as its ligaments. Fortunately an active stabilizing system also exists, in the form of muscles. Their stabilizing role is important at most joints, but crucial in relation to unstable joints such as the shoulder or ankle complex, or joints which may be subjected to gross leverage during soccer, such as the knee. In addition to their stabilizing roles, muscles are also responsible for moving joints.

2.2 Skeletal muscle

Muscles constitute approximately 40–50% of total body weight and even more in soccer players. The main function of skeletal (or striated) muscle is to control the movement of body segments, by a series of patterns of contractions and relaxations, which are under conscious (voluntary) control.

Muscle has four main characteristics: excitability, contractility, extensibility and elasticity. *Excitability* is a muscle's ability to produce a wave of electrical activity (an action potential) in response to chemicals (neurotransmitters) transmitted from a motor nerve. *Contractility*, in turn, is the ability to contract and generate tension in response to an action potential. *Extensibility* is the ability of a muscle to extend without sustaining damage. *Elasticity* is the ability of a muscle to return to its original size and shape after shortening or being stretched.

2.2.1 Gross structure

Skeletal muscle is comprised of long thin muscle fibres which are grouped together to form bundles of various sizes (fasciculi). The spaces between muscle fibres are filled by connective tissue (endomysium) and each fascicle is surrounded by a connective tissue sheath (perimysium). Finally there is an outermost protective layer, the epimysium (Figure 2.2). All three layers of connective tissue extend beyond the muscle fibres to form tendons that attach to other muscles or to bone.

2.2.2 Ultrastructure

The contractile elements of muscle fibres are the myofibrils which are arranged in parallel in small compartments called *sarcomeres*. The two contractile proteins are *actin* and *myosin* and it is the arrangement of these protein chains that give skeletal muscle its striated appearance under the microscope.

2.2.2.1 Muscle contraction: the 'sliding-filament theory'

Skeletal muscles are controlled by means of *motor neurones* (a bundle of which constitute a *motor nerve*). Each such neurone is able to transmit stimuli from the brain towards

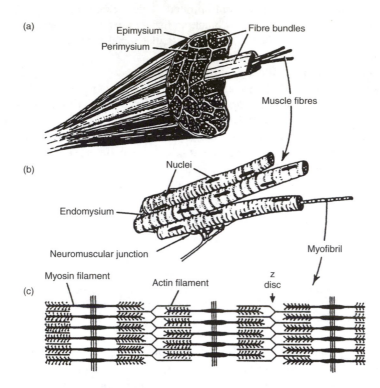

Figure 2.2 The structure of skeletal muscle as revealed by the light microscope. (a) Whole muscle in transverse section showing organization of connective tissue and muscle fibres. (b) Individual muscle fibres revealing characteristic cross-striations and nuclei. (c) Ultrastructure of muscle fibril showing individual protein filaments. (From *Sports Fitness and Sports Injuries*. Reilly, T. (ed.) 1981 Faber & Faber.)

a population of muscle fibres in the form of electrical signals (or *action potentials*). The electrical signals cannot pass directly from motor neurone to muscle fibre, however, because these structures are separated by a small gap. Instead, on arriving at the end of the motor neurone (motor end plate) the action potential causes the release of a neurotransmitter, acteylcholine. This neurotransmitter bridges the gap and stimulates the muscle fibre membrane (sarcolemma), in turn, to produce an action potential of its own. This secondary wave of electrical excitation is able to penetrate deeply into the fibre, reaching all of the myofibrils, via folds in the sarcolemma. It triggers an interaction between the two contractile proteins, which slide over each other and increasingly overlap, pulling the two ends of each sarcomere together.

2.2.3 Muscle fibre types

Muscle fibres have been classified as type I, slow oxidative (SO), type IIb, fast glycolytic (FG) and type IIa, fast oxidative glycolytic (FOG). Type I fibres have a high oxidative capacity (i.e. ability to oxidize carbohydrates, yielding the greatest possible amount of energy) and are fatigue resistant. Type IIb fibres have a low oxidative capacity and a low fatigue resistance. Type IIa fibres possess intermediate properties. Most skeletal muscles are a mixture of the three types of fibres but the relative proportion of each varies according to the regular action of the muscle. Muscles which function in a predominantly ballistic manner, for example, *gastrocnemius* to plantarflex the ankle joint during jumping, contain a high proportion of fast type II fibres. Conversely, muscles that have a predominantly postural function, for example, *soleus* which plantarflexes the ankle during walking, contain a high proportion of slow type I fibres.

The proportion of fibre types within a muscle is not constant. It is affected by the ageing process, injury, denervation and specific training methods, either by voluntary contractions or electrical stimulation. This property of skeletal muscle tissue is known as *plasticity*. The muscles of soccer players are composed of a mixture of the fibre types (Bangsbo, 1993). This mixture reflects the nature of the game which has periods of relative inactivity interspersed with periods of high intensity activity.

2.2.4 The motor unit

A motor neurone and all the muscle fibres that it supplies (innervates) is termed a *motor unit*. The constituent fibres of individual motor units are found to be distributed throughout the muscle and are homogenous, possessing identical structural, metabolic and contractile properties. The number of muscle fibres innervated by a single motor neurone varies from muscle to muscle. Those muscles that control precise movements such as the eye have few muscle fibres per motor unit. Conversely muscles such as the *quadriceps*, which are responsible for powerful gross movements of body segments, have approximately 2000 muscle fibres per motor unit. When a motor neurone is stimulated all the muscle fibres which it innervates contract simultaneously. Motor units are progressively recruited and 'decruited' in order of size so that force may be modulated in a step-wise manner. Small fatigue-resistant motor units (type I) are recruited, most frequently providing control at the low forces required to perform everyday activities (walking, standing from sitting) and maintain posture. Larger motor units (type II) are recruited when strong and fast contractions are required (running, jumping and kicking a ball).

2.2.5 Movement

Muscles span joints and it is the contraction, or shortening, of the muscle belly that produces movement at a joint. Common terms for types of movement produced about a joint are *flexion* (usually a reduction of the joint angle) and *extension* (usually increasing the joint angle), *abduction* (movement away from the body), *adduction* (movement towards the body), *medial rotation* (towards the body) and *lateral rotation* (away from the body). Common movements at the ankle joint are *dorsiflexion* (movement of the foot towards the shin), *plantarflexion* (pointing of the foot downwards), inversion (turning the foot inwards) and eversion (turning the foot outwards). The type of movement produced when muscles contract depends on the attachments of the muscle and the architecture of the joint itself.

Muscles initiating or maintaining a movement are termed *prime movers* or *agonists*, for example, the *quadriceps* extending the knee joint during kicking. Muscles that oppose this movement, or initiate and maintain its converse, are termed *antagonists*, for example, the *hamstrings* opposing knee extension during kicking. When prime movers and antagonists contract together as *fixators* they stabilize a joint, for example, the *quadriceps* and *hamstrings* to stabilize the knee joint during a block tackle.

The contraction of a prime mover that acts across a single joint may produce the desired movement. Many muscles cross more than one joint, for example, *rectus femoris* is an extensor of the knee and a flexor of the hip, and many joints are multiaxial. The unrestrained contraction of such muscles may produce additional and unnecessary movements. The restriction of unwanted movements occurs by the contraction of a partial antagonist muscle acting as a *synergist*. An example of this is powerful flexion of the fingers by the long flexors, for example, during a throw-in. Unrestricted contraction of these muscles would also produce flexion of the wrist joint and would thus reduce function. Function is improved by contraction of the long extensors acting as synergists to prevent wrist flexion.

2.2.6   *Types of muscle action*

During an *isometric* action, the muscle actively develops tension but no movement occurs. It therefore does not perform any external work as the limb is moved no distance about the joint, for example, sustaining a squatting position. Both slow and fast twitch fibres are equally involved in the development of isometric tension.

Dynamic actions occur when a muscle actively develops tension and movement does occur. Such actions may be either *concentric* or *eccentric*. During a concentric action, for example, the action of the *quadriceps* during jumping, the muscular tension rises and the muscle shortens. This action approximates the attachments of the muscle, extending the knee joint and propelling the body upwards. During an eccentric action, for example, *quadriceps* during landing from a jump, the distance between the origin and insertion of the muscle is increased and flexion of the knee is increased. In this case the tension developed within the muscle serves to *regulate* the speed of motion.

Isokinetic movements incorporate dynamic actions through a range of movement at a constant angular limb velocity. These movements are not natural movements but are produced using accommodating resistance devices which load the muscle with maximal force at every point throughout the range of movement. Such computer-controlled dynamometers, for example, Kin-Com, Biodex, Cybex, are used to test muscle performance and for training purposes during rehabilitation programmes.

2.2.7   *Flexibility and strength*

Flexibility refers to the range of movement about a joint. The flexibility of a joint is dependent upon the extensibility and elasticity of the structures surrounding it. These structures include muscles, the fibrous joint capsule and ligaments. Flexibility reduces the risk of injury sustained during overstretching, for example, attempting to reach the ball when intercepting a pass. Conversely, hypermobility may be disadvantageous as joints become unstable and prone to injury.

Increased force generating capacity of a muscle (strength) is accompanied by an increase in cross-sectional area and a corresponding decrease in length. Shortening of a muscle leads to a reduction in joint flexibility and therefore increases the risk of injury. It is therefore essential that strength training regimens are accompanied by flexibility programmes.

When muscle strength is tested, the dominant limb is usually slightly stronger than the non-dominant limb. This phenomenon is not as apparent in muscles of the lower limb than the upper limb and is less so in highly trained individuals. Often a ratio of strength exists between agonists and antagonist muscle groups, for example, the strength of the *hamstrings* is around 60% that of the *quadriceps*. Knowledge of such ratios is important when designing strength training programmes and when diagnosing, and attempting to prevent, injuries.

2.3   Specific muscles (Figure 2.3)

2.3.1   *The muscles of the back*

The term *erector spinae* refers to a group of muscles that surround the spine and run from the sacrum to the head. These muscles act to extend, flex laterally and rotate the trunk and the neck.

2.3.2   *The muscles of the abdominal wall*

The abdominals (*rectus abdominus*, and the *internal* and *external obliques*) run from the ribs to the pelvis. They act to flex, laterally flex and rotate the trunk, for example, during sit-ups.

2.3.3   *The gluteal muscles*

The gluteal muscles (*gluteus maximus, gluteus minimus* and *gluteus medius*) all abduct and laterally rotate the hip joint. Gluteus maximus is also a strong extensor of the hip.

2.3.4   *Main flexors of the thigh*

These muscles (*psoas* and *iliacus*) arise in the abdomen from the pelvis and the front of the spine and insert on the front of the femur. When the lower limb is fixed these muscles will flex the trunk on the thigh, for example, during a sit-up exercise.

2.3.5   *Muscles of the thigh*

The *quadriceps* group actually comprises four muscles (*rectus femoris, vastus medialis, vastus lateralis* and *vastus intermedius*) and is a powerful extensor of the knee joint. *Rectus femoris* also flexes the hip joint. The *hamstring* muscles (*biceps femoris, semimembranosus and semitendinosus*) are powerful flexors of the knee joint. They also extend the hip joint.

Adduction of the thigh is brough about by a group of five muscles, the adductors (groin muscles); some of this group also act as weak rotators and flexors of the thigh. They bring the abducted lower limb back into a neutral position, for example, when running sideways, and may even continue this movement to cross the leg over the other, for example, when kicking with the inside of the foot. These muscles are often tight in soccer players and therefore groin strains are a common injury.

2.3.6   *Muscles of the lower leg*

The anterior tibials (*tibialis anterior, extensor digitorum longus* and *extensor hallucis*) dorsiflex the ankle (pull the foot towards the shin). The posterior tibials (*gastrocnemius, soleus* and *tibialis posterior*) all plantarflex the ankle (point the foot downwards). *Tibialis posterior* also inverts the foot (turns it inwards). The peronei (*peroneus longus, peroneus brevis, peroneus tertius*) initiate eversion (turning outwards) of the foot and plantarflex the ankle, except *peroneus tertius* which dorsiflexes the ankle.

2.4   Muscle actions during soccer skills

2.4.1   *Running*

Running is an integral part of soccer. Indeed soccer players may cover approximately 10–13 km during a single game. The running action may be divided into two stages, *swing*

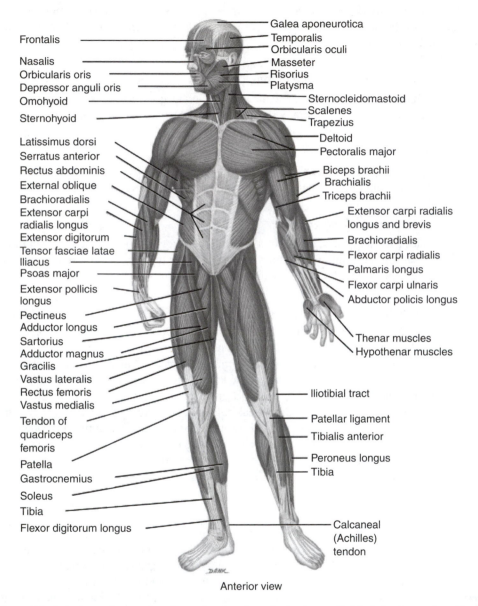

Frontalis

Nasalis
Orbicularis oris
Depressor anguli oris
Omohyoid
Sternohyoid

Latissimus dorsi
Serratus anterior
Rectus abdominis
External oblique
Brachioradialis
Extensor carpi
radialis longus
Extensor digitorum
Tensor fasciae latae
Iliacus
Psoas major
Extensor pollicis
longus
Pectineus
Adductor longus
Sartorius
Adductor magnus
Gracilis
Vastus lateralis
Rectus femoris
Vastus medialis
Tendon of
quadriceps
femoris
Patella
Gastrocnemius
Soleus
Tibia
Flexor digitorum longus

Galea aponeurotica
Temporalis
Orbicularis oculi
Masseter
Risorius
Platysma
Sternocleidomastoid
Scalenes
Trapezius
Deltoid
Pectoralis major
Biceps brachii
Brachialis
Triceps brachii
Extensor carpi radialis
longus and brevis
Brachioradialis
Flexor carpi radialis
Palmaris longus
Flexor carpi ulnaris
Abductor policis longus

Thenar muscles
Hypothenar muscles

Iliotibial tract

Patellar ligament
Tibialis anterior
Peroneus longus
Tibia

Calcaneal
(Achilles)
tendon

Anterior view

Figure 2.3 Anterior and posterior views showing the principal superficial skeletal muscles. (From *Principles of Anatomy and Physiology*, Tortora, G.T. and Grabowski, S.R. © 1993 Harper Collins. This material is used by permission of John Wiley & Sons Inc.)

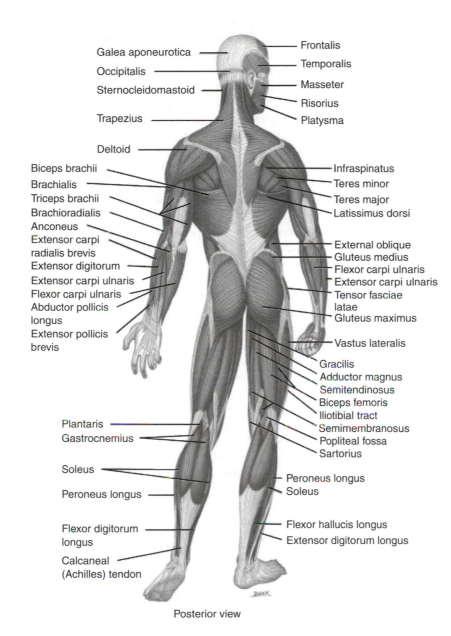

Galea aponeurotica —
Occipitalis —
Sternocleidomastoid —
Trapezius —
Deltoid —
Biceps brachii —
Brachialis —
Triceps brachii —
Brachioradialis —
Anconeus —
Extensor carpi radialis brevis —
Extensor digitorum —
Extensor carpi ulnaris —
Flexor carpi ulnaris —
Abductor pollicis longus —
Extensor pollicis brevis —

Frontalis
Temporalis
Masseter
Risorius
Platysma

Infraspinatus
Teres minor
Teres major
Latissimus dorsi

External oblique
Gluteus medius
Flexor carpi ulnaris
Extensor carpi ulnaris
Tensor fasciae latae
Gluteus maximus

Vastus lateralis
Gracilis
Adductor magnus
Semitendinosus
Biceps femoris
Iliotibial tract
Semimembranosus
Popliteal fossa
Sartorius

Plantaris —
Gastrocnemius —
Soleus —
Peroneus longus —
Flexor digitorum longus —
Calcaneal (Achilles) tendon —

Peroneus longus
Soleus

Flexor hallucis longus
Extensor digitorum longus

Posterior view

Figure 2.3 (Continued).

and *support*. Support begins at the point the foot makes contact with the ground (*foot strike*) and ends at the point the foot leaves contact with the ground (*toe-off*). The swing phase begins at toe-off and ends at foot strike.

At toe-off, the swing leg is in a position of extension of the hip, extension of the knee and plantarflexion of the ankle. The *gluteals* and *hamstrings* are still acting to extend the hip and the *gastrocnemius* to plantarflex the ankle and give a good push off. The actions of *psoas* and *iliacus* flex the hip, the *hamstrings* flex the knee and the *anterior tibials* dorsiflex the ankle. The hip continues to flex and the ankle to dorsiflex to bring the leg forwards in front of the support leg; the *adductors* act to prevent the thigh from swinging outwards. The *quadriceps* then begin to extend the knee in preparation for foot strike.

When foot strike occurs, the hip is in flexion, the knee is in slight flexion and the ankle is normally dorsiflexed and slightly inverted. At this point the weight of the body must be controlled as it hits the ground. The *gluteals* contract to extend the hip, the *quadriceps* and *hamstrings* contract to stabilize the knee joint, and the *adductors* to stabilize the hip. The *anterior tibials* work eccentrically and the *gastrocnemius* concentrically to control the foot as it strikes the ground. The momentum of the body carries it forwards over the ankle joint which acts as a rocker as the foot becomes flat to the ground and then toe-off occurs.

As the speed of running increases longer strides are taken. In this instance, the swing phase involves greater knee flexion and hip extension (the heel almost touching the buttock) and greater hip flexion in the later part of the phase.

When running with a ball, much shorter strides are taken as the player must be ready to change direction and speed. At the toe-off phase the leg may not be as extended. Heel strike may not be as pronounced, instead the foot may land in a more neutral position or be plantarflexed.

The muscles of the arms and trunk also play an important role during running. They act to maintain balance and to counterbalance the rotation of the body when the pelvis rotates.

2.4.2   *Kicking a ball*

There are many different types of kicks in soccer, for example, running kick, volley and push pass (Pronk, 1991). Skilled players can also impose spin on the ball and cause it to dip quickly in flight. In such cases the kicking action is quite complex. For the purposes of this text the kick is simplified into that of movement in one plane. This action may also be divided into four phases; phase one, priming the thigh and leg during backswing; phase two, rotation of the thigh and leg laterally and flexion of the hip; phase three, deceleration of the thigh and acceleration of the leg; and finally stage four, the follow through.

During *phase one*, the hip of the kicking leg is rapidly extended by the action of the *gluteals* and the pelvis is rotated backwards. The knee is flexed by the *hamstrings* and the *anterior tibials* dorsiflex the ankle. These actions are limited by the hip flexors and the adductors which often become over-stretched in many players. The harder the subsequent kick the further the stretch on these muscles. During *phase two*, the *psoas* and *iliacus* contract and the hip flexes to move the thigh and leg forwards and the pelvis rotates forwards. *Phase three* involves the *hamstrings* acting to decelerate the thigh and the *quadriceps* rapidly extend the knee joint. The position of the ankle joint during ball strike is dependent upon the type of kick performed. In addition, the *adductors* will contract to pull the leg towards the body. This is especially relevant during a side kick or push pass. *Phase four* begins after the ball has lost contact with the foot. The leg and thigh will follow through due to the momentum of the thigh, leg and foot. This causes a stretch on the muscles opposing these actions, especially the *hamstrings* as they pass over two joints (De Proft *et al.*, 1988).

The muscles of the non-kicking leg act in a similar fashion to their behaviour during the stance phase of running. However, they act mainly to stabilize the body to provide a stable platform on which the kicking leg may act. This leg is usually abducted and rotated. Again the muscles of the arms and trunk work to maintain poise and balance and to provide a counterbalance to the kicking leg, thus providing more control and speed.

2.4.3   Jumping and heading

Jumping to control the ball in the air is of major importance in soccer. Jumping can occur from a standing position or from a run-up. Take off from a standing jump is usually performed from both feet and from one leg when using a run-up. When performing a standing jump the player will sink down into a position of flexion. The trunk, hips and knees will flex and the ankle will dorsiflex under the action of body weight and gravity but controlled by the agonists to these movements acting eccentrically (*erector spinae, gluteals, hamstrings, quadriceps* and *plantarflexors*). The elbows will flex and the shoulders will be extended. In this position, the body is almost spring-like; the prime movers of the jumping action are on a stretch, storing potential energy ready to be released at the appropriate moment. When the jump itself begins, the prime movers act to launch the body weight up in the air. This is achieved by rapid and powerful contractions of the *erector spinae, gluteals, hamstrings, quadriceps* and *plantar flexors* to produce extension of the trunk, hips, knees and plantar-flexion of the ankles. The arms are also moved rapidly forwards and upwards by flexion of the shoulders and extension of the elbows. When the spine becomes extended during the jumping action a severe stretch may be placed on the abdominal muscles and the hip flexors and injury to these muscles may occur.

Landing from a jump is just as important as the jump itself, as the weight of the body must be controlled as it hits the ground. Basically, it is a reverse of the jumping action. However, this time the muscles of jumping act eccentrically to control joint movement and decelerate the action, thereby increasing shock absorption and decreasing the risk of injury.

The primary aim of most jumps in soccer is to head the ball, but heading may also occur from a standing position. As a player jumps the neck becomes extended partly from the effects of gravity and partly due to the action of the *erector spinae* muscles. As players attempt to make contact with the ball they will aim their head at it. This may involve a combination of movements. Flexion of the neck is the most powerful action but this may be combined with rotation or lateral flexion to direct the ball.

2.4.4   Throwing a football

Throw-ins are usually taken from a short run-up and a two-footed stance. With both feet level, the *erector spinae, gluteals and the hamstrings* contract to extend the spine and the hips. The dorsiflexors act eccentrically to allow the ankles to move into a small degree of plantarflexion without losing balance. The ball is held in both hands and the two arms are held up above the head. The shoulders are moved into full flexion and the elbows also are now fully flexed. This creates full stretch on the antagonist groups and potential energy is stored. As the throw begins, these muscles now become prime movers which contract from a stretched position. The elbows become extended, the shoulders become more extended. The contraction of the *abdominals, psoas* and *iliacus* causes the spine and hips to flex. Dorsiflexion of the ankles is controlled by the eccentric action of *gastrocnemius* and *soleus*.

Summary

An understanding of the importance of joint flexibility and muscle strength combined with a basic knowledge of anatomy and physiology and muscle actions during soccer skills will aid the coach, trainer and medical staff in the design of appropriate training programmes for soccer players.

References

Bangsbo, J. (1993). *The Physiology of Soccer*. August Krogh Institute, University of Copenhagen, pp. 79–82.

De Proft, E., Clarys, J.P., Bollens, E., Cabri, J. and Dufour, W. (1988). Muscle activity in the soccer kick, in *Science and Football* (eds T. Reilly, A. Lees, K. Davids and W.J. Murphy), E & FN Spon, London, pp. 434–40.

Moore, K.L. and Dalley, A.F. II. (1999). *Clinically Oriented Anatomy*, 4th edn, Lippincott, Williams & Wilkins, Philadelphia, PA.

Pronk, N.P. (1991). The soccer push pass. *National Strength and Conditioning Journal*, **13**(2), 6–81.

Williams, P.L., Bannister, L., Berry, M., Collins, P., Dyson, M. and Ferguson. M.W.J., eds (1995). *Gray's Anatomy*, 38th edn, Churchill Livingstone, Edinburgh.

3 Fitness assessment

Thomas Reilly and Dominic Doran

Introduction

The game of soccer places varying physiological demands on performers. Fitness in a soccer-specific context refers to a range of individual characteristics that is a composite of many attributes and competencies. Such competencies by definition include physical, physiological and psychomotor factors. Such qualities are needed in contesting and retaining possession of the ball, maintaining a high work-rate for 90 min of play, reacting quickly and appropriately as opportunities arise and regulating mental attributes before and during match-play.

The balance between these components of soccer-related fitness depends upon participants' level of performance, positional role and the team's style of play. Other inputs to this mix include age, gender, stage of season, environmental influences, history of injury and nutritional status. Re-acquiring desirable fitness levels is especially important after injury and prior to returning to competitive play. Otherwise the individual is vulnerable to re-injury if uncorrected weaknesses, in muscle strength for example, are carried into a game which may exacerbate the condition and lead to sub-optimal personal and team performance.

The possession of these fitness characteristics in isolation does not predispose to successful soccer performance by themselves but must be synergistic with skill development and the acquisition of tactical knowledge. The success of the team depends on how individuals are blended into an effective playing unit. When teams roughly equal in skill and tactical knowledge meet, the one with the higher overall fitness level will have the advantage of being more able to cope with a fast pace of play. Coaches, trainers and sports scientists acknowledge that preparation for competitive match-play calls for a systematic approach. Consideration of individual fitness profiles and the contribution these make to the team must be an integral part of this systematic approach. Attention to fitness profiles is relevant not just in the build-up towards key matches and tournaments but also prior to and throughout the competitive season. The application of a systematic approach to training and match preparation is apparent in the academy and senior team structures of many of the top clubs worldwide. Fitness profiling is achieved by means of a battery of tests. The test items may either be part of a comprehensive physiological assessment or be dedicated solely to assessing performance in soccer. The fitness profiles generated from such batteries have some value in allowing comparisons between individuals and, with the use of normative ranges, individual weaknesses may be identified and remedial training prescribed. Repeated fitness assessment is of further value in that changes in fitness profiles within individuals and throughout the team as a whole can be measured. In this way, the appropriateness and progression of the training regimen applied to the players and the squad can be gauged not only in determining adaptation, but also failure to adapt and maladaptation (over-training).

In this chapter, fitness tests commonly employed in sports science laboratories are described. The application to soccer is highlighted, as are some of their limitations. Field tests are included as a complement to laboratory-based tests, and in some circumstances may be an alternative. Descriptive profiles are provided to examine the variability between individuals, especially at the elite level of soccer play.

3.1   Anthropometry

The biographical and anthropometric profile of soccer players can be characterized by the heterogeneity expressed within teams, between players' competitive levels, and across nationalities (Table 3.1). The average age of top players is 25–27 years with a standard deviation of about 2 years. However, to suggest this constrains the age at which players can compete at the highest levels would be erroneous. Players who do not fall within this age distribution are found to perform at elite levels of play. Players of teenage years do feature in all top club teams, previously Manchester United and currently Leeds United squads are characterized by a high proportion of young players in the first team. This may reflect the development of young talent within the clubs' football academy and progression through the club structure. The expression that 'young teams don't win trophies and championships' is not supported by recent history. Although, these young players tend to be part of experienced squads, age need not be a barrier to performance excellence at the highest level of soccer. Although performance excellence is evident from a younger age, many talented players reach the pinnacle of their own playing careers at a later age, around 25–27 years, which is evidenced by many members of the current England national team. The majority of professional players are in their twenties and traditionally there was a reluctance of managers at top club level to retain the professional services of players once they were into their thirties. A loss of motivation to train may have contributed also to an earlier than necessary retirement from playing professional soccer. Active athletes can maintain fitness levels well into their thirties before physiological functions begin to show signs of deterioration.

Nowadays, professional players do seem prepared to or are able to stay in the 'game' for longer than was traditional. This is probably due to the commercial attractions of maintaining one's playing career as long as possible. Development of appropriate sports medicine and responsive injury management strategies within clubs as well as improved orthopaedic procedures for repair of tissue damage, that in previous decades might have halted a player's career, may also contribute to the trend of professionals staying active for longer than previously. An important adjunct to this increased career longevity has been the emphasis placed on developing a balanced fitness profile with appropriate warm up and warm down techniques that also help in injury prevention. It is noticeable how many players into their thirties are still key members of club and national teams. Goalkeepers seem to have longer playing careers than outfield players and it is not unusual to have players at top level who are well into their thirties (Peter Schmeichel, Manchester City; David Seaman, Arsenal). Indeed there are examples of players who have represented their countries at major international events in their forties. These include Dino Zoff (Italy), Peter Shilton (England) and Pat Jennings (Northern Ireland). This fact may be related to the special requirements of the position and players maturing in tactical judgement with experience in the game. It may also be related to a lower incidence of chronic injuries and degenerative trauma in goalkeeping compared to outfield positions.

The average male soccer player's biographical, anthropometric and performance characteristics are presented although these can vary widely (Table 3.1). Lack of height is

Table 3.1 Biographical, anthropometric and performance characteristics of professional and elite soccer players reported in literature

Source	Nationality	Level	n	Age (years)	Height (cm)	Mass (kg)	Body fat (%)	Somatotype	$\dot{V}O_2\,max$	CMJ (cm)
Faina et al. (1988)	Italian	Professional	27	26.0 ± 4.8	177.2 ± 4.5	74.4 ± 5.8	—	—	58.9 ± 6.1	43.5 ± 4.9
White et al. (1988)	English	Professional D1	17	23.3 ± 0.9	180.4 ± 1.7	76.7 ± 1.5	19.3 ± 0.6	2.6–4.2–2.7	49.6 ± 1.2	59.8 ± 1.3
Togari et al. (1988)	Japanese	National	20	24.2 ± 2.48	175.3 ± 5.8	69.7 ± 5.0	—	—	—	—
Chin et al. (1992)	Hong Kong	Professional	24	26.3 ± 4.2	173.4 ± 4.6	67.7 ± 5.0	7.3 ± ?	—	—	—
Puga et al. (1993)	Portuguese	National	21	27.6 ± ?	178.1 ± ?	73.8 ± ?	11 ± ?	—	—	—
Dunbar and Power (1995)	English	Professional PL	18	22.5 ± 3.6	—	77.7 ± 7.6	12.6 ± 2.9	—	60.7 ± 2.9	—
Tiryaki et al. (1995)	Turkish	Professional D1	16	18–30	178.8 ± 3.8	74.8 ± 6.6	7.6 ± 0.7	—	51.6 ± 3.1	64.8 ± 4.6
Tiryaki et al. (1995)	Turkish	Professional D2	16	18–30	177.7 ± 3.4	69.6 ± 4.1	7.1 ± 0.4	—	51.1 ± 2.0	54.1 ± 5.7
Tiryaki et al. (1995)	Turkish	Professional D3	16	18–30	178.8 ± 5.9	72.7 ± 6.5	7.2 ± 0.4	—	51.3 ± 2.1	57.0 ± 7.5
Mercer et al. (1995)	English	Professional D1	15	24.7 ± 3.8	179.0 ± 8.0	77.6 ± 9.2	16.2 ± 3.4	—	62.6 ± 3.8	44.8 ± 6.8
Raastad et al. (1997)	Norway	Professional	28	23.5 ± 3.0	—	78.9 ± 7.8	—	—	62.8 ± 4.1	—
Bury et al. (1998)	Belgium	Professional D1	15	24.2 ± 2.6	180.7 ± 5.2	76.8 ± 5.2	14.1 ± 1.1	—	62.8 ± 4.0	—
Di-Salvo et al. (1998)	Italian	Professional	44	17.8 ± 0.6	181.3 ± 4.4	72.6 ± 4.7	—	—	—	—
Rico-Sanz et al. (1998)	Puerto Rico	Olympic	8	17.0 ± 2.0	169.8 ± 6.5	63.4 ± 3.1	7.6 ± 3.1	—	69.2 ± ?	—
Rienzi et al. (1998)	South American	Professional	110	26.1 ± 4.0	177.0 ± 6.0	76.4 ± 7.0	10.6 ± 2.6	—	—	—
Wisloff et al. (1998)	Norwegian	Professional D1	14	23.8 ± 3.8	181.1 ± 4.8	76.9 ± 6.3	—	—	67.6 ± 4.0	56.7 ± 5.6
Wisloff et al. (1998)	Norwegian	Professional D1	15	23.8 ± 3.9	180.8 ± 4.9	76.8 ± 7.4	—	—	59.9 ± 4.1	53.1 ± 4.0
Mujika et al. (2000)	Spanish	Professional	17	20.3 ± 1.4	179.9 ± 5.5	74.8 ± 5.5	7.9 ± 1.6	—	—	47.4 ± 5.0
Rico-Sanz et al. (1999a)	Swiss	Professional	17	17.5 ± 1.0	177.3 ± 5.3	69.4 ± 6.4	—	—	—	—
Aziz et al. (2000)	Singaporean	Elite National	23	21.9 ± 3.6	175.0 ± 6.0	65.6 ± 6.1	—	—	58.2 ± 3.7	—
Rienzi et al. (2000)	South American	Professional	11	26.1 ± 4.0	177.0 ± 6.0	76.4 ± 7.0	10.6 ± 2.6	2.2–5.4–2.2	—	—
Sözen et al. (2000)	Turkey	Professional	83	25.5 ± 4.0	177.8 ± 5.5	73.6 ± 8.5	—	—	—	—
Al-Hazzaa et al. (2001)	Saudi Arabian	Professional	154	25.2 ± 3.3	177.2 ± 5.9	73.1 ± 6.8	12.3 ± 2.7	—	56.8 ± 4.8	41.4 ± 2.7
Casajús (2001)	Spanish	Professional	15	26.3 ± 3.1	180.0 ± 7.0	78.5 ± 6.4	8.2 ± 0.91	2.6–4.9–2.3	66.4 ± 7.6	41.6 ± 4.2
Cometti et al. (2001)	France	Professional D1	29	26.1 ± 4.3	179.8 ± 4.4	74.5 ± 6.2	—	—	—	39.7 ± 5.6
Cometti et al. (2001)	France	Professional D2	32	23.2 ± 5.6	178.0 ± 5.8	73.5 ± 14.7	—	—	—	54.7 ± 3.8
Helgerud et al. (2001)	Norwegian	Professional D1	19	18.1 ± 0.8	181.3 ± 5.6	72.2 ± 11.1	—	—	64.3 ± 3.9	—
Craven et al. (2002)	English	Professional D1	14	23	181.0 ± 6.0	80.1 ± 9.2	—	—	—	—
Dowson et al. (2002)	New Zealand	National	21	Senior	178.0 ± 6.8	78.9 ± 6.0	17.4 mm	—	60.5 ± 2.6	48.0 ± 4.6
Strudwick et al. (2002)	English	Professional PL	19	22.0 ± 2.0	177.0 ± 5.9	77.9 ± 8.9	12.3 ± 2.9	—	59.4 ± 6.2	—

not in itself a bar to success in soccer, the possession of particular height characteristics can predispose a player to be placed or migrate toward a certain playing position. Goalkeepers, centre-backs and forward players used as a 'target' for winning possession of the ball with the head tend to be taller than players found in other positions (Bangsbo, 1994c; Wisløff *et al.*, 1998; Sözen *et al.*, 2000; Al-Hazzaa *et al.*, 2001). In contrast, the players deployed in midfield, full-back and on the wings tend to be smaller in size than those in other positional roles. These generic values have but a limited use for comparative purposes when the variability within a given soccer team is large. A coach may modify his team configuration and playing style to accommodate individuals without the expected physical attributes of conventional playing roles, provided they compensate by superior skills and motivation. The average body size noted may also represent ethnic or racial influences. Where national squads have been assessed, this distribution of goalkeepers, centre-backs and target forwards being taller than midfield players and wingers is maintained (Al-Hazzaa *et al.*, 2001). Many top European and South American teams contain players of different racial backgrounds and this can make interpretation of anthropometric profiles more complex (Rienzi *et al.*, 2000).

A particular body size may encourage acquisition of certain skills and force a gravitation toward a specific playing position. This is likely to occur before maturity so that the individual will tend to favour one positional role before playing at senior level. Under-age soccer is organized according to chronological rather than biological age. Advantages bestowed by body size in adolescent and youth soccer may disappear as the late maturers catch up and the gap to the early maturers is narrowed. However, the predisposition toward successful soccer performance is subject to multifactorial inputs of which the anthropometric component is but one (Reilly *et al.*, 2000a). Such observations are also reflected in senior soccer performance (Reilly *et al.*, 2000b).

Physique represents body shape rather than body size and its measurement is referred to as somatotyping. Somatotype is represented on three dimensions – endomorphy, mesomorphy and ectomorphy. Endomorphy reflects roundness and adiposity, mesomorphy indicates muscularity whilst ectomorphy suggests a tendency towards linearity. The somatotype is calculated from a number of limb girth, bone diameter and skinfold thickness measures, in conjunction with height and body mass (Eston and Reilly, 1996). Typical somatotype values for soccer players reported in the early 1990s was 3–5–3, reflecting a trend towards mesomorphy (Reilly, 1990). Recent data reflect this movement in somatotype toward a more mesomorphic component although variability according to both nationality, and level of play is evident. Rienzi *et al.* (2000) reported the somatotype characteristics of elite 'Copa America' players to be 2.2–5.4–2.2 (±0.7, ±1, ±0.6). In line with these observations Casajús (2001) found elite Spanish first division players reflected this mesomorphic dominance, with an average somatotype of 2.4–4.8–2.3 (±0.52, ±0.88, ±0.73). The possession of a strong muscular make-up would be of benefit in game contexts such as tackling, shielding the ball, contesting possession, turning, accelerating, kicking and so on. Given the needs of soccer, muscular development is more pronounced in the thigh and calf compared to the upper body in top soccer players.

Body composition is an important aspect of fitness for soccer as superfluous adipose tissue acts as dead weight in activities where body mass must be lifted repeatedly against gravity. This applies to locomotion during play and in jumping for the ball. The most commonly used model of body composition divides the body into two compartments – fat and fat-free mass. An alternative is to estimate muscle mass from anthropometric measures using the equation of Martin *et al.* (1990), although this may overestimate muscle mass somewhat (Catrysse *et al.*, 1999). Generally, such estimates confirm the tendency towards a muscular make-up among soccer players. The amount of fat in the adult male in his mid-twenties is about 16.5% of body weight. A comparable figure for the adult female is 26%. Lowest values for percentage body fat among athletic groups are found in distance runners, with mean values as low as 4–7% in men. Figures for soccer players are higher than this and reports for average team values have ranged from 9 to 16% (Reilly, 1990), further values are presented in Table 3.1. Higher values are found in goalkeepers than in outfield players, probably because of the higher metabolic loading imposed by match-play and training on outfield players. Soccer players accumulate body fat in the off-season and lose weight more during pre-season training time than in other periods. They may also put on weight when they are recovering from injury and unable to train strenuously, unless they modify their intake of food. Thus, the habitual activity of players at the time of measurement, their diet and the stage of the competitive season should be considered when body composition is evaluated.

The method of estimating or measuring adiposity or percentage body fat should also be considered when interpreting observations presented in the literature. Body fat (or body adiposity) is determined indirectly in live subjects. It may be estimated from chemical measures such as total body water or total body potassium. These methods are not accessible for routine use with soccer players and facilities are not generally available to sports science support groups. Similarly, medical imaging techniques are impractical with athletic groups and the radiographic dose with computerized tomography makes it unsuitable for repeated application. Magnetic resonance imaging (MRI) overcomes this problem but its expense makes it unrealistic for routine use. Portable devices, such as bioelectric impedance analysis and infrared interactance, have not been sufficiently validated for universal use. The scientifically accepted reference method is underwater (hydrostatic) weighing but it is doubtful whether this is actually true at present for all purposes; the assumptions made with regard to body density are not transferable to all highly trained athletes. Besides, it cannot be prescribed for testing squads of soccer players, particularly in field conditions. Air, rather than water, displacement represents an acceptable alternative which is used in some professional clubs.

The most accessible method for obtaining data on body composition is assessment of skinfold thickness by means of calipers. An appropriately trained individual, identifying the correct sites for measurement of skinfolds and using the proper equipment, must make the measurements. Since 1996, the Australian Laboratory Assistance Scheme recommends that seven anatomical sites are measured when assessing adiposity which is in line with the recommendations of the International Society for Advancement of Kinanthropometry (ISAK) (Norton *et al.*, 2000). These sites are – biceps, triceps, sub-scapula and supraspinale, abdominal, front thigh and medial calf. The seven skinfolds should be summed and the resultant value used as an index of subcutaneous adiposity since the skinfold thickness data themselves provide indications of changes in body composition. A percentage body fat value can be calculated in order to provide a target if weight control is desirable. The expression of percentage body fat is widely adopted in literature although more appropriately the dual expression of the summed skinfolds and body fat percentage should be made.

3.2   Muscle function

3.2.1   *Muscle strength*

Various tests of muscle strength and power have been employed for assessment of soccer players. These have ranged from performance tests and measurement of isometric strength to contemporary dynamic measures using computer-linked isokinetic equipment. Tests of anaerobic power output have also evolved as well as short-term jumping performance on the force platform, which have relevance to soccer play. Strength in the lower limbs is of obvious concern in soccer: the quadriceps, hamstrings and triceps surae groups must generate high forces for jumping, kicking, tackling, turning and changing pace. The ability to sustain forceful contractions is also important in maintaining balance and control especially when being challenged for possession. Isometric strength is also an important contributory factor in maintaining a player's balance on a slippery pitch and in ball control. For a goalkeeper almost all the body's muscle groups are important for executing the skills of this positional role. For outfield players the lower part of the trunk, the hip flexors and the plantarflexors and dorsiflexors of the ankle are used most. Upper body strength is employed in throw-ins and the strength of the neck flexors could be important in forcefully heading of the ball. Given the physical nature of the contemporary game, at least a moderate level of upper body strength should prove helpful in preventing being knocked off the ball. High levels of muscular strength are also important in reducing the risk of injury.

Soccer players are generally found to possess only a little above average isometric muscle strength which may reflect inadequate attention to resistance training in their habitual programme. Besides, isometric strength may not truly reflect the ability to exert force in dynamic conditions. It may also be a poor predictor of muscle performance in the game (Reilly, 1994). However, isometric activity is still important in stabilizing the trunk and providing a platform for more dynamic muscular activity of the lower body to take place. Increasingly recognized is the concept of core stability training. Screening and development of isometric muscle activity are integral to most rehabilitation and strength conditioning regimens, although normative data are lacking. Some studies have indicated a relationship between dynamic muscle performance as measurable in laboratory and field contexts. Asami and Togari (1968) reported a significant correlation between knee extension power and ball speed during instep kicking, both factors increasing with experience in the game. Cabri *et al.* (1988) also reported a significant relation between leg strength, measured as peak torque during an isokinetic movement, and kick performance indicated by the distance the ball travelled. The relationship was significant for both eccentric and concentric contractions of hip and knee joints in flexion and extension. Whilst these relationships have been observed in male players, Reilly and Drust (1997) also observed high correlations between peak muscle torque and angular velocity (at a range of velocities up to $6.98 \, \mathrm{rad \, s^{-1}}$) in female soccer players.

The relation between leg strength and kick performance implies that strength training could be effective in improving the kicking abilities of soccer players. Given a certain level of technique, it seems that strength training when added to the normal soccer training improves both muscular strength and kick performance (De Proft *et al.*, 1988). However, such reports are not always mirrored in recent literature. For example, Aagaard *et al.* (1996) have shown that a mix of generic and movement-specific resistance training techniques to improve knee flexor strength produced no concomitant change in ball velocity after 12 weeks of training. Helgerud *et al.* (2001) reported no significant changes in ball velocity after an 8-week programme of training while Cometti *et al.* (2001) reported that higher levels of isokinetic knee extensor strength between professional and amateur players were not reflected in ball velocity.

The relationship between dynamic muscle strength of the knee extensors and kick performance may therefore be dependent on the level of skill already acquired. Trolle *et al.* (1993) measured isokinetic strength of the leg extensors in skilled soccer players at angular velocities between 0 and $4.18 \, \text{rad s}^{-1}$. No relationship was found between these measures and ball velocity recorded during a standardized indoor soccer kick; ball velocity was again unchanged after 12 weeks of strength training. Such findings are not surprising, as average velocities about the knee during a soccer kick range between 13.5 and $18.5 \, \text{rad s}^{-1}$ dependent upon whether kicking is for accuracy or speed (Lees and Nolan, 2002). In the context of isokinetic strength assessment and training, the maximum angular velocity achievable with current technology is about $5.2 \, \text{rad s}^{-1}$. Training at these velocities may negate the training specificity required (in terms of the speed of limb movement) to see any possible changes. The strength of the knee extensors in isolation is not the sole determinant of ball velocity. Kinematic analysis of kicking actions highlights the complex synergy between the hip and lower limb movement patterns initiated during the striking of the ball (Nunome *et al.*, 2002). Where this synergy is disrupted due either to non-specific methods of strength training or assessment, any potential relationship inherent between lower limb strength development and ball velocity will be disrupted. Where the relationship is maintained, improvements in lower limb strength and ball velocity are expressed (Dutta and Subramanium, 2002). Improved rates of force development and improved co-ordination are potential antecedents to improve the development of kicking velocity (Almåsbakk and Hoff, 1996) as such motor control factors may override muscular strength in well-trained soccer players and obscure potential relationships.

The shoulder and trunk muscles are engaged in throwing in the ball from the sidelines and a long throw into the opponents' penalty area can be a rich source of scoring opportunities. The throw-in distance of soccer players has been reported to be related to pull-over strength and trunk flexion strength. Training methods, using a medicine ball, increased strength measures but without a corresponding increase in throw distance (Togari and Asami, 1972). This demonstrates a degree of specificity in the throwing skill and suggests that individual players should be pre-selected to take tactical long throws. Since fitness requirements tend to vary with positional roles, muscle strength values may depend on the player's position. Goalkeepers and defenders were found by Oberg *et al.* (1984) to have higher knee extension torque at $0.52 \, \text{rad s}^{-1}$ than midfield players and forwards. The result was attributable to differences in body size since correction for body surface area removed the positional effect. A similar observation was made by Togari *et al.* (1988) in their tests on Japanese Soccer League players. The goalkeepers were significantly stronger than forwards at slow ($1.05 \, \text{rad s}^{-1}$) speeds of movement, midfield players being intermediate. The differences tended to disappear when the angular velocity was raised to $3.14 \, \text{rad s}^{-1}$. Wisløff *et al.* (1998) have demonstrated that differences in isokinetic strength between playing positions are removed when data are scaled to body mass ($m_b^{0.67}$). Comparative assessment in future should scale the values when attempting to examine the role of positional difference in strength development.

It is now common to monitor the muscle strength of soccer players using isokinetic apparatus such as Cybex, Bio-Dex or Lido (Figure 3.1) systems. These machines offer facilities for determining torque–velocity curves in isokinetic movements and joint-angle curves in a series of isometric contractions. The more complex systems allow for measurement of muscle actions in eccentric as well as concentric and isometric modes. In eccentric actions the limb musculature resists a force exerted by the machine: it is lengthened in the process and hence produces an eccentric contraction.

Isokinetic tests of soccer players have concentrated almost exclusively on lower limb muscle groups and on concentric contractions. Whilst knee extension strength in concentric contractions is correlated with kick performance, an even higher correlation has been

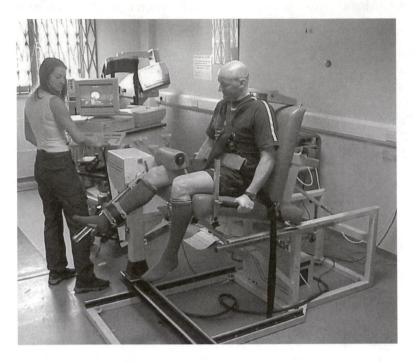

Figure 3.1 Isokinetic dynamometer (Lido) for measurement of muscular function in isometric, concentric or eccentric modes.

reported for knee flexion strength in eccentric actions (Cabri *et al.*, 1988). Although widely used in a strength assessment, some authors view isokinetic dynamometry as not reflecting the specificity of limb movement noted during soccer performance, advocating instead the use of functional tests in performance assessment (Wisløff *et al.*, 1998). Whilst debate exists on the efficacy of isokinetic versus functional testing in relation to performance assessment, the efficacy of isokinetic dynamometry in assessing deficits and imbalances in muscle strength are not disputed (Östenberg *et al.*, 1998; Cometti *et al.*, 2001). Given that approximately 76% of soccer injuries are lower extremity related (Morgan and Oberlander, 2001), high levels of muscular strength in relation to the hamstring to quadriceps ratio would seem important in stabilizing the knee and reducing the risk of injury (Fleck and Falkel, 1986; Fried and Lloyd, 1992). The possession of strong hamstrings, particularly in eccentric modes, is an important requirement for playing soccer. An improper balance between hamstrings and quadriceps strength may predispose towards injury in soccer players. At slow speeds and under isometric conditions, a knee flexor–extensor ratio of 60–65% is recommended (Oberg *et al.*, 1986). This ratio is increased at the higher angular velocities of commercially available apparatus, although the reliability of measurement is reduced at fast speeds. During kicking actions Graham-Smith and Lees (2002) have applied a model of risk assessment for hamstring injury termed the dynamic control ratio (DCR) which is determined using isokinetic dynamometry. Expressed as the eccentric hamstring : concentric quadriceps ratio, it provides a functional assessment of potential injury risk; ideally the DCR ratio should be equivalent to 1.0 although a ratio of less than 0.75 in place kickers has been suggested as a threshold of potential injury risk. Isokinetic strength testing also allows identification of asymmetries in lower limb muscular strength; generally where imbalances are present the weaker limb is the one most liable to injury (Fowler and Reilly, 1993; Östenberg *et al.*, 1998). In soccer, where rapid accelerations, decelerations, cutting and side-stepping manoeuvres can apply substantial mechanical loadings to the knee joint, any inter- or intra-limb asymmetries in knee extensor strength can predispose toward injury particularly where they do not adequately stabilize the knee (Besier *et al.*, 2001). Isokinetic test profiles are also important in monitoring muscle strength gains during rehabilitation using the uninjured side as reference. These comparisons to identify asymmetry or weakness within an individual player may be more important than comparison between teams or between team members. This reservation applies especially to comparisons with data from other laboratories using alternative test protocols.

3.2.2   Anaerobic power

Soccer players are frequently required to produce high power output and sometimes to maintain or repeat it with only a brief period for recovery. The splitting of high-energy intra-muscular phosphagens contributes along with anaerobic glycolysis to the maximal power a player can develop. These substrates (ATP, creatine phosphate and glycogen) may be used for combustion by muscle at the onset of exercise and result in a high anaerobic work production. A number of methodological techniques have been utilized to assess maximal intensity exercise performance in soccer players; these estimates are taken to reflect the re-synthesis of ATP via anaerobic energy systems. The maximum power output can be calculated from performance on the stair-run test of Margaria *et al.* (1966). Measurement is made of the time taken for the player to run between two steps on the stairs, the vertical distance between which is known.

The maximal ability to generate muscular power can be measured as a response to jumping on a force platform. This test requires expensive and complex equipment, which is not available for routine assessments. Power output in vertical jumping can be calculated, knowing the player's body mass, the vertical distance through which body mass is moved and the flight time. The vertical distance itself is a good measure of muscular performance, that is, mechanical work done. This value can be recorded using a digital system attached to the subject's waist and based on the extension of a cord, which is pulled as the individual jumps vertically (Figure 3.2). This method has now replaced the classical Sargent jump technique for measuring vertical jump. It is preferable to the standing broad jump which is influenced by leg length and which does not permit calculation of power output. Performance of soccer players in such tests of jumping ability tends to show up positional influences. Generally, goalkeepers, defenders and forwards perform better than midfield players on counter-movement jump (vertical jump) performance (Di-Salvo and Pigozzi, 1998; Wisløff *et al.*, 1998; Al-Hazzaa *et al.*, 2001).

Bosco *et al.* (1983) described another method for measuring mechanical power output in jumping. It requires jumping repeatedly for a given period, usually 60 s, the higher time and jumping frequency being recorded. The jumps are performed on a touch-sensitive mat, which is connected to a timer. Power output can be estimated knowing the subject's body mass and the time between contacts on the mat. Performance at various parts of the 1-min test can be compared, the tolerance to fatigue as the test progresses being indicative the anaerobic glycolytic capacity.

As soccer players must be prepared to repeat fast bursts of activity supported by anaerobic glycolysis, the high anaerobic capacity should be important to play well. The Wingate

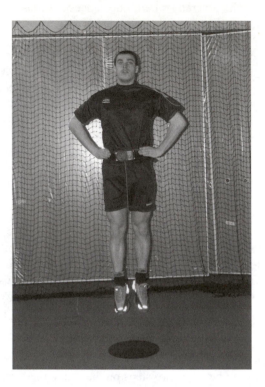

Figure 3.2 Measurement of counter-movement jumping ability: power output is calculated from the height of the jump and body mass if flight time is measured.

test which entails a 30-s all-out effort on a cycle ergometer has been widely adopted as a test of anaerobic capacity. Al-Hazzaa *et al*. (2001) have presented normative data for the Saudi Arabian national team with peak power relative to body mass ranging from 11.31 to 13.50 W kg^{-1} dependent upon positional role. The test duration of 30 s is too short to tax anaerobic capacity completely, more correctly it represents an estimate of anaerobic power. A significant aerobic contribution to Wingate test performance of up to 40% has been suggested (Medbø and Tabata, 1986). Such findings are in line with the low-moderate correlations found between $\dot{V}O_2$ max and anaerobic sprint performance during the Wingate test and repeated sprints in soccer players (Aziz *et al*., 2000; Al-Hazzaa *et al*., 2001). Another major limitation of such anaerobic capacity profiles is the mode of exercise. Measurement of power production and anaerobic capacity on a treadmill is more appropriate for soccer players than using the traditional Wingate test. Power output may be measured whilst the player runs as fast as possible on a 'self-powered' treadmill (Figure 3.3). The speed of the belt is determined by the effort of the subject. The horizontal forces produced can be determined from a load cell attached to the individual by means of a harness worn around the waist (Lakomy, 1984). Repeated bursts of exercise, such as 6 s in duration may be performed and power profiles determined at different recovery periods.

An anaerobic capacity test of potential relevance to soccer players is that used by Medbø *et al*. (1988). It employs measurement of the maximum accumulated oxygen deficit (MAOD) based on the close relationship between the observed oxygen deficit during intense exercise and the anaerobic energy production. Odetoyinbo and Ramsbottom (1995)

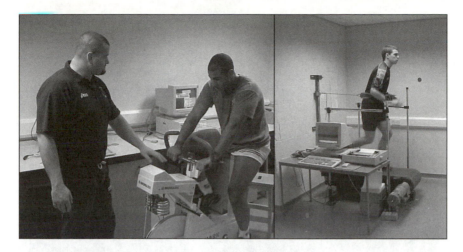

Figure 3.3 Laboratory equipment for measuring anaerobic power output in both soccer non-specific and specific modes.

have utilized a high intensity 20-m shuttle test to estimate MAOD in soccer players using this proposed relationship. The test requires participants to complete as many 20-m shuttles as possible at a speed pre-determined by the final velocity achieved on a previously performed 20-m shuttle test. The authors suggest the test to be both highly specific to soccer performance and sensitive to changes in aerobic and anaerobic (MAOD) metabolism with training. However, it is thought by some that the energy demand during exercise of higher intensities than the maximal aerobic power is underestimated and may not adequately represent anaerobic energy production (Bangsbo *et al.*, 1993).

Ability to perform high-intensity exercise in soccer training contexts is usually recorded by coaching staff from time trials over short distances. The systematic application of such tests in both laboratory and field settings is common place and has proved valuable as both a research and performance evaluation tool (Amigó *et al.*, 1998; Aziz *et al.*, 2000; Helgerud *et al.*, 2001). Repeated sprint tests over 5, 10 and 20 m are an integral part of the Australian Institute of Sport's protocols for soccer performance assessment (Tumilty, 2000). Where a systematic and careful approach is applied to such tests with adequate familiarization and preparation of the players, they are a valuable tool in assessing sprinting ability of soccer players.

3.2.3 *Muscle fibre types*

Muscle performance characteristics of soccer players in many respects are determined by their distribution of fibre types. Traditionally needle biopsy techniques have been the method of choice for muscle fibre typing (Figure 3.4), although Rico-Sanz *et al.* (1999a) have reported that it may be possible to differentiate non-invasively between muscle fibre types using magnetic resonance spectroscopy, based on the relative concentrations of muscle metabolites in different fibres. Muscle fibre types are categorized as fast twitch (FT) or slow twitch (ST) based on the speed of response when stimulated. An alternative classification is based on the histochemical characteristics of the different motor units. This classification distinguishes between slow oxidative (SO or type I), fast glycolytic (FG or type IIb), and fast

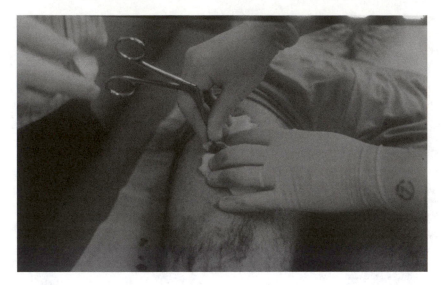

Figure 3.4 Muscle biopsy taken from the vastus lateralis: the technique shown here is the 'conchotome technique'.

oxidative glycolytic (FOG or type IIa). The functions of the different fibre types and their recruitment during exercise have been reviewed by Williams (1990).

Analysis of work rate of soccer players during match-play provides an overview that suggests that an ability to sustain physical effort over 90 min, albeit discontinuously at different exercise intensities varying from low to high intensity is important (Drust *et al.*, 1998; Rienzi *et al.*, 2000). Such intermittent activity profiles are compatible with the development of both ST and FT muscle fibre characteristics; a balanced combination of muscle fibre types would be expected in top players. The muscle fibres in the vastus lateralis of Swedish professional club players were found to be 59.8 (±s.d. 10.6)% FT. The percentage FT area was 65.6 (±10.6)% depicting a FT/ST mean fibre area of 1.28 (±0.22) (Jacobs *et al.*, 1982). These figures suggest that the fibre types of soccer players are closer to sprinters than to endurance athletes in make-up. However, a large variability was observed within the squad, the number of FT fibres ranging from 40.8 to 79.1%. It would be expected that the fibre types of the goalkeeper and central defenders, in which an anaerobic profile of physical performance dominates, would be biased towards FT fibres. Amigó *et al.* (1998) reported a predominance of FT fibres in some forward players. Examination of the oxidative enzymes of soccer players' gastrocnemius muscles provides a different perspective (Bangsbo and Mizuno, 1988). The relative occurrence of ST (type I), FTa (type IIa) and FTb (type IIb) fibres in four Danish professionals was 55.9 (range 48–63.6)%, 39.8 (33–46.5)% and 4.4 (3.0–5.5)%, respectively (Bangsbo and Mizuno, 1988). A reduction in muscle fibre area with 3 weeks of detraining was observed only in the FTa fibres and the decrease was small (7%). The number of capillaries around the fibres was also reduced with detraining but only in the ST fibres. At the time of full training mitochondrial activities of oxidative enzymes were similar to those noted for cross-country skiers in the case of 3-hydroacyl coenzyme A dehydrogenase (HAD). Values for citrate synthase were intermediate between middle-distance runners and cyclists (Bangsbo *et al.*, 1993). Similar detraining effects on muscle fibre area and enzymatic characteristics have been shown in trained adolescent soccer players (Amigó *et al.*, 1998).

Later studies from the same laboratory showed a fibre type distribution of 48.5 (34.6–61.0)% type I, 44.1 (32.0–65.4)% type IIa and 7.4 (0.3–18.7)% type IIb in the gastrocnemius of eight Danish professional club members (Bangsbo and Mizuno, 1988). A recent examination of fibre type distribution in the gastrocnemius of 15 elite soccer players, showed that type I fibres were predominant 60.2 (2.9)%, with type IIa and type IIb equivalent to 33.7 (3.1)% and 6.7 (1.5)%, respectively (Bangsbo *et al.*, 1993). The mean number of capillaries around each fibre type was approximately 4.67, 4.73 and 4.39 per fibre, respectively (Bangsbo and Lindquist, 1992) while a mean of 2.26 (0.09) per fibre has also been noted (Bangsbo *et al.*, 1993).

The biochemical properties of fibre types may be affected by the nature and intensity of training. Andersen *et al.* (1994) studied nine national level Danish players during a 3-month strength-training programme for the quadriceps muscles. A decrease in type IIa fibres from 35.4 to 26.7% was linked to a corresponding significant increase in type IIb fibres from 5.5 to 14.9%. An inverse correlation was noted between percentage type I fibres and the number of knee extensions performed over 50 s in an all-out test. The results demonstrated that well-trained soccer players could increase short-term muscle performance by strength training but that changes other than those in fibre type are responsible for the progress created by the training.

Eight Finnish second division players were examined by Smaros (1980). Biopsies taken from the vastus lateralis muscle showed an average fibre type distribution of 47% FT and 53% ST. Moreover, muscle biopsies taken at the end of a game showed that the reduction in glycogen stores occurred mainly in the ST fibres. Depletion of muscle glycogen is well noted after simulated and actual match play (Balsom *et al.*, 1999; Rico-Sanz *et al.*, 1999a,b). Japanese university players displayed fibre type percentages very close to those values found in the Finnish study (Ryushi *et al.*, 1979). Ryushi *et al.* reported no relation between percentage fibre area and isometric strength but the maximal power of the knee extensors per kilogram body mass was highly correlated ($r = 0.734$) with the percentage fibre area.

Amigó *et al.* (1998) reported that fibre type distribution alters with ageing. In a group of 14–16-year-old adolescent players the proportion of type I fibres decreased from 54, 53 and 51% while type II increased to 46, 47 and 49%, respectively across 14, 15 and 16-year-olds, respectively. While such alterations may be the result of biological ageing, it may also reflect adherence to habitual soccer-specific training programmes. Kuzon *et al.* (1990) compared the skeletal muscle characteristics of the vastus lateralis of 11 top Canadian junior (mean age 18.1 ± 1.3 years) players to those of age-matched controls. Percentage values for the controls were $51.4 \pm 12.5\%$ type I (ST), $29.5 \pm 7.3\%$ type IIa (FTa) and $19.1 \pm 13.1\%$ type IIb (FTb), similar figures being reported for the soccer players. The footballers had the greater capillary supply, characterized by a significantly greater number of capillaries around each muscle fibre (5.7 ± 0.9 versus 4.9 ± 0.4), a significantly larger capillary density (282.7 ± 42 versus 220 ± 38.1 mm^{-1}) and a significantly higher capillary to fibre ratio (2.2 ± 0.6 versus 1.7 ± 0.1). The findings were taken to indicate that soccer might induce simultaneous adaptations in both muscle fibre types.

Any inferences about fitness levels, muscle fibre types and elite soccer play must be tentative. Fibre type distributions could reflect positional role and so a team may be relatively heterogeneous in muscle composition. Parente *et al.* (1992) showed that there was a higher percentage of type I fibres (67%) among the midfield players than in defenders (44%) and forwards (38%) that they studied. The defenders had more type IIb fibres (49%) than the

Table 3.2 Muscle fibre composition of soccer players from various published sources[a]

Players	n	Muscle	Fibre type		$\dot{V}O_2 max$ (ml kg^{-1} min^{-1})	References
			FT	ST		
Malmo FC	19	Vastus lateralis	59.8 ± 10.6	40.2		Jacobs et al. (1982)
Danish professionals	4	Gastrocnemius (IIa, IIb)	44	55.9	66.2	Bangsbo and Mizuno (1988)
Danish professionals	8	Gastrocnemius (IIa, IIb)	51.5	48.5	60.4 ± 3.1	Bangsbo and Lindquist (1992)
Finnish Second Division	8	Vastus lateralis (IIa, IIb)	53	47 ± 13.3	63.6 ± 6.6	Smaros (1980)
Spanish and Italian semi-professionals	12	Quadriceps (undefined)	61.2 (type IIa, IIb)	38.8 (type I, IIc)	—	Montanari et al. (1990)
Japanese university players	12	Vastus lateralis	55.4 (type IIa, IIb)	44.6	—	Ryushi et al. (1979)
Canadian national junior team	11	Vastus lateralis	47.1 (type IIa, IIb)	52.9 ± 18.8	—	Kuzon et al. (1990)
Italian players (unspecified)		Quadriceps (undefined)				Parente et al. (1992)
Defenders	10		56	44		
Midfield	10		33	67		
Forwards	10		62 (type IIa, IIb)	38 (type I)		
Spanish Junior Amateur		Vastus lateralis				Amigó et al. (1998)
14 years	14	Vastus lateralis	54	46	—	
15 years	16	Vastus lateralis	53	47	—	
16 years	7	Vastus lateralis	51	49	—	

Note

a Values of $\dot{V}O_2$max (±s.d.) are also included where they were reported or could be calculated.

forwards (40%) or the midfield players (17%) These characteristics correspond to the demands imposed by the playing role. Fibre type distributions also vary between skeletal muscles. The gastrocnemius has an important function in locomotion whereas the quadriceps comprises the important muscle group in powerful kicking. The observations on soccer players may therefore reflect a relatively higher FT proportion in vastus lateralis than in the gastrocnemius compared to the ratio noted in other athletic populations. Table 3.2 summarizes a sample of the fibre type characteristics presented in the literature.

3.3   Aerobic fitness

The aerobic system is the main source of energy provision during soccer match play (Bangsbo, 1994a). This fact is indicated both by measurement of physiological responses during games and by the metabolic characteristics of soccer players' muscles. The maximum oxygen uptake or $\dot{V}O_2$ max indicates the upper limit of the body's ability to consume oxygen. The $\dot{V}O_2$ max represents an integrated physiological function with contributions from lung, heart, blood and active muscles. Pulmonary function is not normally a limiting factor in maximal aerobic performance and the main use of single-breath spirometry lies in screening for any impairment or lung restruction. The oxygen transport system is influenced by the O_2-carrying capacity of the blood. Along with the maximal cardiac output, this determines the amount of oxygen delivered to the active muscle cells. The amount of oxygen delivered is important in soccer because of the large contribution the aerobic system provides towards energy production (Bangsbo, 1994b). The oxygen-carrying capacity is determined by the concentration of haemoglobin in the blood, which affects the binding of O_2 in red blood cells, or the blood volume. Thus, total body haemoglobin is highly correlated with the maximal oxygen uptake. Blood volume and total body haemoglobin tend to be about 20% higher in endurance trained athletes than in non-athletes. Haemoglobin concentration and haematocrit (the percentage of blood volume occupied by red blood cells) of soccer players generally lie within the normal range. Blood tests tend to have most value in screening for anaemia or deficient iron stores in cases of players whose exercise performances fall below expectations.

The amount of blood delivered to the active muscles during strenuous exercise depends on the cardiac output. This measure is an indication of the stroke volume and maximal heart rate. The maximal heart rate is not increased as a result of training and is not itself an indicator of fitness. The heart responds to strenuous training by becoming larger and more effective as a pump. The chambers (particularly the left ventricle) increase in volume from a repetitive overload stimulus such as endurance running whilst the walls of the heart thicken and may increase its force development capability as a result of a pressure stimulus. Soccer has been classified by Michel *et al.* (1994) as a highly dynamic activity with low static component and as such these adaptations are to be expected. Hypertrophy of cardiac muscle is reflected in a greater stroke volume and a larger left ventricular size, which enables more blood to fill the chamber before the heart contracts. Sözen *et al.* (2000) have examined cardiac adaptations in a cross-section of elite Turkish soccer players. As a result of participation in soccer, a higher degree of cardiac hypertrophy was apparent in the soccer players compared to an active control group. No identifiable positional effects were apparent, this homogeneity in adaptive response being attributed to the commonality of squad training practice. Positive adaptations in ventricular volume and strength of ventricle contractions are manifest in a lower heart rate at rest. This effect is apparent in observations on well-trained professional soccer players. Resting heart rates of 48 ($\pm$1) beats min^{-1} and 54–59 beats min^{-1}

have been reported for English League players and elite Turkish players (Sözen *et al.*, 2000). The slower than normal heart rate allows extended relaxation time during diastole for the pressure to drop below the normal level of about 80 mmHg. The pulse pressure, the difference between systolic and diastolic pressures, with a value of 50 mmHg for the English League players, was superior to the normal 40 mmHg (Reilly, 1979).

The $\dot{V}O_2$ max may be affected by pulmonary ventilation, pulmonary diffusion, the O_2-carrying capacity of the blood, the cardiac output and the arteriovenous difference in O_2 concentrations. It is measured in a progressive exercise test to voluntary exhaustion (Figure 3.5). A motorized treadmill provides the most appropriate mode of exercise for testing soccer players. Expired air is analyzed for its O_2 and CO_2 content and the minute ventilation ($\dot{V}E$) is also measured. The attainment of maximal oxygen uptake ($\dot{V}O_2$ max) is indicated by a plateau in $\dot{V}O_2$ near exhaustion, a rise in the respiratory exchange ratio above 1.1 ($\dot{V}CO_2/\dot{V}O_2$), the elevation of heart rate to its age-predicted maximum or the blood lactate concentration reflecting anaerobic metabolism. Although providing the gold standard, such laboratory tests are regularly supplemented in club settings with the use of the 20-m shuttle test (Ramsbottom *et al.*, 1988).

The average values of $\dot{V}O_2$ max for top-level soccer players tend to be high, supporting the belief that there is a large contribution from aerobic power to playing the game (Bangsbo, 1994a). Nevertheless, $\dot{V}O_2$ max values do not reach the same levels as in specialist endurance sports such as cross-country running and skiing, distance running or orienteering

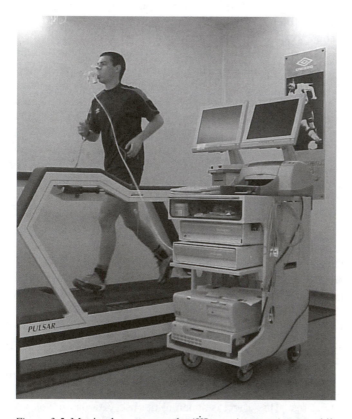

Figure 3.5 Maximal oxygen uptake ($\dot{V}O_2$ max) measurement whilst a player runs to exhaustion.

where values frequently exceed $80 \, \text{ml} \, \text{kg}^{-1} \, \text{min}^{-1}$. Values reported for elite players lie in the region 55–$70 \, \text{ml} \, \text{kg}^{-1} \, \text{min}^{-1}$, the higher figures tending to be found at the top level of soccer and when players are at peak fitness.

Whilst $\dot{V}O_2$ max values may be influenced by differences in standards of play and training regimens, the stage of the season should also he considered. The $\dot{V}O_2$ max of professional soccer players improves significantly in the pre-season period when there is an emphasis on aerobic training (Reilly, 1979; Bangsbo, 1994c; Helgerud *et al.*, 2001). However, not all recent reports demonstrate increased $\dot{V}O_2$ max with either pre-season training or progression through the season (Raastad *et al.*, 1997; Bury *et al.*, 1998; Wisløff *et al.*, 1998; Casajús, 2001). Explanation for the attenuation in $\dot{V}O_2$ max increases as the season progresses may lie in an extension of the playing season and number of games played, implementation of more systematic and appropriate training regimens within clubs and a reduced time available for detraining to occur in the off-season. Further emphasis on improving the $\dot{V}O_2$ max adds little to the quality of play and fails to improve the quality of ball passing although significant increases in work-rate are apparent (Helgerud *et al.*, 2001). When two teams of equal skill meet, the one with superior aerobic fitness would have the edge, being able to play the game at a faster pace throughout (Helgerud *et al.*, 2001). A high rank order correlation between mean $\dot{V}O_2$ max of Hungarian soccer teams and their finishing position in the Hungarian First Division Championship has been demonstrated (Apor, 1988). Mean $\dot{V}O_2$ max values for the first, second, third and fourth teams were 66.6, 64.3, 63.3 and $58.1 \, \text{ml} \, \text{kg}^{-1} \, \text{min}^{-1}$, respectively. Common factors such as stability in the team, avoidance of injury and other similar factors help to maintain both $\dot{V}O_2$ max and team performance independently. Reilly *et al.* (2000b) argued that while $\dot{V}O_2$ max alone does not predispose toward success in soccer, a minimum threshold of approximately $60 \, \text{ml} \, \text{kg}^{-1} \, \text{min}^{-1}$ is apparent where players falling below it may fail to perform with success.

The $\dot{V}O_2$ max varies with positional role, when such roles can be clearly differentiated. Sözen *et al.* (2000) suggested that irrespective of playing formation, squad-based training can obscure any specific positional adaptations in cardiovascular structure or aerobic capacity. When English League players were subdivided into positions according to 4–3–3 and 4–4–2 configurations, the midfielders had significantly higher aerobic power values than those in the other positions. Central defenders have significantly lower relative values than the other outfield players while the fullbacks and strikers have values that are intermediate (Reilly, 1979; Bangsbo, 1994c; Wisløff *et al.*, 1998; Al-Hazzaa *et al.*, 2001). Positional difference in $\dot{V}O_2$ max are most obvious when data are scaled relative to body mass. Expressing data as $\text{ml} \, \text{kg}^{-1} \, \text{min}^{-1}$ implies a linear response between oxygen uptake and body mass when none in reality exists. This means individuals with low body mass are overestimated and heavier players are underestimated (Wisløff *et al.*, 1998). When data are expressed relative to body mass raised to the power of body mass 0.75, such problems are avoided and a more meaningful expression of maximal aerobic capacity is evident (Wisløff *et al.*, 1998). Positional differences in $\dot{V}O_2$ max may be removed with this method (Wisløff *et al.*, 1998; Al-Hazzaa *et al.*, 2001). The significant correlation between $\dot{V}O_2$ max and distance covered in a game ($r = 0.67$) underlines the need for a high work-rate in midfield players who link between defense and attack. Goalkeepers have lower values than centre-backs, an observation confirmed by various researchers and reinforced by the highest values for adiposity among goalkeepers. Four goalkeepers in the German national team had values of $56.2 \, (\pm 1.2) \, \text{ml} \, \text{kg}^{-1} \, \text{min}^{-1}$ compared to $67.0 \, (\pm 4.5) \, \text{ml} \, \text{kg}^{-1} \, \text{min}^{-1}$ for the squad as a whole (Hollmann *et al.*, 1981). The $\dot{V}O_2$ max of 19 professional players in the First Division of the Portuguese League was $59.6 \, (\pm 7.7) \, \text{ml} \, \text{kg}^{-1} \, \text{min}^{-1}$; the average values for goalkeepers and central defenders were below, whilst midfield players and forwards were above $60\,\mathrm{ml\,kg^{-1}\,min^{-1}}$ (Puga *et al.*, 1993). Al-Hazzaa *et al.* (2001), reported $\dot{V}O_2$ max values for elite Saudi Arabian players of $56.8 \pm 4.8\,\mathrm{ml\,kg^{-1}\,min^{-1}}$ which are similar to Singaporean national squad's average of $58.2 \pm 3.7\,\mathrm{ml\,kg^{-1}\,min^{-1}}$ (Aziz *et al.*, 2000). These values are generally toward the lower end of $\dot{V}O_2$ max data reported in literature. At the other end of the spectrum Spanish First Division players' maximal oxygen uptake were reported as $66.4 \pm 7.6\,\mathrm{ml\,kg^{-1}\,min^{-1}}$ (Casajús, 2001), while elite Norwegian players demonstrated similar level of $\dot{V}O_2$ max at $67.68 \pm 4.0\,\mathrm{ml\,kg^{-1}\,min^{-1}}$ (Wisløff *et al.*, 1998). In contrast a sample of players drawn from Norwegian division 1–3 clubs indicated average $\dot{V}O_2$ max to be 62.8 ± 4.1 (Raastad *et al.*, 1997). English Premier League players had $\dot{V}O_2$ max estimated at $59.4 \pm 6.2\,\mathrm{ml\,kg^{-1}\,min^{-1}}$ (Strudwick *et al.*, 2002). This heterogeneity in $\dot{V}O_2$ max values across nationality and similar competitive levels suggests that, although important in maintaining work-rate, $\dot{V}O_2$ max may not discriminate between average and exceptional players (Reilly *et al.*, 2000b).

Whilst the $\dot{V}O_2$ max indicates the maximal ability to consume oxygen in strenuous exercise, it is not possible to sustain exercise for very long at an intensity that elicits $\dot{V}O_2$ max. The upper level at which exercise can be sustained for a prolonged period is thought to be indicated by the so-called 'anaerobic threshold': this is usually expressed as the work-rate corresponding to a blood lactate concentration of $4\,\mathrm{mmol\,l^{-1}}$, the onset of accumulation of lactate in the blood (OBLA) or as a deflection in the relation between ventilation and oxygen consumption with incremental exercise (the ventilatory threshold). The inflection point in blood lactate response to incremental exercise represented 83.9% $\dot{V}O_2$ max in top Finnish players studied by Rahkila and Luhtanen (1991). The $\dot{V}O_2$ max corresponding to a blood lactate concentration of $3\,\mathrm{mmol\,l^{-1}}$ was about 80% of $\dot{V}O_2$ max for both a continuous and an interval test on Danish players running on a treadmill (Bangsbo and Lindquist, 1992). This reference lactate level for the continuous test was significantly correlated with distance covered in a game. Elite Norwegian players had lactate thresholds estimated as occurring between 80 and 86% of $\dot{V}O_2$ max (Raastad *et al.*, 1997; Helgerud *et al.*, 2001). The ventilatory threshold has been measured at 77, 79 and 76% of $\dot{V}O_2$ max in English League First Division, Spanish La Liga players and Saudi Arabian national squad members, respectively (White *et al.*, 1988; Al-Hazzaa *et al.*, 2001; Casajús, 2001). Positional variation in lactate threshold (Tlac) is also evident. Bangsbo (1994a) showed in elite Danish players that midfield and full-backs had higher lactate thresholds than goalkeepers and central defenders. The intermittent nature of soccer means that frequently players operate at or above this intensity although the average fractional utilization of $\dot{V}O_2$ max is deemed to be 75–80% $\dot{V}O_2$ max (Reilly, 1990). The lactate and ventilatory thresholds also provide a sensitive sub-maximal indicator of physiological adaptation; as such it represents a means of quantifying a change in performance through a season (Al-Hazzaa *et al.*, 2001; Casajús, 2001).

3.4   Field tests

Soccer coaches are continually on the lookout for appropriate tests which allow them to assess fitness of players in field conditions. A convenient practical test for estimating $\dot{V}O_2$ max is the multi-stage shuttle run (Leger *et al.*, 1988) and its subsequent modifications (Ramsbottom *et al.*, 1988). The speed is dictated by a rhythm on a tape recorder, the individual response to each running intensity being monitored. This test is now accepted as a valid method of indirectly estimating the maximal oxygen uptake. It is suitable for testing of soccer squads as it has a reasonable fidelity to movements in the game. The ability to recover quickly from strenuous exercise may be important in soccer, which involves intermittent efforts interspersed with short rests. The Harvard Step Test designed initially for college men and later used in testing of military conscripts provides a fitness index which is based on the recovery of pulse rates over 3.5 min after a standard work-rate. The test has a history of use in the fitness assessment of soccer players (Reilly, 1990). It lost favour in recent years due to the availability of alternatives such as the 20-m shuttle run and the greater specificity associated with these other tests.

Bangsbo (1994c) described various running tests specifically designed for assessing soccer performance. They included a sprint test performed seven times over a short course of about 35 m with 25-s rest between sprints. The duration of each sprint is recorded and a fatigue index is obtained by comparing the fastest and slowest sprints. Blood lactate concentrations (between 9 and 14 mmol l^{-1}) attest to large anaerobic involvement in the activity. Aerobic mechanisms are stressed in an intermittent field test conducted on a space equal to the penalty area and over a running course incorporating ziz-zag, backwards, sideways and forwards running. The relevance of the intermittent endurance test was shown by the correlation between test performance and distance covered by the player in competitive matches.

A recent development in performance monitoring has been the establishment of protocols that simulate the soccer activity performance in laboratory and/or field settings. These various tests are generally derivatives of the shuttle running tests that already exist for assessing $\dot{V}O_2$ max. Rico-Sanz *et al.* (1999a) validated an intermittent shuttle test (JRS fatigue test) that consist of shuttling running at three different velocities which are designed to be representative of the work-rate occurring during match-play. The test is exhaustive and designed to evaluate muscle glycogen depletion during soccer play. The Loughborough Intermittent Shuttle Test (LIST) is another shuttle-based protocol designed to replicate the work-rate activity pattern noted during soccer. It comprises a two-part test with an initial fixed period of shuttle running at different intensities over 20 m, while the second part comprises continuos 20-m shuttle running alternating between 55 and 95% of $\dot{V}O_2$ max to exhaustion (Nicholas *et al.*, 2000).

Bangsbo (1994c) has developed two-field tests also based upon the shuttle running concept; the YO-YO intermittent endurance test and the YO-YO intermittent recovery test. In both tests, players perform 20-m shuttle runs with a short active recovery phase before recommencing the test at a higher running speed, the aim being to complete as many shuttles as possible. As such it assesses soccer specific endurance. A departure from the shuttle running of these tests is that presented by Drust *et al.* (2000) who used a treadmill-based intermittent protocol that alternates periods of high- and low-intensity activity to simulate the activity patterns noted during soccer performance. A variety of field and laboratory assessment methodologies exists to assess both physiological function and to simulate soccer-specific performance, their roles as modes of identifying young talent and monitoring senior players are the subject of current research (Reilly *et al.*, 2000a,b).

3.5   Agility and flexibility

The dynamic nature of soccer requires the possession of not only speed but also agility. Agility refers to the capability to change the direction of the body abruptly. The ability to turn quickly, dodge and side-step calls for good motor coordination and is reflected in a standardized agility run test. Dallas Tornado players were found to have average times on the Illinois Agility Run above the 99.95 percentile for the test norms (Raven *et al.*, 1976). The test distinguished the soccer players as a group from the normal population better than

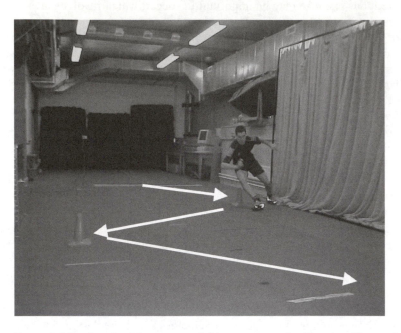

Figure 3.6 Dual setting laboratory or field assessment of agility via a 20-m sprint.

any field test used for strength, power and flexibility. This finding is understandable, since soccer players have to be capable of dodging and weaving past opponents. A 40-m sprint fatigue test that has an agility component has been incorporated into a battery of fitness assessment tasks for soccer although differentiation of the speed and agility components has not been assessed (Williams *et al.*, 1997). A 20-m soccer-specific agility test which requires participants to sprint a zig-zag course around four cones which deviated to the left by 4 m then to the right by 4 m and is repeated four times has been reported by Buttifant *et al.* (2002) (Figure 3.6). An agility test that incorporated dribbling a football was described by Reilly and Holmes (1983) for use with young players.

Joint flexibility is an important factor in soccer. Testing for limitation in the range of motion at a joint can be of benefit in screening for injury predisposition. Extensive techniques for determining joint range of motion were presented by Borms and Van Roy (1996). Factor analysis of a number of fitness tests on English games players showed that flexibility in a range of movements at the hip joint afforded protection against injury (Reilly and Stirling, 1993). Muscle tightness, particularly in the hamstring and adductor groups, has been linked with increased risk of muscle injury in Swedish professionals (Ekstrand, 1982). Graham-Smith and Lees (2002) investigated potential for hamstring injury during place kicking tasks in relation to muscle length and range of motion. They suggested that the dynamic action of place kicking places an additional stretch of 10% on the biceps femoris than can be applied during a maximal static stretch. Although this may not predispose a player to increased injury risk in itself, a combination of fatigue and imbalances in muscle strength as is evident in many players may sow the seed of future injury. Two-thirds of the players had flexibility values poorer than non-players. This may be an adaptation to soccer performance, but it could also reflect a lack of attention to flexibility practices in training. Limited range of motion has also been noted at the ankle joint in Japanese (Haltori and Ohta, 1986) and English League (Reilly, 1979) players, although the goalkeepers were exceptions among the English professionals. The Japanese players were less flexible than a reference group in inversion, eversion, plantarflexion and dorsiflexion. This may reflect an adaptive response of soft tissue around the ankle, which improves stability at the joint.

In games, the player has to move the entire body quickly, rather than one segment. Whole-body reaction time (WRT) of soccer players was studied by Togari and Takahashi (1977). No differences were found in simple WRT between regular and substitute players but the regular players had the faster choice WRTs. No differences were observed between any of the various playing positions, although goalkeepers were generally faster to react in choice WRT. Fast diving movements are particularly relevant to goal keeping skills (Suzuki *et al.*, 1988). This superiority in WRT is likely to be largely a product of training specific to that position.

Summary

Fitness for soccer cannot be determined by a single parameter since the game demands a large ensemble of physical, physiological and psychological capabilities. Fitness profiles are likely to vary within the playing season and the emphasis placed on different components of fitness will change with the stage of the season. Variables linked with fitness are influenced not just by training regimens but also by the stimulus provided by competing regularly. Successful play at top level in contemporary soccer depends on how individuals are knitted together into a competent unit and so the combination of physiological characteristics may vary from player to player. Nevertheless, it is possible to generalize on physiological characteristics of specialists in this sport. Anthropometric factors can determine the positional role most appropriate for the player. Body adiposity is highest among players pre-season but players at major international tournaments (with the exception of the goalkeeper) tend to have very little surplus fat. The physique of players generally shows muscular development, reflected in a high mesomorphy and low ectomorphy somatotype profile. Muscular strength, particularly in fast isokinetic movements, does seem to favour game-related performance. Anaerobic power of soccer players tends towards the profile of sprinters for the goalkeeper and central defenders whereas anaerobic capacity would seem to be more important in the other positions. Imbalanced muscular development (inappropriate flexor–extensor ratios, unilateral weakness, muscle tightness) may predispose towards injury. Leg muscle composition is not extreme, the fibre type distribution favouring fast movements but demonstrating histochemical properties of aerobically trained athletes. The moderately large heart volumes and aerobic power of players complement this aspect. Values for $\dot{V}O_2$ max above $60\,\mathrm{ml\,kg^{-1}\,min^{-1}}$ would seem desirable for outfield players. Sensory physiological mechanisms are also relevant considerations in the make-up of soccer players. It is likely that central factors in deciding the timing of game-related movements, supported by sufficiently well-developed muscular strength, motor coordination and oxygen transport mechanisms to implement the decisions, are the keys that open up opportunities for success in soccer play.

References

Aagaard, P., Simonsen, E.B., Trolle, M., Bangsbo, J. and Klausen, K. (1996). Specificity of training velocity and training loads on gains in isokinetic knee joint strength. *Acta Physiologica Scandanavica*, **156**, 123–9.

Al-Hazzaa, H.M., Alumuzaini, K.S., Al-Rafaee, A., Sulaiman, M.A., Dafterdar, M.Y., Al-Ghamedi, A. and Khuraiji, K.N. (2001). Aerobic and anaerobic power characteristics of Saudi elite soccer players. *Journal of Sports Medicine and Physical Fitness*, **41**, 54–61.

Almåsbakk, B. and Hoff, J. (1996). Co-ordination the determinant of velocity specificity? *Journal of Applied Physiology*, **80**, 2046–52.

Amigó, N., Cadefau, J.A., Ferrer, I., Tarrados, N. and Cussó, R. (1998). Effect of summer intermission on skeletal muscle of adolescent soccer players. *Journal of Sports Medicine and Physical Fitness*, **38**, 298–304.

Andersen, J.L., Bangsbo, J., Klitgaard, H. and Saltin, B. (1992). Changes in short-term performance and muscle fibre-type composition by strength training of elite soccer players. *Journal of Sports Sciences*, **10**, 62–3.

Andersen, J.L., Klitgaard, H., Bangsbo, J. and Saltin, B. (1994). Myosin heavy chain isoforms in single fibres from m.vastus lateralis of soccer players: effects of strength training. *Acta Physiologica Scandinavica*, **150**, 21–6.

Apor, P. (1988). Successful formulae for fitness training, in *Science and Football* (eds T. Reilly, A. Lees, K. Davids and W.J. Murphy), E & FN Spon, London, pp. 95–107.

Asami, T. and Togari, H. (1968). Studies on the kicking ability in soccer. *Research Journal of Physical Education*, **12**, 267–72.

Aziz, A.R., Chia, M. and Teh, K.C. (2000). The relationship between maximal oxygen uptake and repeated sprint performance indices in field hockey and soccer players. *Journal of Sports Medicine and Physical Fitness*, **40**, 195–200.

Balsom, P.D., Wood, K., Olsson, P. and Ekblom, B. (1999). Carbohydrate intake and multiple sprint sports: with special reference to football (soccer). *International Journal of Sports Medicine*, **20**, 48–52.

Bangsbo, J. (1994a). Energy demands in competitive soccer. *Journal of Sports Sciences*, **12**, S5–12.

Bangsbo, J. (1994b). The physiology of soccer—with special reference to intense intermittent exercise. *Acta Physiologica Scandinavica*, **619**(suppl), 1–155.

Bangsbo, J. (1994c). *Fitness Training in Football—A Scientific Approach*, HO & Storm, Bagsvaerd.

Bangsbo, I. and Lindquist, F. (1992). Comparison of various exercise tests with endurance performance during soccer in professional players. *International Journal of Sports Medicine*, **13**, 125–32.

Bangsbo, J. and Mizuno, M. (1988). Morphological and metabolic alterations in soccer players with detraining and retraining and their relation to performance, in *Science and Football* (eds T. Reilly, A. Lees, K. Davids and W.J. Murphy), E & FN Spon, London, pp. 114–24.

Bangsbo, J., Michalsik, L. and Petersen, A. (1993). Accumulated O_2 deficit during intense exercise and muscle characteristics of elite athletes. *International Journal of Sports Medicine*, **14**, 207–13.

Besier, T.F., Llyod, D.G., Cochrane, J.L. and Ackland, T.R. (2001). External loading of the knee joint during running and cutting maneuvers. *Medicine and Science in Sports and Exercise*, **33**, 1168–75.

Borms, J. and Van Roy, P. (1996). Flexibility, in *Kinanthropometry and Exercise Physiology Laboratory Manual*, (eds R. G. Eston and T. Reilly) E & FN Spon, London, pp. 115–44.

Bosco, C.P., Luhtanen, P. and Komi, P. (1983). A simple method for measurement of mechanical power in jumping. *European Journal of Applied Physiology*, **50**, 273–82.

Bury, T., Marechal, R., Mahieu, P. and Pirnay, F. (1998). Immunological status of competitive football players during the training season. *International Journal of Sports Medicine*, **19**, 364–68.

Buttifant, D., Graham, K. and Cross, K. (2002). Agility and speed in soccer players are two different performance parameters, in *Science and Football IV* (eds W. Spinks, T. Reilly and A. Murphy), Routledge, London, pp. 329–32.

Cabri, J., De Proft, F., Dufour, W. and Clarys, J.P. (1988). The relation between muscular strength and kick performance, in *Science and Football* (eds T. Reilly, A. Lees, K. Davids and W.J. Murphy), E & FN Spon, London, pp. 186–93.

Casajús, J.A. (2001). Seasonal variation in fitness variables in professional soccer players. *Journal of Sports Medicine and Physical Fitness*, **41**, 463–7.

Catrysse, E., Zinzen, E., Cuboor, D., Verlinden, M., Van Roy, P., Duquet, W. and Clarys, J.P. (1999). A revision of the anthropometric four compartment individual classification model for use in in vitro studies. *Journal of Sports Sciences*, **17**, 910.

Chin, M.K., Lo, Y.S.A., Li, C.T. and So, C.H. (1992). Physiological profiles of Hong Kong elite soccer players. *British Journal of Sports Medicine*, **26**, 262–6.

Cometti, G., Maffiuletti, N.A., Pousson, M., Chatard, J.C. and Maffulli, N. (2001). Isokinetic strength and anaerobic power of elite sub-elite and amateur French soccer players. *International Journal of Sports Medicine*, **22**, 45–51.

Craven, R., Bulter, M., Dickenson, L., Kinch, R. and Ramsbottom, R. (2002). Dietary analysis of a group of English first division players, in *Science and Football IV* (eds W. Spinks, T. Reilly and A. Murphy), Routledge, London, pp. 230–3.

De Proft, F., Cabri, J., Dufour, W. and Clarys, J.P. (1988). Strength training and kick performance in soccer players, in *Science and Football* (eds T. Reilly, A. Lees, K. Davids and W.J. Murphy), E & FN Spon, London, pp. 108–13.

Di-Salvo, V. and Pigozzi, F. (1998). Physical training of football players based on their positional rules (sic) in the team. *The Journal of Sports Medicine and Physical Fitness*, **38**, 294–7.

Drust, B., Reilly, T. and Rienzi, E. (1998). Analysis of work rate in soccer. *Sports Exercise and Injury*, **4**, 151–5.

Drust, B., Reilly, T. and Cable, N.T. (2000). Physiological responses to laboratory based soccer specific intermittent and continuous exercise. *Journal of Sports Sciences*, **18**, 885–92.

Dowson, M.N., Cronin, J.B. and Presland, J.D. (2002). Anthropometric and physiological differences between gender and age groups of New Zealand national soccer players, in *Science and Football IV* (eds W. Spinks, T. Reilly and A. Murphy), Routledge, London, pp. 63–71.

Dunbar and Power (1995). Fitness profiles of English professional and semi-professional soccer players using a battery of field tests. *Journal of Sports Sciences*, **13**, 501–2.

Dutta, P. and Subramanium, S. (2002). Effects of six weeks of isokinetic strength training combined with skills training on football kicking performance, in *Science and Football IV* (eds W. Spinks, T. Reilly and A. Murphy), Routledge, London, pp. 333–9.

Ekstrand, J. (1982). Soccer injuries and their prevention. Doctoral thesis, Linkoping University.

Eston, R.G. and Reilly, T. (eds) (1996). *Kinanthropometry and Exercise Physiology Laboratory Manual*, E & FN Spon, London.

Faina, M., Gallozzi, C., Lupo, S. *et al.* (1988). Definition of the physiological profile of the soccer player, in *Science and Football* (eds T. Reilly, A. Lees, K. Davids and W.J. Murphy), E & FN Spon, London, pp. 158–63.

Fleck, S.J. and Falkel, J.E. (1986). Value of resistance training for the reduction of sports injuries. *Sports Medicine*, **3**, 61–8.

Fowler, N. and Reilly, T. (1993). Assessment of muscle strength asymmetry in soccer players, in *Contemporary Ergonomics* (ed. E.J. Lovesey), Taylor and Francis, London, pp. 327–32.

Fried, T. and Lloyd, G.J. (1992). An overview of common soccer injuries. Management and prevention. *Sports Medicine*, **14**, 269–75.

Graham-Smith, P. and Lees, A. (2002). Risk assessment of hamstring injury in rugby union place kicking, in *Science and Football IV* (eds W. Spinks, T. Reilly and A. Murphy), Routledge, London, pp. 183–189.

Helgerud, J., Engen, L.C., Wisløff, U. and Hoff, J. (2001). Aerobic training improves soccer performance. *Medicine and Science in Sports and Exercise*, **33**, 1925–31.

Haltori, K. and Ohta, S. (1986). Ankle joint flexibility in college soccer players. *Journal of Human Ergology*, **15**, 85–9.

Hollmann, W., Liesen, H., Mader, A. *et al.* (1981). Zur Hochstund Dauer leistungsfahigkeit der deutschen Fussball-Spitzenspieler. *Deutsch Zeitschrift für Sportmedizin*, **32**, 113–20.

Jacobs, I., Westlin, N., Rasmusson, M. and Houghton, B. (1982). Muscle glycogen and diet in elite soccer players. *European Journal of Applied Physiology*, **48**, 297–302.

Kuzon, W.M. Jr, Rosenblatt, J.D., Huebel, S.C., Leatt, P., Plyley, M.J., McKee, N. and Jacobs, I. (1990). Skeletal muscle fibre type, fibre size and capillary supply in elite soccer players. *International Journal of Sports Medicine*, **11**, 99–102.

Lakomy, H. (1984). An ergometer for measuring the power generated during sprinting. *Journal of Physiology*, **354**, 33P.

Lees, A. and Nolan. L. (2002). Three-dimensional kinematic analysis of the instep kick under speed and accuracy conditions, in *Science and Football IV* (eds W. Spinks, T. Reilly and A. Murphy), Routledge, London, pp. 16–21.

Leger, L.A., Mercier, D., Gadoury, C. and Lambert, J. (1988). The multistage 20 metre shuttle run test for aerobic fitness. *Journal of Sports Sciences*, **6**, 93–101.

Margaria, R., Aghemo, P. and Rovelli, E. (1966). Measurement of muscular power (anaerobic) in man. *Journal of Applied Physiology*, **21**, 1661–4.

Martin, A.D., Spenst, L.F., Drinkwater, D.T. and Clarys, J.P. (1990). Anthropometric estimates of muscle mass in men. *Medicine and Science in Sports and Exercise*, **22**, 729–33.

Medbø, J. and Tabata, I. (1986). Relative importance of aerobic and anaerobic energy release during short-lasting exhausting bicycle exercise. *Journal of Applied Physiology*, **67**, 1881–6.

Medbø, J., Mohn, A., Tabata, I., Bahr, R. and Sejersted, G. (1988). Anaerobic capacity determined by the maximal accumulated oxygen deficit. *Journal of Applied Physiology*, **64**, 50–60.

Mercer, T.H., Gleeson, N. and Mitchell, J. (1995). Fitness profiles of professional soccer players before and after pre-season conditioning, in *Science and Football III* (eds T. Reilly, J. Bangsbo and M. Hughes), E & FN Spon, London, pp. 112–17.

Michel, J.H., Haskell, W.L. and Raven, P. (1994). Classification of sports. *Journal of the American College of Cardiology*, **24**, 864–6.

Montanari, G., Vecchiet, L. and Recoy-Campo, G.L. (1990). Structural adaptations to the muscle of soccer players, in *Sports Medicine Applied to Football* (ed. G. Santilli), CONI, Rome, pp. 169–78.

Morgan, B.E. and Oberlander, M.A. (2001). An examination of injuries in major league soccer. *The American Journal of Sports Medicine*, **29**, 426–30.

Mujika, I., Padilla, S., Ibanez, J., Izquierdo, M. and Gorostiaga, E. (2000). Creatine supplementation and sprint performance in soccer players. *Medicine and Science in Sports and Exercise*, **32**, 518–25.

Nicholas, C.W., Nuttall, F.E. and Williams, C. (2000). The Loughborough intermittent shuttle test: a field test that simulates the activity pattern of soccer. *Journal of Sports Sciences*, **18**, 97–104.

Norton, K., Marfell-Jones, M., Whittingham, N., Kerr, D., Carter, L., Saddington, K. and Gore, C. (2000). Anthropometric assessment protocols, in *Physiological Tests for Elite Athletes*, Australian Sports Commission (ed. C. Gore), Human Kinetic Publishers, Champaign, IL, pp. 66–85.

Nunome, H., Ikegami, Y., Asai, T. and Sato, Y. (2002). Three dimensional kinetics analysis of in-side and instep soccer kicks, in *Science and Football IV* (eds W. Spinks, T. Reilly and A. Murphy), Routledge, London, pp. 27–31.

Oberg, B., Eskstrand, J., Moller, M. and Gillquist, J. (1984). Muscle strength and flexibility in different positions of soccer players. *International Journal of Sports Medicine*, **5**, 213–16.

Oberg, B., Moller, M., Gillquist, J. and Ekstrand, J. (1986). Isokinetic torque levels in soccer players. *International Journal of Sports Medicine*, **7**, 50–3.

Odetoyinbo, K. and Ramsbottom, R. (1995). Aerobic and anaerobic field testing of soccer players. *Journal of Sports Sciences*, **13**, 506.

Östenberg, A., Roos, E., Ekdahl, C. and Roos, H. (1998). Isokinetic extensor strength and functional performance in healthy female soccer players. *Scandinavian Journal of Medicine and Science in Sports*, **8**, 257–64.

Parente, C., Montagnari, S., De Nicola, A. and Tajana, O.F. (1992). Anthropometric and morphological characteristics of soccer players according to positional role. *Journal of Sports Sciences*, **10**, 155.

Puga, N., Ramos, L., Agostinho, J. *et al.* (1993). Physical profile of a First Division Portuguese professional football team, in *Science and Football II* (eds T. Reilly, J. Clarys and A. Stibbe), E & FN Spon, London, pp. 40–2.

Rahkila, P. and Luhtanen, P. (1991). Physical fitness profile of Finnish national soccer teams candidates. *Science and Football*, **5**, 30–3.

Ramsbottom, R., Brewer, J. and Williams, C. (1988). A progressive shuttle test run test to estimate maximal oxygen uptake. *British Journal of Sports Medicine*, **22**, 141–4.

Raastad, T., Høstmark, A.T. and Strømme, S.B. (1997). Omega-3 fatty acid supplementation does not improve maximal aerobic power, anaerobic threshold and running performance in well-trained soccer players. *Scandinavian Journal of Medicine and Science in Sports*, **7**, 25–31.

Raven, P., Gcttman, L., Pollock, M. and Cooper, K. (1976). A physiological evaluation of professional soccer players. *British Journal of Sports Medicine*, **109**, 209–16.

Reilly, T. (1979). *What Research Tells the Coach about Soccer*, American Alliance for Health, Physical Education, Recreation and Dance, Washington, DC.

Reilly, T. (1990). Football, in *Physiology of Sports* (eds T. Reilly, N. Secher, P. Snell and C. Williams), E & FN Spon, London, pp. 371–425.

Reilly, T. (1994). Physiological profile of the player, in *Football* (*Soccer*) (ed B. Ekblom), Blackwell Scientific, London, pp. 78–94.

Reilly, T. and Drust, B. (1997). The isokinetic muscle strength of female soccer players and *Coaching and Sport Science Journal*, **2**, 12–17.

Reilly, T. and Holmes, M. (1983). A preliminary analysis of selected soccer skills. *Physical Education Review*, **6**, 64–71.

Reilly, T. and Stirling, A. (1993). Flexibility, warm-up and injuries in mature games players, in *Kinanthropometry IV* (eds W. Duquet and J.A.P. Day), E & FN Spon, London, pp. 119–23.

Reilly, T., Williams, A.M., Nevill, A. and Franks, A. (2000a). A multidisciplinary approach to talent identification in soccer. *Journal of Sports Sciences*, **18**, 695–702.

Reilly, T., Bangsbo, J. and Franks, A. (2000b). Anthropometric and physiological predispositions for elite soccer. *Journal of Sports Sciences*, **18**, 669–83.

Rienzi, E., Drust, B., Reilly, T, Carter, J.E.L. and Martin, A (2000). Investigation of the anthropometric and work rate profiles of elite South American international soccer players. *Journal of Sports Medicine and Physical Fitness*, **40**, 162–9.

Rico-Sanz, J., Frontera, W.R., Molé, P.A., Rivera, M.A., Rivera-Brown, A. and Meredith, C.N. (1998). Dietary and performance assessment of elite soccer players during a period of intense training. *International Journal of Sports Nutrition*, **8**, 230–40.

Rico-Sanz, J., Zehnder, M., Buchli, R., Kuhne, G. and Boutellier, U. (1999a). Non-invasive measurement of muscle high-energy phosphates and glycogen concentrations in elite soccer players by ^{31}P- and ^{13}C-MRS. *Medicine and Science in Sports and Exercise*, **31**, 1580–6.

Rico-Sanz, J., Zehnder, M., Buchli, R., Dambach, M. and Boutellier, U. (1999b). Muscle glycogen degradation during simulation of a fatiguing soccer match in elite soccer players examined noninvasively by ^{13}C-MRS. *Medicine and Science in Sports and Exercise*, **31**, 1587–93.

Ryushi, T., Asami, T. and Togari, H. (1979). The effect of muscle fibre composition on the maximal power and the maximal isometric strength of the leg extensor muscle. *Proceedings of the Department of Physical Education*, (College of General Education, University of Tokyo), **13**, 11–45.

Smaros, G. (1980). Energy usage during a football match, in *Proceedings of the 1st International Congress of Sports Medicine Applied to Football* (ed. L.Vecchiet), Vol II, D. Guanillo, Rome, pp. 795–801.

Sözen, A.B., Akkaya, V., Demirel, S., Kudat, H., Tükek, T., Ünal, M., Beyaz, M., Güven, Ö. and Korkut, F. (2000). Echocardiographic findings in professional league soccer players. *Journal of Sports Medicine and Physical Fitness*, **40**, 150–5.

Strudwick, A., Reilly, T., Doran, D. (2002). Anthropometric and fitness characteristics of elite players in two football codes. *Journal of Sports Medicine and Physical Fitness*, **42**, 239–42.

Suzuki, S., Togari, H., Lsokawa, M. *et al.* (1988). Analysis of the goalkeeper's diving motion, in *Science and Football* (eds T. Reilly, A. Lees, K. Davids and W.J. Murphy), E & FN Spon, London, pp. 468–75.

Tíryaki, G., Tuncel, F., Yamer, F., Agaoglu, S.A. and Gümüdad, H. (1995). Comparison of the physiological characteristics of the first, second and third league Turkish soccer players, in *Science and Football III* (eds T. Reilly, J. Bangsbo and M. Hughes), E & FN Spon, London, pp. 32–6.

Togari, H. and Asami, T. (1972). A study of throw-in training in soccer. *Proceedings of the Department of Physical Education* (College of General Education, University of Tokyo), **6**, 33–8.

Togari, H., Ohashi, J. and Ohgushi, T. (1988). Isokinetic muscle strength of soccer players, in *Science and Football* (eds T. Reilly, A. Lees, K. Davids and W.J. Murphy), E & FN Spon, London, pp. 181–5.

Togari, M. and Takahashi, K. (1977). Study of 'whole-body reaction' in soccer players. *Proceedings of the Department of Physical Education* (College of General Education, University of Tokyo), **11**, 35–41.

Trolle, M., Aagard, P., Simonsen, P., Bangsbo, J. and Klausen, K. (1993). Effects of strength training on kicking performance in soccer, in *Science and Football II* (eds T. Reilly, J. Clarys and A. Stibbe), E & FN Spon, London, pp. 95–7.

Tumilty, D. (2000). Protocols for the physiological assessment of male and female soccer players, in *Physiological Tests for Elite Athletes* (ed. C.J. Gore), Human Kinetics, Champaign IL, pp. 356–62.

White, J.E., Emery, T.M., Kane, J.E., Groves, R. and Risman, A.B. (1988). Pre-season fitness profiles of professional soccer players, in *Science and Football* (eds T. Reilly, A. Lees, K. Davids and W.J. Murphy), E & FN Spon, London, pp. 164–71.

Williams, C. (1990). Metabolic aspects of exercise, in *Physiology of Sports* (eds T. Reilly, N. Secher, P. Snell and C. Williams), E & FN Spon, London, pp. 3–67.

Williams, C., Reid, R.M. and Coutts, L. (1973). Observations on the aerobic power of university rugby players and professional soccer players. *British Journal of Sports Medicine*, **7**, 390–1.

Williams, A.M., Borrie, A., Cable, T., Gilbourne, D., Lees, A., MacLaren, D. and Reilly, T. (1997). *Umbro: Conditioning for Football*. TSL Publishing, London.

Wisløff, U., Helgerud, J. and Hoff, J. (1998). Strength and endurance of elite soccer players. *Medicine and Science in Sports and Exercise*, **30**, 462–7.

4 Physiology of training

Jens Bangsbo

Introduction

Soccer players need a high level of fitness to cope with the physical demands of a game (see Chapter 3) and to allow for their technical skills to be utilized throughout a match. Therefore, fitness training is an important part of the overall training programme. Common to all types of fitness training in soccer is that the exercise performed should resemble match-play as closely as possible. In this chapter the various categories of fitness training in soccer are described and specific examples of training drills are outlined.

4.1 Components of fitness training

Fitness training has to be multifactorial in order to cover the different aspects of physical performance in soccer. Thus, the training can be divided into a number of components based on the different types of physical demands during a match (see Figure 4.1). The terms aerobic and anaerobic training are based on the energy pathway that dominates during the activity periods of the training session. Aerobic and anaerobic training represent exercise intensities below and above the maximum oxygen uptake, respectively. However, during a training game, the exercise intensity for a player varies continuously, and some overlap exists between the two categories of training. Figure 4.2 shows examples of exercise intensities during games and drills within aerobic and anaerobic training.

The separate components within fitness training are briefly described in this chapter. They include aerobic, anaerobic and specific muscle training. For further discussion of the practical aspects of fitness training in soccer and for additional suggestions on activities which can be used in soccer, the reader is referred to Bangsbo (1994).

4.1.1 Aerobic training

Aerobic training causes changes in central factors such as the heart and blood volume, which result in a higher maximum oxygen uptake (Ekblom, 1969). A significant number of peripheral adaptations also occur with this type of training (Henriksson and Hickner, 1996). The training leads among other things to a proliferation of capillaries and an elevation of the content of mitochondrial enzymes, as well as the activity of lactate dehydrogenase 1–2 isozymes (LDH_{1-2}). Furthermore, the mitochondrial volume and the capacity of one of the shuttle systems for NADH are elevated (Schantz and Sjøberg, 1985). These changes cause marked alterations in muscle metabolism. The overall effects are an enhanced oxidation of lipids and sparing of glycogen, as well as a lowered lactate production, both at a given and at the same relative work-rate (Henriksson and Hickner, 1996).

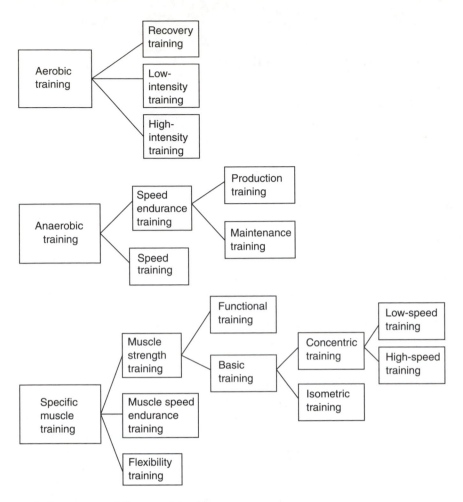

Figure 4.1 Components of fitness training in soccer.

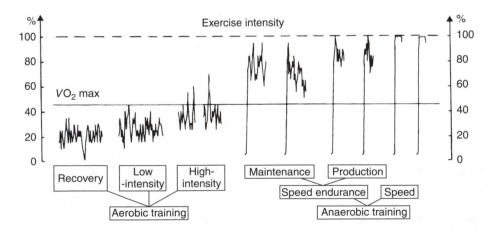

Figure 4.2 Examples of exercise intensities of a player during games within aerobic and anaerobic training, expressed in relation to maximal intensity (100%). The exercise intensity eliciting maximum oxygen uptake and maximal intensity of the player are represented by the lower and upper dotted horizontal lines, respectively.

The optimal way to train the central versus the peripheral factors is not the same. Maximum oxygen uptake is most effectively elevated by exercise intensities around 80–100% of $\dot{V}O_2$ max (20–40% of maximal exercise intensity; Figure 4.2). For muscle adaptation to occur, an extended period of training appears to be essential, and therefore, the mean intensity has once in a while to be below 80% of $\dot{V}O_2$ max. This does not imply that high-intensity training does not elevate the number of capillaries and mitochondrial volume in the muscles engaged in the training, but the duration of this type of training is often too short to obtain optimal adaptations at a local level.

The dissociation between changes in $\dot{V}O_2$ max and muscle adaptation by means of training and detraining is illustrated by results from a study on top-class players (Bangsbo and Mizuno, 1988). The players abstained from training for 3 weeks. It was found that $\dot{V}O_2$ max was unaltered, whereas performance in a field test was lowered by 8%, and there was a reduction in oxidative enzymes of 20–30% (Figure 4.3).

The recovery processes from intense exercise are related both to the oxidative potential and to the number of capillaries in the muscles (Tesch and Wright, 1983). Thus, aerobic training not only improves endurance performance of an athlete, but also appears to influence an athlete's ability to perform maximal efforts repeatedly. The overall aim of aerobic training is to increase the work-rate during competition, and to minimize a decrease in technical performance as well as lapses in concentration induced by fatigue towards the end of a game. The specific aims of aerobic training are as follows.

- To improve the capacity of the cardiovascular system to transport oxygen. Thus, a larger percentage of the energy required for intense exercise can be supplied aerobically, allowing a player to work at a higher exercise intensity for prolonged periods of time.
- To improve the capacity of muscles specifically used in soccer to utilize oxygen and to oxidize fat during prolonged periods of exercise. Thereby, the limited store of muscle glycogen is spared and a player can exercise at a higher intensity towards the end of a game.
- To improve the ability to recover after a period of high-intensity exercise. As a result, a player requires less time to recover before being able to perform in a subsequent period of high-intensity exercise.

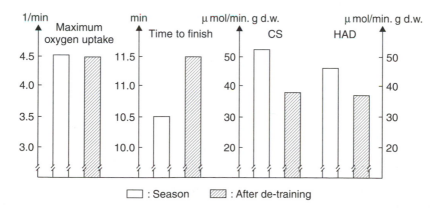

Figure 4.3 Maximum oxygen uptake, performance in a field test, activities of oxidative enzymes citrate synthase (CS) and b-hydroxy-CoA-dehydrogenase (HAD; involved in fat oxidation) of Danish top-class soccer players during the season and after 3 weeks of holiday.

Table 4.1 Principles of aerobic training

	Heart rate				Oxygen uptake	
	% of HR$_{max}$		*Beats min^{-1}*		*% of $\dot{V}O_2$ max*	
	Mean	*Range*	*Mean*[a]	*Range*[a]	*Mean*	*Range*
Low-intensity training	65	50–80	130	80–160	55	20–70
Moderate-intensity training	80	70–90	160	140–180	70	60–85
High-intensity training	90	80–100	180	160–200	85	70–100

Note
a If HR$_{max}$ is 200 beats min^{-1}.

4.1.1.1   Components of aerobic training

Aerobic training can be divided into three overlapping components: *aerobic low-intensity training* (Aerobic$_{LI}$), *aerobic moderate-intensity training* (Aerobic$_{MO}$) and *aerobic high-intensity training* (Aerobic$_{HI}$) (see Figures 4.1 and 4.2). Table 4.1 illustrates the principles behind the various categories of aerobic training, which take into account that the training may be performed as a game, and thus, the heart rate (HR) of a player will frequently alternate between categories during the training.

During Aerobic$_{LI}$ the players perform light physical activities, such as jogging and low-intensity games. This type of training may be carried out the day after a match or the day after a hard training session to help a player recover to a normal physical state. Aerobic$_{LI}$ may also be used to avoid the players getting into a condition known as 'overtraining' in periods involving frequent training sessions (maybe even twice a day) and a busy competitive schedule of matches.

The purpose of Aerobic$_{MO}$ training is to elevate the capillarization and the oxidative potential in the muscle (peripheral factors). The functional significance is an optimization of the substrate utilization and thereby an improvement in endurance capacity. The main aim of Aerobic$_{HI}$ training is to improve central factors such as the pump capacity of the heart, which is closely related to $\dot{V}O_2$ max. These improvements increase a player's capability to exercise repeatedly at high intensities for prolonged periods of time during a match.

4.1.2   Anaerobic training

During a match, a player frequently performs activities that require rapid development of force, such as sprinting or quickly changing direction. Furthermore, findings of high blood lactate concentrations in top-class players during match-play indicate that the lactate-producing energy system (glycolysis) is highly stimulated during periods of a game. Therefore, the capacity to perform high-intensity exercise repeatedly should be specifically trained. This can be achieved through anaerobic training.

Anaerobic training results in an increase in the activity of creatine kinase (CK) and glycolytic enzymes, such an increase implies that a certain change in an activator results in a higher rate of energy production of the anaerobic pathways. Intense training does not appear to influence the total creatine phosphate (CP) pool, but it leads to elevated muscle glycogen levels, which is of importance for performance during repeated high-intensity exercise (Reilly and Bangsbo, 1998). The capacity of the muscles to release and neutralize H^+ (buffer capacity) is also increased after a period of anaerobic training (Pilegaard *et al.*, 1999). This will lead to a lower reduction in pH for a similar amount of lactate produced during high-intensity exercise. Therefore, the inhibitory effects of H^+ within the muscle cell are smaller, which may be one of the reasons for a better performance in high-intensity tests after a period of anaerobic training. Another important effect of anaerobic training is an increased activity of the muscle Na^+/K^+ pumps resulting in a reduced net loss of potassium from the contracting muscles during exercise, which may also lead to increased performance (Bangsbo, 1997).

The overall aim of anaerobic training is to increase an athlete's potential to perform high-intensity exercise. The specific aims of anaerobic training are summarized below. These are:

- To improve the ability to act quickly and to produce power rapidly. Thus, a player reduces the time required to react and elevates performance of sprinting.
- To improve the capacity to produce power and energy continuously via the anaerobic energy-producing pathways. Thereby, a player elevates the ability to perform high-intensity exercise for longer periods of time.
- To improve the ability to recover after a period of high-intensity exercise, which is particularly important in soccer. As a result, a player requires less time before being able to perform maximally in a subsequent period of exercise, and the player will be able to perform high-intensity exercise more frequently during a match.

4.1.2.1 Components of anaerobic training

Anaerobic training can be divided into *speed training* and *speed endurance training* (see Figure 4.1). The aim of speed training is to improve a player's ability to act quickly in situations where speed is essential. Speed endurance training can be separated into two categories: *production training* and *tolerance training*. The purpose of production training is to improve the ability to perform maximally for a relatively short period of time, whereas the aim of tolerance training is to increase the ability to sustain exercise at a high intensity. Table 4.2 illustrates the principles of the various categories of anaerobic training.

Table 4.2 Principles of anaerobic training

		Duration		Intensity	Number of repetitions
		Exercise	Rest		
Speed training		2–10	>10 times the exercise duration	Maximal	2–10
Speed-endurance training	Production	15–40	>5 times the exercise duration	Almost maximal	2–10
	Maintenance (tolerance)	20–90	1–3-fold exercise duration	High	2–10

Anaerobic training must be performed according to an interval principle. During speed training the players should perform maximally for a short period of time (<10 s). The periods between the exercise bouts should be long enough for the muscles to recover to near resting conditions, so as to enable a player to perform maximally in a subsequent exercise bout. In soccer, speed is not merely dependent on physical factors. It also involves rapid decision-making, which must then be translated into quick movements. Therefore, speed training should mainly be performed with a ball. Speed drills can be designed to promote a player's ability to sense and predict situations, and the ability to decide on the opponents' responses in advance.

By speed endurance training, the creatine kinase and glycolytic pathways are highly stimulated. The exercise intensity should be almost maximal to elicit major adaptations in the enzymes associated with anaerobic metabolism. In *production training*, the duration of

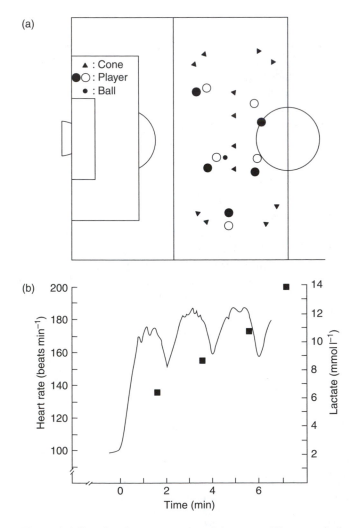

Figure 4.4 Speed endurance soccer training game. The game is described in the boxed text and the playing field is illustrated in (a). (b) Blood lactate (filled squares) and heart rate (line) of one player performing the game.

the exercise bouts should be relatively short (15–40 s), and the rest periods in between the exercise bouts should be comparatively long (2–4 min) in order to maintain a very high intensity during the exercise periods throughout an interval training session. In *tolerance training* the exercise periods should be 20–90 s and the duration of the rest periods should be 1–3-fold longer than the exercise periods, to allow the players to become progressively fatigued.

The adaptations caused by speed endurance training are mostly localized to the exercising muscles. Thus, it is important that a player performs movements in a manner similar to match-play. This can be obtained with high-intensity games or drills with a ball. Figure 4.4 illustrates a game within the tolerance category of speed endurance training. It also shows HR and blood lactate values for a player during the game illustrating that the game fulfil the criteria for speed endurance training.

Speed endurance training is both physically and mentally demanding for the players. Therefore, it is recommended that this type of training is only used by top-class players. When there is a limited time amount of time available for training, time can be better utilized on other forms of training.

4.1.3   Specific muscle training

Specific muscle training involves training of muscles in isolated movements. The aim of this type of training is to increase performance of a muscle to a higher level than can be attained just by playing soccer. Specific muscle training can be divided into *muscle strength, muscle speed endurance* and *flexibility* training (Figure 4.1). The effect of this form of training is specific to the muscle groups that are engaged, and the adaptation within the muscle is limited to the kind of training performed.

A brief description of muscle strength training is given below. Further information about strength training as well as an overview of muscle endurance and flexibility training for soccer players can be obtained from Bangsbo (1994).

Area of field: One-third of a soccer field

Number of players: Ten

Organization: A team consists of five players. The teams perform man-to-man marking. Six small 'goals' (pairs of cones) are placed at various positions on the field.

Description: Ordinary soccer play. The players must try to play the ball through the goals to a team-mate.

Rule: The players are not allowed to run through the goals.

Intervals: Exercise periods of 1 min and rest periods of 1 min.

Scoring: A point is scored for passing the ball through one of the goals to a team-mate.

Comments: The players should be motivated to exercise continuously at a high intensity. If one player cannot cope with the marking of an opponent, the intensity of the other players can be affected. It is therefore important to have players of equal ability marking each other. In order to avoid delays extra balls should be placed around the field. The exercise demands can be controlled by changing the number or the width of the goals.

4.1.3.1   Strength training

Many activities in soccer are forceful and explosive, for example, jumping, kicking, tackling and turning. The power output during such activities is related to the strength of the muscles involved in the movements. Thus, it is beneficial for a player to have a high level of muscular strength, which can be obtained by strength training.

Strength training can result in hypertrophy of the muscle, partly through an enlargement of muscle fibres. In addition, training with high resistance can change the fibre type distribution in the direction of fast twitch fibres (Andersen *et al.*, 1994). There is also a neuromotor effect of strength training and part of the increase in muscle strength can be attributed to changes in the nervous system. Improvements in muscular strength during isolated movements seem closely related to training speeds. However, significant increases in force development at very high speeds (10–$18\,\mathrm{rad\,s}^{-1}$) have also been observed with slow-speed high resistance training (Aagaard *et al.*, 1994).

One essential function of the muscles is to protect and stabilize joints of the skeletal system. Hence, strength training is of importance also in preventing injuries as well as re-occurrence of injuries. A prolonged period of inactivity, for example, during recovery from an injury, will considerably weaken the muscle. Thus, before a player returns to training after an injury, a period of strength training is needed. The length of time required to regain strength depends on the duration of the inactivity period but generally several months are needed. For a group of soccer players observed 2 years after a knee operation, it was found that the average strength of the quadriceps muscle of the injured leg was only 75% of the strength in the other leg (Ekstrand, 1982).

The overall aim of muscle strength training is to develop an athlete's muscular make-up. The specific aims of muscle strength training are as follows:

- To increase muscle power output during explosive activities such as jumping and accelerating.
- To prevent injuries.
- To regain strength after an injury.

4.1.3.2   Components of strength training

Strength training can be divided into *functional strength training* and *basic strength training* (Figure 4.1). In functional strength training, movements related to the sport are used. The training can consist of activities in which typical movements are performed under conditions that are physically more stressful than normal. During basic strength training, muscle groups are trained in isolated movements. For this training, different types of conventional strength training machines and free weights can be used, but body weight may also be used as resistance. Strength training should be carried out in a manner that resembles activities and movements specific to the sport. Based on the separate muscle actions the basic strength training can be divided into isometric, concentric and eccentric muscle strength training (Figure 4.1). Several principles can be used in concentric strength training. Table 4.3 illustrates a principle which is based on determinations of five-repetition maximum (5-RM) and which allows for muscle groups to be trained at both slow and fast speeds.

Common to the two types of strength training is that the exercise should be performed with a maximum effort. After each repetition the player should rest a few seconds to allow for a higher force production in the subsequent muscle contraction. The number of repetitions in a set should not exceed 15. During each training session two to four sets should be

Table 4.3 Principles of muscle strength training

	Work-load	Number of repetitions	Rest between repetitions	Number of sets
Concentric				
Low-speed	5-RM	5	2–5	2–4
High-speed	50% of 5-RM	15	1–3	2–4
Isometric	85–100% of max maintained for 5–15 s	5–10	5–15	2–4

Note
RM: repetition maximum.

performed with each muscle group, and rest periods between sets should be longer that 5 min. During this time, the athletes can exercise with other muscle groups.

4.1.4   Training methods

Fitness training in soccer should be performed closely related to the activities in soccer, since this ensures that the specific muscle groups used in the sport are trained. Second, the players will develop technical and tactical skills under conditions similar to those encountered during a match. Third, this form of training usually provides greater motivation for the players compared to training not focused on the sport.

Individual physical demands must be considered when planning fitness training and a part of the fitness training may be performed on an individual basis. The training should be focused on improving both the player's strong and weak abilities. It is important to be aware of the fact that, due to hereditary differences, there will always be differences in the physical capacity of players, irrespective of training programmes.

4.2   Planning fitness training

The time course of adaptations in the various tissues should be taken into account when planning fitness training. A change in heart size is rather slow, and there is a need for training over a long period of time (years) to improve the pump capacity of the heart significantly. Blood volume changes more quickly than the heart size, but this adaptation is optimal first after a dimensional development of the cardiovascular system has occurred. The content of oxidative enzymes in a tissue and the degree of capillarization of skeletal muscle change more rapidly than the volume of a tissue, for example, the heart, but months of regular training are needed to obtain considerable increases in muscle capillaries and oxidative enzymes. On the other hand, a reduction in these parameters can occur with a time constant of weeks (see Figure 4.3). The changes in glycolytic enzymes are rapid and they can be markedly elevated within a month of appropriate training (Houston *et al.*, 1979; Bangsbo *et al.*, 1993).

When planning fitness training in soccer the year may be divided into a season and an off-season. Table 4.4 shows how the different types of fitness training can be structured through the two periods of the year. The higher the number given, the more important is the form of training. The scheme is based on an 8-month season and a 4-month off-season of which the last 8 weeks before the season are spent in the club. The scheme is only a guideline, since there are difference in duration of the season and off-season from country to country, and

Table 4.4 Priority of fitness training through a year

	Off-season		Pre-season		Season							
Aerobic training												
Moderate-intensity	3344	4445	5555	4433	4343	4343	4343	4343	4343	4343	4343	4343
High-intensity	2223	3234	4445	4555	5545	5545	5545	5545	5545	5545	5545	5444
Anaerobic training												
Speed endurance	1111	1111	2234	4555	4353	4353	4353	4353	4353	4353	4353	3453
Speed	1111	1111	2234	4555	5555	5555	5555	5555	5555	5555	5555	5554
Muscle strength training												
Basic	3334	5555	5543	2323	2323	2323	2323	2323	2323	2323	2323	2222
Functional	2222	3333	3344	4343	4343	4343	4343	4343	4343	4343	4343	4322
Muscle speed training	1111	1112	3333	3333	3333	3333	3333	3333	3333	3333	3333	3333
Flexibility training	3232	3434	4444	4444	4444	4444	4444	4444	4444	4444	4444	4444

Note
Each single number represents a week. For practical purposes each month is given 4 weeks. The values represent the following priorities: 1, very low priority; 2, low priority; 3, moderate priority; 4, high priority; 5, very high priority.

some countries have a mid-season break. It should also be emphasized that there may be deviation in the priority of the various aspects of fitness training due to specific demands of a team.

During the off-season, the players should regularly perform sessions with Aerobic$_{MO}$ training, since the oxidative enzyme capacity is rapidly lost with inactivity and it takes months to restore the enzyme levels. The training will reduce the decrease in fitness level, which always occurs on cessation of normal training and competition. A gradual transition between the various phases of the off-season also keeps the risk of injuries low and leaves time for other types of football training, such as tactical and technical training.

In the first part of the off-season, it is reasonable to emphasize basic strength training, since the adaptations from this type of training can be easily maintained. As the start of the season approaches, the amount of basic strength training should be reduced and more time should be allocated to functional strength training and playing soccer.

During the last 6 weeks or so of the off-season, the players should frequently perform sessions of Aerobic$_{HI}$ training, speed training and, for elite players, also speed endurance training. Such training should be supplemented by regular matches at a high competitive level.

During a season, Aerobic$_{HI}$ training should be given a high priority (see Table 4.4). Speed training and, for top-class players, speed endurance training should also be performed regularly. Endurance capacity may be maintained by frequently including prolonged training sessions with only short rest periods. The extent of strength training during the season should be determined by the total training time available.

Summary

With appropriate training, performance of a player during a match can be increased and the risk of injury can be reduced. In order to design an efficient training programme it is important to be aware of the different components of fitness training in soccer. Aerobic training increases the ability to exercise at an overall higher intensity during a match, and minimizes a decrease in technical performance induced by fatigue towards the end of a game. Anaerobic training elevates a player's potential to perform high-intensity exercise during a game. Muscle strength training, combined with technical training, improves a player's power output during explosive activities in a match.

Fitness training should mainly be performed with a ball. This ensures that the specific muscles used within soccer are trained. Equally important is that players should develop their technical skills under conditions similar to those encountered during competition.

References

Aagaard, P., Trolle, M., Simonsen, E.B., Klausen, K. and Bangsbo, J. (1994). Moment and power generation during maximal knee extension performed at low and high speed. *European Journal of Applied Physiology*, **69**, 376–81.

Andersen, J.L., Klitgaard, H., Bangsbo, J. and Saltin, B. (1994). Myosin heavy chain isoform in single fibres from m. vastus lateralis of soccer players: effects of strength-training. *Acta Physiologica Scandinavica*, **150**, 21–6.

Bangsbo, J. (1994). *Fitness Training in Football – A Scientific Approach*, HO & Storm, Bagsvaerd.

Bangsbo, J. (1997). Physiology of muscle fatigue during intense exercise, in *Clinical Pharmacology of Sport and Exercise* (eds T. Reilly and M. Orme), Elsevier, Amsterdam, pp. 123–31.

Bangsbo, J. and Mizuno, M. (1988). Morphological and metabolic alterations in soccer players with detraining and retraining and their relation to performance, in *Science and Football* (eds T. Reilly, A. Lees, K. Davids and W.J. Murphy), E & FN Spon, London, pp. 114–24.

Bangsbo, J., Petersen, A. and Michalsik, L. (1993). Accumulated O_2 deficit during intense exercise and muscle characteristics of elite athletes. *International Journal of Sports Medicine*, **14**, 207–13.

Ekblom, B. (1969). Effect of physical training on the oxygen transport system in man. *Acta Physiologica Scandinavica Supplement*, **328**, 5–45.

Ekstrand, J. (1982). Soccer injuries and their prevention. Thesis, Linköping University Medical Dissertation 130.

Henriksson, J. and Hickner, R.C. (1996). Skeletal muscle adaptation to endurance training, in *Intermittent High Intensity Exercise* (eds D.A.D. MacLeod, R.J. Maughan, C. Williams, C.R. Madeley, J.C.M. Sharp and R. W. Nutton), E & FN Spon, London, pp. 5–26.

Houston, M.E., Bentzen, H. and Larsen, H. (1979). Interrelationships between skeletal muscle adaptations and performance as studied by detraining and retraining. *Acta Physiologica Scandinavica*, **105**, 163–70.

Pilegaard, H., Domino, K., Noland, T., Juel, C., Hellsten, Y., Halestrap, A.P. and Bangsbo, J. (1999). Effect of high intensity exercise training on lactate/H^+ transport capacity in human skeletal muscle. *American Journal of Physiology*, **276**, E255–61.

Reilly, T. and Bangsbo, J. (1998). Anaerobic and aerobic training, in *Applied Sport Science: Training in Sport* (ed. B. Elliott), John Wiley, Chichester, pp. 351–409.

Schantz, P. and Sjøberg, B. (1985). Malate-aspartate and alpha-glycerophosphate shuttle enzyme levels in untrained and endurance trained human skeletal muscle. *Acta Physiologica Scandinavica* **123**, 12A.

Tesch, P.A. and Wright, J.E. (1983). Recovery from short-term intense exercise: its relation to capillary supply and blood lactate concentration. *European Journal of Applied Physiology*, **52**, 98–103.

5 Motion analysis and physiological demands

Thomas Reilly

Introduction

The physiological demands of soccer play are indicated by the exercise intensities at which the many different activities during match-play are performed. There are implications not only for fitness assessment and selection of players but also for their training regimes. Since the training and competitive schedules of players comprise their occupational roles, there are consequences too for their habitual activities, daily energy requirements and energy expenditures. Finally, there are repercussions for the prevention of injuries as far as possible.

The exercise intensity during competitive soccer can be indicated by the overall distance covered. This represents a global measure of work-rate which can be broken down into the discrete actions of an individual player for a whole game. The actions or activities can be classified according to type, intensity (or quality), duration (or distance) and frequency. The activity may be juxtaposed on a time-base so that the average exercise-to-rest ratios can be calculated. These ratios can then be used in physiological studies designed to represent the demands of soccer and also in conditioning elements of the soccer players' training programmes. These work-rate profiles can be complemented by monitoring physiological responses where possible.

Various aspects of the exercise intensities in soccer are examined in this chapter. Work-rate during play and factors influencing work-rate profiles are considered prior to a review of physiological responses to playing. These are restricted to heart-rate and metabolic measures. The compatibility between the demands of play, training stimuli and fitness measures is addressed.

5.1 Motion analysis

In the early applications of motion analysis to professional soccer, it was presumed that work-rate could be expressed as distance covered in a game, since this determines the energy expenditure. Activities were coded according to intensity of movements, the main categories being walking, jogging, cruising, sprinting, whilst other game-related activities such as backing, playing the ball and so on were investigated. The observer utilized a learnt map of pitch markings in conjunction with visual cues around the pitch boundaries and spoke into a tape recorder. The method of monitoring activity was checked for reliability, objectivity and validity (Reilly and Thomas, 1976), and is still considered to be the most appropriate way of monitoring one player per game (Reilly, 1990, 1993, 1994a, 1997; Reilly *et al.*, 2000).

Coded commentary of activities on to a tape recorder by a trained observer has been correlated with measurements taken from video recordings. The latter method entails establishing stride characteristics for each subject according to the various exercise intensities. The data can then be translated into distances or velocities of movements. With careful checks on procedure the two methods are in good agreement (Reilly, 1994a). Video recording has been employed in studies of referees during matches where heart-rate can be measured throughout the game (Catterall *et al.*, 1993). Stride frequencies can be counted on playback of the video; these can be expressed as distance for each discrete event, provided the stride length for each activity is determined separately for the individual concerned.

An alternative appropriate for data collection is to set the activity profile alongside a time-base. This permits establishment of exercise-to-rest ratios; these can be useful both in designing training drills and in interpreting physiological stresses. This approach is nowadays straightforward once video systems are linked with computerized methods of handling the observations.

A comprehensive review of methods of motion analysis in soccer has been published elsewhere (Reilly, 1994a). The various methods used have incorporated cine-film of samples of individuals to eventually encompass the whole team, overhead cine-views of the pitch for computer-linked analysis of movements, and synchronized cameras for calculation of activities using trigonometry. Hand-notation methods of recording activities on paper have also been used whilst computerized notation analysis, currently utilized for analyzing patterns of play, has great potential for producing work-rate information. More recently, the use of six synchronized cameras linked to a computer has enabled the collection of movement and behavioural information on all 22 players on the pitch. Whatever method is adopted must comply with quality control specifications.

A summary of overall work-rate reported in the literature (Table 5.1) indicates that outfield players should be able to cover 8–13 km during the course of a match. This is done more or less continuously. The overall distance covered during a game is only a crude measure of work-rate due to the frequent changes in activities. These amount to approaching 1000 different activities in a game, or a break in the level or type of activity once every 6 s (Reilly and Thomas, 1976). The changes embrace alterations in pace and direction of movement, execution of game skills and tracking opponents' movements.

Table 5.1 Mean distance covered per game according to various sources[a]

Source	n	Distance covered (m)	Method
Finnish	7	7100	TV cameras (2)
South American	18	8638	Videotape
English	40	8680	Tape-recorder
Japanese	2	9845	Trigonometry (2 cameras)
Swedish	10	9800	Hand notation
Japanese	—	9971	Trigonometry
English Premier	6	10 104	Videotape
Belgian	7	10 245	Cine-film
Danish	14	10 800	Video (24 cameras)
Swedish	9	10 900	Cine-film
Czech	1	11 500	Undisclosed
Australian	20	11 527	Videotape
Japanese	50	11 529	Trigonometry

Note
a Sources of the data are cited in Reilly (1994b) and Rienzi *et al.* (2000).

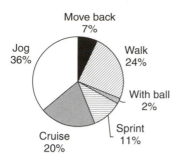

Figure 5.1 Relative distances covered in different categories of activity for outfield players during soccer match-play. (Data from Reilly and Thomas, 1976.)

The overall distance covered by outfield players during a match consists of 24% walking, 36% jogging, 20% cruising submaximally (striding), 11% sprinting, 7% moving backwards and 2% moving in possession of the ball. Masked withing the broad categories are sideways and diagonal movements. These figures (Figure 5.1) are fairly representative of contemporary play in the English top divisions and seemingly are indicative also of other major national leagues in Europe and at top level in Japan (Reilly, 1994b; Williams *et al.*, 1999).

The categories of cruising and sprinting can be combined to represent high-intensity activity in soccer. The ratio of low- to high-intensity exercise is then found to be about 2.2 to 1 in terms of distance covered (Reilly and Thomas, 1976). In terms of time, this ratio is about 7 to 1, denoting a predominantly aerobic outlay of energy. On average each outfield player has a short rest pause of only 3 s every 2 min, though rest breaks are longer and occur more frequently than this at lower levels of play where players are more reluctant to run to support a colleague in possession of the ball. Generally, less than 2% of the total distance covered by top players is with the ball. The vast majority of actions are 'off the ball', either in running to contest possession, support team-mates, track opposing players, execute decoy runs, counter-runs by marking a player, jump for the ball or tackle an opponent, or play the ball with one touch only.

Most activity during a game at top level is at a low or submaximal level of exertion but the importance of high-intensity efforts cannot be over-emphasized. Players generally have to run with effort (cruise) or sprint every 30 s but sprint all-out only once every 90 s. The timing of these anaerobic efforts, whether in possession of the ball or without, is crucial since the success of their deployment plays a dominant role in the outcome of the game. Although work-rate profiles are relatively consistent for players from game to game, it is the high-intensity exercise which is the most constant feature (Bangsbo, 1994).

The work-rate in women's soccer seems to be at the same relative intensity but, overall the average distance covered is less than in the men's game. Nevertheless some international female midfield players cover distances overlapping the work-rates of their male counterparts. Miyamura *et al.* (1997) analysed a series of matches played by national women's teams and concluded that, compared to men, the women took longer rest periods most notably before restarting play after the ball went out of play.

5.2 Factors affecting work-rates

The work-rate is determined to a large extent by the positional role of the player. The greatest distances are covered by midfield players who have to act as links between defence and

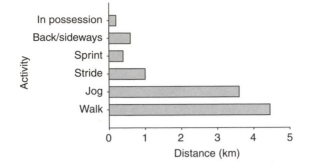

Figure 5.2 Mean distances covered (km) during a soccer match by an international striker playing in an English League match.

attack. This factor has been noted in English (Reilly and Thomas, 1976), Swedish (Ekblom, 1986) and Danish (Bangsbo *et al.*, 1991) League matches. In the studies of English League players, the full-backs showed the greatest versatility: although they covered more overall distance than the centre-backs, they covered less distance sprinting. The greatest distances covered sprinting were found in the strikers and midfield players. A characteristic work-rate profile of an English League striker who represented his country in the World Cup finals of 1990 and 1994 is shown in Figure 5.2.

The greatest overall distance covered by the Danish midfield players was due to more running at low speeds. This denotes an aerobic type of activity profile for the midfield players in particular. A more anaerobic type of profile is found in the centre-back and sweeper or libero. The pace of walking was found to be slower in centre-backs than for any other outfield position. Centre-backs and strikers have to jump more frequently than full-backs or midfield players (Reilly and Thomas, 1976). The frequency of once every 5–6 min denotes that while jump endurance may not be as important in soccer as in basketball and volleyball, anaerobic power output and the ability to jump well vertically are requirements for play in central defence and in attack as a 'target player'.

The goalkeeper has been observed to cover about 4 km during a match. Time spent standing still is much greater than for outfield players. The work-rate profile emphasizes anaerobic efforts of brief duration when the goalkeeper is involved directly in play. The goalkeeper is engaged in play more than any of the outfield players though the extent of this has been reduced by the rule-changes introduced in 1992 prohibiting back-passes from defenders. In a comparison of matches from the 1991–92 season with the 1997–98 season, goalkeepers were less frequently involved in the 'modern game'. This rule has only had a marginal effect on the activities of outfield players. Later measures to reduce the length of rest pauses and maintain the pace of play have increased the activity of outfield players and raised the tempo of the game (Williams *et al.*, 1999).

The ability to sustain prolonged exercise is dependent on a high maximal aerobic power ($\dot{V}O_2$ max), but the upper limit at which continuous exercise can be maintained is influenced by the so-called 'anaerobic threshold' and a high fractional utilization of $\dot{V}O_2$ max. Soccer play calls for an oxygen uptake corresponding roughly to 75% $\dot{V}O_2$ max (Reilly, 1990), a value likely to be close to the 'anaerobic threshold' of top soccer players. It has been shown that midfield players in the English League have higher $\dot{V}O_2$ max values than players in other outfield positions. The $\dot{V}O_2$ max was found to be correlated significantly with the distance covered in a game, underlining the need for a high work-rate and a high aerobic fitness level, particularly in midfield players (Reilly, 1993). Smaros (1980) confirmed this strong relation between $\dot{V}O_2$ max and distance covered in a game, but noted also that the $\dot{V}O_2$ max influenced the number of sprints the players attempted. Bangsbo and Lindquist (1992) showed that the distance covered was correlated with performance in a continuous field test over 2.16 km with the maximal oxygen uptake, and with the oxygen uptake corresponding to 3 mmol l^{-1} blood lactate level. It seems that work-rate in soccer matches depends on physiological indicators of aerobic fitness as found in distance runners (Jacobs, 1981).

The style of play may influence the work-rates of players. Emphasis on retaining possession, slowing the pace of the game and delaying attacking moves until opportunities to penetrate defensive line-ups are presented place emphasis on speed of movement in such critical phases of the game. Conversely the direct method of play, characteristic of some English clubs in the 1990s, contrast with the more methodical build-up of offensive plays in European and South American teams. The direct method (known also as 'Route 1'), used by the Republic of Ireland team in the 1988 European Championship and 1990 and 1994 World Cups, raises the pace of the game at all times. The main elements are faster transfer of the ball from defence to attack to create scoring opportunities, use of long passes rather than a sequence of short passes, exploitation of defensive errors, harrying opponents into mistakes when in possession of the ball and midfield players taking turns to support the strikers when on the offensive (Reilly *et al.*, 1991). This style of play has a levelling effect on the work-rate of outfield players since all players are expected to exercise at a high intensity 'off-the-ball'. A similar equalization of aerobic fitness demands applies to the 'total football' style of play as exhibited first by the Netherlands national side in 1974 and characteristic of many top European club sides today. The South American style is more rhythmic and the overall distance covered in a game is about 1.5 km less than in the English Premier League (Rienzi *et al.*, 2000).

Whilst the 'direct method' tends to even out differences in work-rate between playing positions, the Irish style of play evolved to accommodate individual differences. For example, team membership for the 1994 World Cup qualifying matches included some players known for exceptionally high work-rates and one player whose training programme was habitually hampered by chronic injury. This acknowledges that the team managers do consider building teams around highly talented players. The direct method was also employed by Norway's national team during the 1990s, although in this instance probably compensated for a lack of exceptionally gifted individuals in the team.

Computerized methods for notating movements during play have been developed for describing playing patterns (Hughes, 1988). Variables associated with each player, the position in the field of play and the player's action are entered into this computer by means of a 'concept' keyboard. This comprises a pad with 128 touch-sensitive cells over which a chart illustrating the pitch is superimposed. Information from viewing a video tape of a game is input by attributing codes to each activity on the ball. A comprehensive profile is of patterns of play and individual players' contributions to it is built up and illustrated by computer graphics at the end of each game. Comparison of England and Ireland international matches shows the superiority of the latter in fewer lost possessions per touch on the ball (particularly when in possession in defence), a feature of the 'direct method' of play (Reilly *et al.*, 1991). Such observations should be interpreted against the basic background of physiological stress imposed on players by the game strategies adopted.

5.3   Fatigue

Fatigue can be defined as a decline in performance due to the necessity to continue performing. In soccer it is manifest in a deterioration in work-rate towards the end of a game. Studies which have compared work-rate between first and second halves of matches have provided evidence of the occurrence of fatigue.

Belgian university players were found to cover on average a distance of 444 m more in the first half than in the second half (Van Gool *et al.*, 1988). Bangsbo *et al.* (1991) reported that the distance covered in the first half was 5% greater than in the second. This decrement does not necessarily occur in all players. Reilly and Thomas (1976) noted an inverse relation between aerobic fitness ($\dot{V}O_2$ max) and decrement in work-rate. The players with the higher $\dot{V}O_2$ max values, those in midfield and full-back positions, did not exhibit a significant drop in distance covered in the second half. In contrast, all the centre-backs and 86% of the strikers had higher figures for the first half, the difference between halves being significant. It does seem that the benefits of a high aerobic fitness level are especially evident in the later stages of a match.

The amount of glycogen stored in the thigh muscles pre-match appears to have an important protective function against fatigue. Swedish club players with low glycogen content in the vastus lateralis muscle were found to cover 25% less overall distance that the other players (Saltin, 1973). A more marked effect was noted for running speed; those with low muscle glycogen stores pre-match covered 50% of the total distance walking and 15% at top speed compared to 27% walking and 24% sprinting for players who started with high muscle glycogen concentrations. Attention to diet and maintaining muscle glycogen stores by not training too severely are recommended in the immediate build-up for competition. These considerations would be most important in deciders where drawn matches are extended into 30-min extra time.

Youth players of a professional club showed positive responses to consuming a maltodextrin solution during training. The subjective assessments of coaches of their players' performance corroborated the judgements of the players (Miles *et al.*, 1992). Dietary advice given to the senior professionals resulted in an alteration of nutritional support. The distribution of macronutrients in the diet of the players also improved (Reilly, 1994b). Manipulation of energy intake by provision of a high carbohydrate diet improved performance in a running test designed to interpret the activity profile of a soccer match (Bangsbo *et al.*, 1992). Whilst goals may be scored at any time during a game, most are scored towards the end of a game. This is exemplified by data from the Scottish League during an extended period of the 1991–92 season (Figure 5.3). A higher than average scoring rate occurred in the final 10 min of play. This cannot be explained simply by a fall in work-rate, as logically this would be balanced out between the two opposing teams. It might be accounted for by the more pronounced deterioration among defenders which gives an advantage to the attackers towards the end of a game. Alternatively it may be linked with 'mental fatigue', lapses in concentration as a consequence of sustained physical effort leading to tactical errors that open up goal-scoring chances. The phenomenon may be a factor inherent in the game, play becoming more urgent towards the end despite the fall in physical capabilities. Indeed, Rahnama *et al.* (2002) noted more critical incidents in contesting possession of the ball in the first 15 min and in the last 15 min of the game. Irrespective of the nature of the fatigue process, a team that is physiologically and tactically prepared to last 90 min of intense play is likely to be an effective unit.

Environmental conditions may also impose a limit on the exercise intensity that can be maintained for the duration of a soccer game or hasten the onset of fatigue during a match.

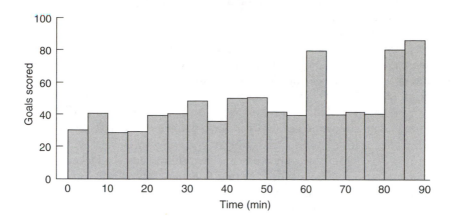

Figure 5.3 Time (min) at which goals were scored in a sample of 283 Scottish League soccer matches played in the 1991–92 season.

Major soccer tournaments, for example, the World Cup finals in Spain in 1982, in Italy in 1990 and in the USA in 1994, have been held in hot countries with ambient temperatures above 30°C. The work-rate is adversely affected when hot conditions are combined with high humidity. Performance is influenced both by the rise in core temperature and dehydration, and sweat production will be ineffective for losing heat when relative humidity is 100%. Cognitive function, akin to the kind of decision-making required during match-play, is better maintained during 90 min of continuous exercise when water is supplied intermittently to subjects compared to a control condition (Reilly and Lewis, 1985). Adequate hydration pre-exercise and during the intermissions is important when players have to play in the heat. The opportunity to acclimatize to heat prior to competing in tournaments in hot climates is an essential element in the systematic preparation for such events. This may be realized by astute location of training camps, a good physiological adaptation being realised within 10–14 days of the initial exposure in hot weather or regular and frequent exposures to heat in an environmental chamber.

The major consequences of playing in cold conditions are likely to be associated with liability to injury. This would be pronounced when playing on icy pitches without facilities for underground heating. Muscle performance deteriorates as muscle temperature falls; therefore a good warm-up prior to playing in cold weather and use of appropriate sportswear to maintain warmth and avoid the deterioration in performance synonymous with fatigue would be important. It is also established that injury is more likely to occur in players if their warm-up routine is inappropriate (Reilly and Stirling, 1993). Therefore, pre-match exercises should engage the muscle groups employed during the game, particularly in executing soccer skills.

The interactions between environmental variables and soccer performer are covered more extensively elsewhere in this volume. A consensus statement of nutritional needs of the soccer player and guidelines for fluid replacement to offset work-rate deterioration are outlined by Ekblom and Williams (1994).

5.4 Physiological responses to match-play

The relative metabolic loading during soccer play could be indicated if direct measurements were available for both energy expenditure during competition and the maximal aerobic power ($\dot{V}O_2$ max). Direct measurement made from collections of expired air in Douglas bags have indicated energy expenditure rates of 22–44 kJ min^{-1} (Covell *et al.*, 1965) and 32.3 kJ min^{-1} (Yamaoka, 1965). These values are likely to be underestimates due to the restrictions placed on the players by the apparatus and also to the low skills of the subjects used in these investigations. Higher values than these were reported by Seliger (1968a,b) for Czech players, mean figures being 54.8 kJ min^{-1} for energy expenditure and 76.1 min^{-1} for minute ventilation. The $\dot{V}O_2$ of 35.5 ml kg^{-1} min^{-1} is in close agreement with figures of 35–38 and 29–30 ml kg^{-1} min^{-1} for two Japanese players (Ogushi *et al.*, 1993). These approaches to data collection are likely to have hampered the activities of the players. A lightweight telemetric system for measuring oxygen concentrations, such as the K2 device (Kawakami *et al.*, 1992), is an improvement. The original devices lacked the facility for recording CO_2 but contemporary designs (e.g. Metamax, Borsdorf) can measure O_2 and CO_2. An alternative research strategy has been to measure heart-rate during match-play and juxtapose the observations on heart rate–$\dot{V}O_2$ regression lines determined during running on a treadmill. The error involved in using this method of estimating energy expenditure is small (Bangsbo, 1994). Allowing for any imperfections in such extrapolations from laboratory to field conditions, the heart rate is a useful indicator of the overall physiological strain during play.

Traditionally, long-range radio telemetry has been employed to monitor heart rate data during friendly matches or simulated competitions. In recent years, the use of short-range radio telemetry (Sport-Tester) has been adopted. Observations generally confirm that the circulatory strain during match-play is relatively high and does not fluctuate greatly during a game. The variability increases in the second half of play at university level (Figure 5.4), as the player takes more rest periods (Florida-James and Reilly, 1995). Rohde and Espersen (1988) reported that the heart rate was about 77% of the heart-rate range (maximal minus resting heart rate) for 66% of the playing time. For the larger part of the remaining time the heart rate was above this level.

The heart rate during soccer varies with the work-rate and so may differ between playing positions and between first and second halves. Van Gool *et al.* (1983) reported mean figures

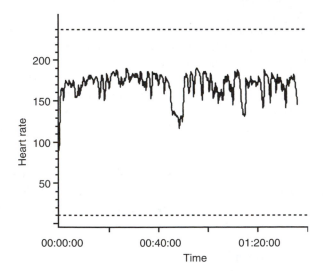

Figure 5.4 Mean heart rate (beats min^{-1}) during a whole game for a university player.

of 155 beats min^{-1} for a centre-back and for a full-back, 170 beats min^{-1} for a midfield player, and 168 and 171 beats min^{-1} for two forwards. This pattern was closely related to the distances covered by the players in a match. The same research group reported mean values for a Belgian university team during a friendly match of 169 beats min^{-1} in the first half and 165 beats min^{-1} in the second half. Again, the physiological responses reflected a drop in the work-rate during the second half. These trends have been confirmed in matches played by English university teams (Florida-James and Reilly, 1995).

The heart rate during soccer has been employed in several reports to estimate the relative metabolic loading during match-play (see Table 5.2). Most estimates are that exercise intensity during soccer is about 75–80% $\dot{V}O_2$ max (Reilly, 1990). Whilst the limitations of extrapolating from laboratory to field data, using heart rate–$\dot{V}O_2$ regression lines, suggest that this figure may represent an overestimate, comprehensive calculations indicate this error is not very large (Bangsbo, 1994).

The severity of exercise can also be indicated by measurements of blood lactate concentrations. Progressively higher lactates have been observed in matches from the fourth to the top division in the Swedish League (Ekblom, 1986). Gerisch *et al.* (1988) demonstrated that higher blood lactate concentrations are associated with man-to-man marking roles compared with a zone-coverage responsibility. Ekblom (1986) claimed that peak values above 12 mmol l^{-1} were frequently measured at the higher levels of soccer play. Activity could not be sustained continuously under such conditions which reflect the intermittent consequences of anaerobic metabolism during competition. Whilst most studies of blood lactate concentration have shown values of 4–6 mmol l^{-1} during play (see Table 5.3), such measures are determined by the activity in the 5 min prior to obtaining the blood samples. Consequently, higher values are generally noted when observations are made at half-time compared to the end of the match.

Table 5.2 Mean values for heart rate (beats min^{-1}) during soccer

Series	Heart rate	Match-play situation
Seliger (1968a)	160	Model 10-min game
Seliger (1968b)	165	Model 10-min match
Reilly (1986)	157	Training matches
Ogushi *et al.* (1993)	161	Friendly match (90 min)
Ali and Farrally (1991)	169	Friendly match (90 min)
Bangsbo (1994)	167	Competitive game (90 min)
Florida-James and Reilly (1995)	161	Competitive game (90 min)

Table 5.3 Mean ($\pm$ s.d.) blood lactate concentrations (mmol l^{-1}) during soccer

First half	Second half	Sources
4.9 ± 1.6	4.1 ± 1.3	Smaros (1980)
5.1 ± 1.6	3.9 ± 1.6	Rohde and Espersen (1988)
5.6 ± 2.0	4.7 ± 2.2	Gerish *et al.* (1988)
4.9	4.4	Bangsbo *et al.* (1991)
4.4 ± 1.2	4.5 ± 2.1	Florida-James and Reilly (1995)

Whilst muscle glycogen is the main fuel for exercise during match-play, muscle triglycerides, liver glycogen and blood-borne free fatty acids (FFA) are also used for oxidative metabolism. The increase in FFA is most pronounced in the last 15 min of a game (Bangsbo, 1994). This corresponds to an elevation in catecholamines which increase lipolysis in adipose tissue and raise the concentrations of fatty acids in the blood. The plasma concentrations of catecholamines are higher in the second half than in the first half of a game and their effects are enhanced by a rise in growth hormone. Without these contributions from fat as a source of fuel, the available glycogen would be depleted before the end of play. It is likely that there is some contribution to energy production from protein sources, particularly branched chain amino acids, but this is probably small (Wagenmakers *et al.*, 1989).

5.5   Game-related activities

The distance covered in a game underrepresents the energy expended because the extra demands of game skills are not accounted for. These include the frequent accelerations and decelerations, angular runs, changes of direction, jumps to contest possession, tackles, avoiding tackles and the many aspects of direct involvement in play. Some attempts have been made to quantify the additional physiological demands of game skills over and above the physiological cost of locomotion.

Dribbling the ball is an example of a game skill amenable to physiological investigation in a laboratory context. Reilly and Ball (1984) examined physiological responses to dribbling a soccer ball on a treadmill at speeds of 9, 10.5, 12 and 13.5 km h^{-1}, each for 5 min. A rebound box at the front of the treadmill returned the ball to the player's feet after each touch forward. The procedure allowed precise control over the player's activity while expired air, blood lactate level and perception of effort were measured. The energy cost of dribbling, which entailed one touch of the ball every two to three full stride cycles was found to increase linearly with the speed of running. The added cost of dribbling was constant at 5.2 kJ min^{-1} (Figure 5.5). This value is likely to vary in field conditions according to the closeness of ball control the player exerts.

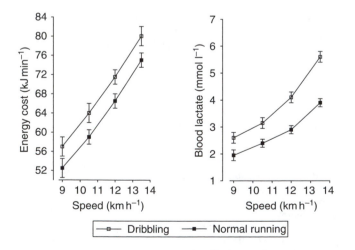

Figure 5.5 Physiological responses to running at different speeds are higher when dribbling a ball than in normal running. (Data from Reilly and Ball, 1984. Reproduced with permission from the American Alliance for Health, Physical Education, Recreation and Dance, Reston, VA 22091.)

When dribbling with tight control of the ball, the stride rate increases and the stride length shortens compared with normal running at the same speed; these changes are likely to contribute to the additional energy cost. Increasing or decreasing the stride length beyond that freely chosen by the individual causes the oxygen consumption for a given speed to increase. The energy cost may be further accentuated in matches as the player changes stride characteristics irregularly or feigns lateral movements whilst in possession of the ball in order to outwit an opponent. A reduction in stride length when dribbling is perhaps needed to effect controlled contact with the ball and propel it forward with the right amount of force by the swinging leg. The muscle activity required for kicking the ball and the action of synergistic and stabilizing muscles to facilitate balance while the kicking action is being executed are also likely to contribute to the added energy cost.

Perceived exertion rises while dribbling in parallel with the elevation in metabolism (Reilly and Ball, 1984). All-out efforts are likely to be limited by attaining the ceiling in perceived exertion and so top running speeds may not be attained in dribbling practices unless the frequency of ball contact is reduced. This in effect is done on occasions when a player kicks the ball ahead to allow himself space to accelerate past a stationary or more slowly moving opponent.

Blood lactate levels are also elevated as a consequence of dribbling the ball, the increased concentrations being disproportionate at the high speed (see Figure 5.5). In the study of Reilly and Ball (1984), the lactate inflection threshold was estimated to occur at $10.7\,\mathrm{km\,h^{-1}}$ for dribbling but not until $11.7\,\mathrm{km\,h^{-1}}$ in normal running. This finding indicates that the metabolic strain of fast dribbling will be underestimated unless the additional anaerobic loading is considered.

Kawakami *et al.* (1992) measured $\dot{V}O_2$ using a telemetric system (K2, Rome) which players wore as a back-pack whilst executing game-related drills. The highest $\dot{V}O_2$ was $4.0\,\mathrm{l\,min^{-1}}$ whilst dribbling, though it varied between 2.0 and $4.0\,\mathrm{l\,min^{-1}}$ for drills such as 1 versus 1 and 3 versus 1 practices.

There are other game-related activities that require unorthodox movements. About 16% of the distance covered by players in a game is in moving backwards or sideways. The percentage is highest in defenders who may, for example, have to back up quickly under high kicks forward from the opposition's half or move sideways in jockeying for position prior to tackling. The added physiological costs of unorthodox directions of movements have been examined by getting nine soccer players to run on a treadmill at speeds of 5, 7 and $9\,\mathrm{km\,h^{-1}}$,

Table 5.4 Mean (±s.d.) for energy expended (kJ min^{-1}) and ratings of exertion at three speeds and three directional modes of motion ($n = 9$)

Speed (km h)$^{-1}$	Forwards	Backwards	Sideways
Energy expended			
5	37.0 ± 2.6	44.8 ± 6.1	46.6 ± 3.2
7	42.3 ± 1.7	53.4 ± 3.5	56.3 ± 6.1
9	50.6 ± 4.9	71.4 ± 7.0	71.0 ± 7.5
Perceived exertion			
5	6.7 ± 0.1	8.6 ± 2.0	8.7 ± 2.0
7	8.0 ± 1.4	11.2 ± 2.9	11.3 ± 3.2
9	10.2 ± 2.1	14.0 ± 2.0	13.8 ± 2.5

Source: Reilly and Bowen (1984).

running normally, running backwards and running sideways (Reilly and Bowen, 1984). The extra energy cost of the unorthodox modes of running increased disproportionately with speed of movement. Running backwards and running sideways did not differ in terms of energy expenditure or ratings of perceived exertion (Table 5.4). Clearly improving the muscular efficiency in these unorthodox modes of movement would benefit the player.

Summary

The physiological responses of soccer players to match-play denote that a combination of demands is imposed on soccer players during competition. The critical phases of play for an individual call for anaerobic efforts but these are superimposed on a background of largely aerobic submaximal activities. The intermittent and acyclical nature of activity during competition means that it is difficult to model game-related protocols in laboratory experiments. It is likely that field studies with a greater specificity to the game will be employed more in future investigations of the physiology of soccer. The work-rate and activity profiles can be used to design appropriate training protocols to optimize fitness and ensure that performance during play is enhanced. Whilst physiological considerations have a place in a systematic preparation for competition, performance ultimately depends on the quality with which individual skill and team tactics are executed.

References

Ali, A. and Farrally, M. (1991). Recording soccer players' heart rates during matches. *Journal of Sports Sciences*, **9**, 183–9.

Bangsbo, J. (1994). The physiology of soccer: with special reference to intense intermittent exercise. *Acta Physiologica Scandinavica*, **150**(Suppl), 619.

Bangsbo, J. and Lindquist, F. (1992). Comparison of various exercise tests with endurance performance during soccer in professional players. *International Journal of Sports Medicine*, **13**, 125–32.

Bangsbo, J., Norregaard, L. and Thorso, F. (1991). Activity profile of professional soccer. *Canadian Journal of Sports Science*, **16**, 110–16.

Bangsbo, J., Norregaard, L. and Thorso, F. (1992). The effect of carbohydrate diet on intermittent exercise performance. *International Journal of Sports Medicine*, **14**, 207–13.

Catterall, C., Reilly, T., Atkinson, G. and Coldwells, A. (1993). Analysis of the work rates and heart rates of association football referees. *British Journal of Sports Medicine*, **27**,153–56.

Covell, B., El Din, N. and Passmore, R. (1965). Energy expenditure of young men during the weekend. *Lancet*, **1**, 727–8.

Ekblom, B. (1986). Applied physiology of soccer. *Sports Medicine*, **3**, 50–60.

Ekblom, B. and Williams, C. (1994). Nutrition and football. *Journal of Sports Sciences*, **12**(Suppl), 1.

Florida-James, G. and Reilly, T. (1995). The physiological demands of Gaelic football. *British Journal of Sports Medicine*, **29**, 41–5.

Gerisch, G., Rutemöller, E. and Weber, K. (1988). Sports medical measurements of performance in soccer, in *Science and Football* (eds T. Reilly, A. Lees, K. Davids and Y.W. Murphy), E & FN Spon, London, pp. 60–7.

Hughes, M. (1988). Computerised notation analysis in field games. *Ergonomics*, **31**, 1585–92.

Jacobs, I. (1981). Lactate, muscle glycogen and exercise performance in man. *Acta Physiologica Scandinavica*, Suppl. 495.

Kawakami, Y., Nozaki, D., Matsuo, A. and Fukunaga, T. (1992). Reliability of measurement of oxygen uptake by a portable telemetric system. *European Journal of Applied Physiology*, **65**, 409–14.

Miles, A., MacLaren, D. and Reilly, T. (1992). The efficacy of a new energy drink: a training study. Communication to the Olympic Scientific Congress, Benalmadena, Spain, 14–19 July.

Miyamura, S., Seto, S. and Kobayashi, H. (1997). A time analysis of men's and women's soccer, in *Science and Football III* (eds T. Reilly, J. Bangsbo and M. Hughes), E & FN Spon, London, pp. 251–57.

Ogushi, T., Ohashi, J., Nagahama, H. *et al.* (1993). Work intensity during soccer match-play (a case study), in *Science and Football II* (eds T. Reilly, J. Clarys and A. Stibbe), E & FN Spon, London, pp. 121–3.

Rahnama, N., Reilly, T. and Lees, A. (2002). Injury risk associated with playing actions during competitive soccer. *British Journal of Sports Medicine*, **36**, 354–9.

Reilly, T. (1986). Fundamental studies in soccer, in *Sportspielforschung: Diagnose Prognose* (eds H. Kasler and R. Andresen), Verlag Ingrid Czwalina, Hamburg, pp. 114–20.

Reilly, T. (1990). Football, in *Physiology of Sports* (eds T. Reilly, N. Secher, P. Snell and C. Williams), E & FN Spon, London, pp. 371–425.

Reilly, T. (1993). Science and football: an introduction, in *Science and Football II* (eds T. Reilly, J. Clarys, and A. Stibbe) E & FN Spon, London, pp. 3–11.

Reilly, T. (1994a). Motion characteristics, in *Football (Soccer)* (ed. B. Ekblom), Blackwell Scientific, London, pp. 31–42.

Reilly, T. (1994b). Physiological aspects of soccer. *Biology and Sport*, **11**, 3–20.

Reilly, T. (1997). Energetics of high intensity exercise (soccer) with particular reference to fatigue. *Journal of Sports Sciences*, **15**, 257–63.

Reilly, T. and Ball, D. (1984). The net physiological cost of dribbling a soccer ball. *Research Quarterly for Exercise and Sport*, **55**, 267–71.

Reilly, T. and Bowen, T. (1984). Exertional cost of changes in directional modes of running. *Perceptual and Motor Skills*, **58**, 49–50.

Reilly, T. and Lewis, W. (1985). Effects of carbohydrate feeding on mental functions during sustained exercise, in *Ergonomics International '85* (eds I.D. Brown, R. Goldsmith, K. Coombes and M.A. Sinclair), Taylor and Francis, London, pp. 700–2.

Reilly, T. and Stirling, A. (1993). Flexibility, warm-up and injuries in mature games players, in *Kinanthropometry IV* (eds W. Duquet and J.A.P. Day), E & FN Spon, London, pp. 119–23.

Reilly, T. and Thomas, V. (1976). A motion analysis of work-rate in different positional roles in professional football match-play. *Journal of Human Movement Studies*, **2**, 87–97.

Reilly, T., Bangsbo, J. and Franks, A. (2000). Anthropometric and physiological predispositions for elite soccer. *Journal of Sports Sciences*, **18**, 669–83.

Reilly, T., Hughes, M. and Yamanaka, K. (1991). Put them under pressure. *Science and Football*, **5**, 6–9.

Rienzi, E., Drust, B., Reilly, T., Carter, J.E.L. and Martin, A. (2000). Investigation of anthropometric and work-rate profiles of elite South American international players. *Journal of Sports Medicine and Physical Fitness*, **40**, 162–9.

Rohde, H.C. and Espersen, T. (1988). Work intensity during soccer training and match-play, in *Science and Football* (eds T. Reilly, A. Lees, K. Davids and W. Murphy), E & FN Spon, London, pp. 68–75.

Saltin, B. (1973). Metabolic fundamentals in exercise. *Medicine and Science in Sport*, **5**, 137–46.

Seliger, V. (1968a). Heart rate as an index of physical load in exercise. *Scripta Medica, Medical Faculty, Brno University*, **41**, 231–40.

Seliger, V. (1968b). Energy metabolism in selected physical exercises. *Internationale Zeitschrift für Angewandte Physiologie*, **25**, 104–20.

Smaros, G. (1980). Energy usage during football match, in *Proceedings of the 1st International Congress on Sports Medicine Applied to Football*, vol. 11 (ed. L. Vecehiet), D. Guanello, Rome, pp. 795–801.

Van Gool, D., Van Gerven, D. and Boutmans, J. (1983). Heart rate telemetry during a soccer game: a new methodology. *Journal of Sports Sciences*, **1**, 154.

Van Gool, D., Van Gerven, D. and Boutmans, J. (1988). The physiological load imposed on soccer players during real match-play, in *Science and Football* (eds T. Reilly, A. Lees, K. Davids and W. Murphy), E & FN Spon, London, pp. 51–9.

72 *Thomas Reilly*

Wagenmakers, A.J.M., Brookes, J.H., Conley, J.H. *et al*. (1989). Exercise-induced activation of the branched-chain 2-oxo acid dehydrogenase in human muscle. *European Journal of Applied Physiology*, **59**, 159–67.

Williams, A.M., Lee, D. and Reilly, T. (1999). *A Quantitative Analysis of Matches Played in the 1991–92 and 1997–98 Seasons*. The Football Association, London.

Yamaoka, S. (1965). Studies on energy metabolism in athletic sports. *Research Journal of Physical Education*, **9**, 28–40.

6 Nutrition

Don MacLaren

Introduction

Soccer may be considered an endurance sport, incorporating periods of intense exercise interspersed with lower levels of activity over 90 min. The estimated energy requirements for a soccer game embracing both casual recreational play and top-class professional games are between 21 and 73 kJ min^{-1} (5–17 kcal min^{-1}) (Reilly, 1990). For a 70-kg player, the result could be the loss of approximately 100–200 g of carbohydrate. Since the body's stores of carbohydrate are limited (approximately 300–400 g), this loss is significant. If muscle stores of carbohydrate are not adequately replenished, subsequent performance will be impaired. The carbohydrate intake of elite soccer players is often inadequate and so the concentrations of carbohydrate in active muscle may become low.

The energy demands of soccer are such that there is likely to be a significant production of heat within the body. Even in cold conditions, considerable amounts of sweat are lost in an attempt to dissipate this heat, thus resulting in a degree of dehydration (Maughan, 1991). A mild degree of dehydration will impair skilled performance and affect strength, stamina and speed. An adequate fluid intake is necessary to offset the effects of dehydration.

Despite the fact that the major causes of fatigue for soccer players are the depletion of muscle glycogen stores and dehydration, players are forever looking for nutritional supplements to help improve their performance and aid recovery. Vitamins and minerals are the legal products that players may consider, although substances such as creatine, sodium bicarbonate, caffeine and alcohol will be briefly discussed.

In this chapter the nutritional requirements of soccer players are considered in terms of energy, carbohydrates, fluid intake and vitamins. A brief examination is also undertaken of some ergogenic substances. Where possible, references will be made to research in soccer, and recommendations stated for use before, during and after training or competition.

6.1 Energy

6.1.1 Sources of muscular energy

Human energy provides the basis for movement in all sports, and any successful performance depends on the ability of the athlete to produce the right amount at the right time. Sports differ in their energy requirements, and thus each sport imposes specific energy demands on the athlete. Soccer entails multiple sprints, yet is an endurance sport. Consequently the ability to generate energy rapidly as for a sprint is as necessary as the ability to generate sustained energy over the 90 min of a match.

Adenosine triphosphate (ATP) is the only usable form of energy in muscle contractions, where the ATP is broken down to ADP by the action of enzymes (ATPases) at the cross-bridge heads of muscle filaments:

$$ATP \rightarrow ADP + Pi + Energy\ (30.5\,kJ\,mol^{-1})$$

Adenosine triphosphate is a high-energy compound which is used as an immediate source of energy for muscle activity. During intense activity such as in a sprint, the ATP stores will be used first. The ATP stores are in very limited supply and are capable of providing energy for only a few seconds. Another related high-energy source, creatine phosphate (CP), is also found in skeletal muscle in small amounts. Although it cannot be used as an immediate source of energy, it can rapidly replenish ATP:

$$CP \rightarrow Cr + Pi + Energy\ (43.1\,kJ\,mol^{-1})$$
$$CP + ADP \rightarrow ATP + Cr + Energy\ (12.6\,kJ\,mol^{-1})$$

Creatine phosphate and ATP are known as the phosphagen stores and are the rapid response energy systems for muscular activity. Together they are capable of supplying sufficient energy for about 8–10 s of high-intensity exercise. They are sometimes referred to as the ATP–CP system.

Because ATP and CP are found in very small amounts within skeletal muscle and can be used up in a matter of seconds, it is necessary to have alternative stores of energy. Carbohydrate, fat and protein comprise the other energy stores and can provide the body with enough ATP to last for many weeks of starvation. Table 6.1 summarizes the amount of energy stores in the body for a 70-kg person.

If the exercise is continued for more than a few seconds, a further energy system is brought into action. This system has been called the lactic acid system because lactic acid is produced as a consequence. Glycogen, from the muscle's store, is used rather than blood glucose because it is more readily available. The glycogen is broken down to release glucose-1-phosphate which is then converted to glucose-6-phosphate before being further broken down via glycolysis to produce pyruvic acid. Under normal steady-state conditions, the pyruvic acid is oxidized to carbon dioxide and water, producing quite a lot of energy in the process. Under conditions of intense exercise, where the rate of pyruvic acid production is greater than its oxidation, or conditions where oxygen is inadequate, the pyruvic acid is converted to lactic acid:

$$Glycogen \rightarrow G\text{-}1\text{-}P \rightarrow G\text{-}6\text{-}P \rightarrow \rightarrow \rightarrow \rightarrow Pyruvic\ acid$$
$$Pyruvic\ acid \rightarrow Lactic\ acid$$

Table 6.1 Energy stores for a 70-kg man (15% body fat)

Energy source	Amount (g)	Energy (kJ)
Fat (adipose tissue)	10 500	378 000
Muscle glycogen	350	5600
Liver glycogen	100	1600
Blood glucose	20	320
Protein	9000	153 000

If the exercise is prolonged and sustained at a steady state, then the complete oxidative breakdown of pyruvic acid is possible:

Pyruvic acid → Carbon dioxide + Water + Energy

The amount of energy produced from the breakdown of a glucose molecule to lactic acid results in the formation of three ATP molecules whereas complete oxidation results in the formation of 39 ATPs. The usefulness of the lactic acid system is that it is switched on very rapidly (within 1 s), so promoting ATP production in times where the phosphagens may be compromised and other energy stores are incapable of being used. Almost 50% of the energy during intense 6-s bouts of sprinting is derived from the breakdown of glycogen and the resultant build-up of lactic acid (Boobis *et al.*, 1987). Of course the end result is an accumulation of lactic acid which decreases the pH of the muscle cell and results in fatigue (MacLaren *et al.*, 1989).

Other sources of energy during prolonged steady-state exercise include glucose from the liver transported via blood, fatty acids from triglyceride stores in adipose tissue, and amino acids from a variety of tissues:

Glucose → Carbon dioxide + Water + Energy
Fatty acid → Carbon dioxide + Water + Energy
Amino acid → Carbon dioxide + Water + Energy
Amino acid → Glucose → Carbon dioxide + Water + Energy

The relative contribution of these sources for energy supply to the muscle during prolonged exercise can be seen in Figure 6.1.

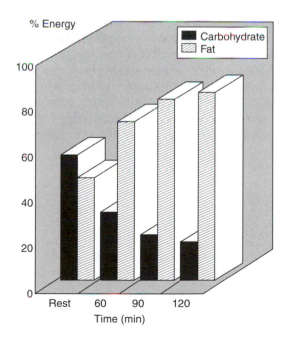

Figure 6.1 Relative contribution of carbohydrate and lipid as energy sources during prolonged moderately intense exercise.

6.1.2   *Nutritional sources*

Carbohydrates and fats are the principal sources of energy for muscular activity in humans, although energy may also be obtained from protein and from alcohol. The amounts of energy contained within 1 g of these nutrients are:

Carbohydrates	16 kJ (3.75 kcal)
Fats	37 kJ (9 kcal)
Protein	17 kJ (4 kcal)
Alcohol	29 kJ (7 kcal)

The data above clearly illustrate the advantages of fats as an energy store compared to carbohydrates. This advantage is reinforced on realising that carbohydrates are stored bound with water whereas fats are not; three molecules of water are stored with every one molecule of glycogen.

Although alcohol appears to be a potentially useful source of energy, it can only be metabolised by the liver and not by muscle. It is therefore of no ergogenic benefit to the soccer player and is more likely to have adverse effects on performance.

6.1.2.1   *Carbohydrates*

Carbohydrates are composed of carbon, hydrogen and oxygen, of which the simplest form are monosaccharides such as glucose, fructose and galactose. They are also known as the simple sugars. Disaccharides are made up from two monosaccharide molecules, for example, sucrose (table sugar) is made from glucose and fructose, maltose is made from two glucose units. The complex carbohydrates are polysaccharides of which starch is the plant storage form and glycogen the animal storage form. All the carbohydrates contain approximately the same amount of energy, that is, $16 \, kJ \, g^{-1}$. Table 6.2 provides examples of the types of food high in carbohydrates which are either complex carbohydrates or simple sugars.

Traditionally, carbohydrate-containing foods have been classified as simple or complex. In this instance foods which are high in simple carbohydrates include glucose, sugar, fruit, jams, sweets and confectionery products, whereas complex carbohydrates include foods high in starch, such as bread, pasta, potato and rice. This classification is an oversimplification since most naturally occurring foods contain a mixture of carbohydrate types, and the issue becomes even more complex in the case of processed or composite foods. For example,

Table 6.2 Foods containing 100 g of either complex carbohydrates or simple sugars

Complex carbohydrates	Simple sugars
400 g white bread	100 g sugar
250 g wholemeal bread	150 g chocolate bar
400 g boiled pasta	130 g honey
400 g baked potato	150 g jam
300 g boiled rice	140 g shortcake biscuit
130 g crackers	170 g fruit cake
120 g cornflakes	1 1 litre coca-cola
150 g weetabix	2 1 litre lemonade
750 g baked beans	

carbohydrate rich dishes such as the various pizzas contain simple and complex carbohydrates. The glycaemic index (GI) has been used to categorize foods according to their ability to raise blood glucose. Since all carbohydrate-containing foods raise blood glucose levels following digestion and absorption, the GI is a useful way of ranking composite and simple carbohydrate sources based on this ability.

The GI is a means of ranking foods according to their ability to elevate blood glucose concentration compared to a reference nutrient such as glucose. In order to determine the GI, subjects have to eat 50 g of the food containing carbohydrate and have their blood glucose concentration measured over a set period of time. The area under the blood glucose concentration graph is then calculated and expressed as a percentage when compared with the area under a curve of blood glucose following ingestion of 50 g of glucose under similar conditions:

$$GI = \frac{\text{blood glucose area after 50 g of 'test' food}}{\text{blood glucose area after 50 g of glucose}} \times 100$$

The GI reflects the rate of digestion and absorption of the food, and so the tests have to be carried out after a 12-h overnight fast.

Tables of the measured GI of various foods have been published (Foster-Powell and Brand Miller, 1995), and examples of foods which have a high, moderate or low GI can be seen in Table 6.3. Examination of the types of foods in the categories show that it is not possible to state simply that simple carbohydrates produce a rapid rise in blood sugar and that complex carbohydrates take longer to do so. For example chocolate (which would be considered as a simple carbohydrate) has a low GI, whereas baked potato (which would be categorized as a source of complex carbohydrate) has a high GI. Quite clearly, rapid digestion and absorption will lead to a short, sharp rise in blood glucose as with simple carbohydrates, whereas a slower, lower but more protracted rise in blood glucose is likely with complex carbohydrates. The net effect may be equal areas under the curves or ones in which there is a diminished or enhanced response. The GI does not equate to a rapid or slow appearance of glucose in the blood, rather it is a measure of the total amount of glucose appearing in the blood.

6.1.2.2 Fats

Fats are composed essentially of carbon, hydrogen and oxygen. Fat is stored in the form of triglycerides, which consist of a glycerol and three fatty acid molecules. They may be found

Table 6.3 Classification of some foods based on their glycaemic index

Low GI foods		Moderate GI foods		High GI foods	
Food	*GI*	*Food*	*GI*	*Food*	*GI*
Fructose	23	Sucrose	65	Glucose	100
All-bran	42	Muesli	68	Cornflakes	84
Apple	36	Mango	55	Watermelon	72
Pear	36	Banana	53	Pineapple	66
Pasta	41	Rice	56	Bread	70
Chocolate	49	Popcorn	55	Jelly beans	80
Peanuts	14	Crisps	54	Corn chips	73
Apple juice	41	Orange juice	57	Sports drinks	95
Yoghurt	33	Ice cream	61	Baked potato	83

in adipose tissue, between muscles, and within muscle cells. During exercise triglycerides in adipose tissue are broken down to glycerol and fatty acids, which are then transported to the muscles for oxidation in the mitochondria. Fatty acids have been classified as being saturated or unsaturated depending on whether the bonds between the carbon atoms are saturated or not. Saturated fatty acids are generally of animal origin and are solid at room temperature whereas unsaturated fatty acids are mainly from plant origin and are liquid at room temperature. From a health perspective it is advisable to consume fats which are unsaturated rather than fats which are saturated (suggestions are that 60% of fat intake should be in the form of unsaturated fatty acids).

6.1.2.3   Protein

Proteins are made up of 20 or so naturally occurring amino acids. These molecules consist of carbon, hydrogen and nitrogen. It is the nitrogen that has to be broken and eliminated as urea if taken in excess. Proteins are essential in the body not only as structural components but also because all enzymes are proteins; many of the hormones that integrate metabolism are proteins, and so are large macromolecules such as haemoglobin. The human body has a minimum requirement of protein in quantity and also in quality. The requirement with regard to quality is concerned with the fact that out of the 20 amino acids which make up all the proteins in our body, eight of them are essential. The eight essential amino acids are needed in small amounts regularly since we are incapable of making them ourselves, and a deficiency will ultimately lead to death. Table 6.4 shows the essential amino acids, and the foods rich in proteins. Essential amino acids can be found in meat, fish and dairy products, as well as in plant products such as cereals, pulses and nuts.

6.1.3   Recommended energy intake

Among the most important questions asked of nutritionists are those concerned with how much food needs to be consumed per day, and what type of food should be eaten in order to remain healthy. The Department of Health and Social Security (DHSS) in the United Kingdom provides information regarding this in its Recommended Intakes of Nutrients. The information presented contains the approximate amounts of energy needed to be consumed in order to balance the energy expended, as well as the protein, mineral and vitamin requirements in order to lead a healthy existence (DHSS, 1979).

The energy requirements are based on the recommendations of a committee whose members examined the average energy expended by individuals of different sexes, varying ages

Table 6.4 Essential amino acids and foods high in protein

Essential amino acids	Foods high in protein
Histidine (children only)	Milk (whole or skimmed)
Isoleucine	Cheese and cottage cheese
Leucine	Yoghurt
Lysine	Meat
Methionine	Fish
Phenylalanine	Eggs
Threonine	Beans/peas/lentils
Tryptophan	Peanuts/soy products
Valine	Broccoli/sprouts

and having different activity levels. For example, a 25-year-old man with a sedentary lifestyle is recommended to consume about 10 500 kJ (2500 kcal), whereas a man of similar age but who is considered very active should consume about 14 000 kJ (3333 kcal). Older people are recommended to have a lower energy intake than younger individuals, and males to have a higher energy intake compared with females of the same age and lifestyle.

The approximate energy intake of a 20–30-year-old elite soccer player is likely to be 8500–16 500 kJ (2033–4000 kcal) per day (Clark, 1994). Lower values may reflect off-season periods whereas higher values are likely to be found during periods of training or match-play. Any significant variations from this resulting in either an excessive intake or reduced intake could lead to overweight or undernutrition, respectively. Clearly there are times when the soccer player may need to increase energy intake, such as during intense pre-season training, or to reduce energy intake, such as during periods of comparative inactivity in recovering from injury or surgery.

General recommendations regarding the contribution of carbohydrates, fats and protein to the energy intake are as follows:

Carbohydrate	55–60%
Fat	25–30%
Protein	10–15%

This means that a soccer player whose energy intake is 16 MJ (16 000 kJ) should be consuming 550–600 g of carbohydrate (55 or 60% of 16 000 ÷ 16), 108–130 g of fat (25 or 30% of 16 000 ÷ 37), and 94–141 g of protein (10 or 15% of 16 000 ÷ 17). In what form should these nutrients be consumed? Recommendations based on health requirements suggest that about 60% of the carbohydrates should be in the form of complex carbohydrates, and 60% of the fats in the form of unsaturated fats. With respect to protein, the requirements for the essential amino acids should be met. This can be achieved by eating a variety of foods containing protein; vegans (who avoid meat, milk and dairy products) need to pay special attention to the correct mix of pulses and legumes in their diet.

Recommendations have been made regarding the minimum amounts of protein to be consumed per day, and these have stated a value amounting to $0.7 \, g \, kg^{-1}$ body weight. For a 70-kg person this would be 49 g of protein. Since the Western diet normally contains 15% energy intake from protein, an individual would have to eat less than 5470 kJ (1300 kcal) per day in order not to meet the recommendations; the value would be less than 8400 kJ (2000 kcal) if 10% of the energy intake was in the form of protein. A contribution of less than 10% energy intake from protein is unlikely except in cases of dieting, starvation or possibly carbohydrate loading.

The recommendations for protein intake of athletic populations is normally $1.0–2.0 \, g \, kg^{-1}$ per day. This higher value reflects the greater protein turnover as a result of increased activity and loss of amino acids in sweat. The extremely high protein intake of some bodybuilders and weight trainers (in excess of $2.0 \, g \, kg^{-1}$ body weight per day) is unlikely to be of any significant benefit unless anabolic steroids are taken in combination. The recommendations for soccer players has been suggested to be in the range $1.4–1.7 \, g \, kg^{-1}$ per day (Lemon, 1994).

Dietary surveys of soccer players have highlighted the inadequacies of their dietary intakes with respect to nutrients (Table 6.5). These findings demonstrate that the diet of elite soccer players tends to be low in carbohydrate and too high in fat. The importance of an adequate carbohydrate intake (i.e. 55–60% of total energy intake) cannot be overstressed.

Table 6.5 Nutrient intake of soccer players

Reference	No. of subjects	Energy (kJ)	Carbohydrates (%)	Fat (%)	Protein (%)
Jacobs *et al.* (1982)	15	20 700	47	39	14
Short and Short (1983)	8	12 500	43	41	16
Maughan (1997)	26	11 000	54.4	31.5	15.9
	25	12 800	48.1	35.0	14.3

The energy cost of playing First Division English League soccer has been estimated as being 69 kJ (16.4 kcal) per minute (Reilly and Thomas, 1979). This represents an energy expenditure of 6210 kJ (1480 kcal) for a 90-min match. If 60% of the energy were to come from carbohydrates, 227 g of carbohydrate would need to be consumed to replace it afterwards. How could this be consumed? Table 6.2 provides possible answers.

6.2   Importance of carbohydrates

6.2.1   Early studies

One of the earliest studies to show the importance of carbohydrates to athletic performance was carried out on marathon runners in the Boston marathons of 1924 and 1925 (Levine *et al.*, 1924; Gordon *et al.*, 1925). After the 1924 marathon, the doctors undertaking the study found that the post-race blood glucose levels of six runners studied were decreased, and that there was a strong relation between their physical condition and their blood glucose concentration. As a result of these findings, the doctors encouraged a group participating in the 1925 marathon to go on a diet high in carbohydrate for 24 h before the race and to eat some sweets after they had run about 24 km. The results proved very encouraging insofar as the runners improved their times, had a higher post-race blood glucose concentration than the previous year and were in better condition.

The next studies reported on humans were performed in Scandinavia in 1939, where it was shown that when exhausted subjects were given 200 g of carbohydrate they were able to continue exercising for a further 40 min (Christensen and Hansen, 1939). The same researchers also showed that when subjects were on a high carbohydrate diet they could exercise for a longer period of time than when on a normal or a low carbohydrate/high fat diet.

6.2.2   Muscle glycogen

The introduction of the muscle biopsy technique to nutritional studies in the 1960s led to the discovery that a high carbohydrate diet undertaken for three days resulted in an elevated muscle glycogen content (Bergstrom *et al.*, 1967). The effect was an increase in time to exhaustion when cycling an ergometer at 75% $\dot{V}O_2$ max compared to a mixed diet, and to a diet high in fat (Table 6.6).

The same research group also observed that a more pronounced effect could be obtained if the muscle glycogen levels were depleted by exercise before undertaking a high carbohydrate regime for 3 days. Under such circumstances, muscle glycogen concentrations of 4 g per 100 g wet muscle were obtained.

Table 6.6 Relationship between carbohydrate intake, muscle glycogen concentration and time to exhaustion

Diet	Muscle glycogen concentration (g per 100 g wet muscle)	Time to exhaustion (min)
Mixed diet (50% carbohydrate)	1.75	115
Low carbohydrate (0% carbohydrate)	0.6	60
High carbohydrate (80% carbohydrate)	3.5	170

Source: Bergstrom *et al.* (1967).

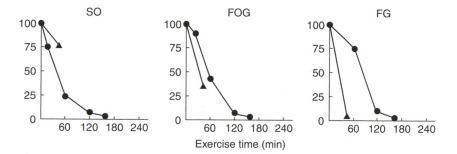

Figure 6.2 Rate of glycogen disappearance from three muscle fibre types during moderate, that is, 65% $\dot{V}O_2$ max (●), and intense, that is, 90% $\dot{V}O_2$ max (▲), exercise. (Data from Vollestad *et al.*, 1984.)

This muscle biopsy technique was used to further demonstrate that muscle glycogen concentrations fell during prolonged exercise, and that the very low levels coincided with the development of fatigue. Thus it was concluded that depletion of muscle glycogen caused fatigue.

The exercise intensity is related to muscle glycogen depletion in specific muscle fibre types (Vollestad *et al.*, 1984; Vollestad and Blom, 1985). Figure 6.2 highlights the effects of exercise intensities of approximately 65 and 90% $\dot{V}O_2$ max on the glycogen depletion of slow oxidative (SO), fast oxidative glycolytic (FG) fibres. It can be seen clearly that at high exercise intensities glycogen stores were depleted in FG fibres, whilst all fibre types had depleted stores at the moderate exercise intensities. It is important to appreciate that repeated bouts of sprint activity (such as all-out sprints in soccer) will lead to glycogen depletion in the FG fibres, and thus impair sprint performance; lower exercise intensities will be possible, but not intense bouts.

Since there is much evidence relating muscle glycogen depletion to fatigue during prolonged and possible to high-intensity exercise, there needs to be a consideration as to how much and when to consume carbohydrates in order to replenish the depleted stores. Costill *et al.* (1981) observed the effects of feeding meals containing 188, 325, 525 or 648 g of carbohydrate in a 24-h period following a 16-km run at approximately 80% $\dot{V}O_2$ max, which was in turn followed by five 1-min sprints corresponding to an exercise intensity of approximately 130% $\dot{V}O_2$ max. The exercise resulted in a 60% decrease in muscle glycogen content, which was only restored when 525–648 g of carbohydrate was consumed; the consumption of 188–325 g of carbohydrate failed to restore muscle glycogen content in a 24-h period. Other studies have reported similar findings insofar as rates of glycogen re-synthesis after depletion were maximal when 0.7–$3.0\,\mathrm{g\,kg^{-1}}$ carbohydrate was consumed every 2 h (Blom *et al.*, 1987; Ivy *et al.*, 1988).

The question regarding timing of the meal following exercise was addressed by Ivy *et al.* (1988). They showed that the highest rates of glycogen re-synthesis occurred when carbohydrates were consumed immediately after prolonged exercise. If the carbohydrate is given two or more hours after the exercise, then restoration takes longer to achieve. This clearly has implications for the soccer player who trains or competes late in the evening and decides not to eat anything until breakfast, or the professional who trains in the morning, skips lunch and eats a hearty meal in the evening. It is unlikely that muscle glycogen stores will be sufficiently replenished. Drinking carbohydrate beverages immediately after a match or training may be desirable under these circumstances, although normal practice may be to delay this for some time. Coaches and soccer players would be well advised to give serious consideration to post-exercise nutrition.

6.2.3   Muscle glycogen and soccer

Is there any evidence that muscle glycogen depletion and/or carbohydrate feedings are pertinent to performance in soccer? Saltin (1973) examined the effects of pre-exercise muscle glycogen levels on work-rates of nine subjects during soccer play. Five of the players had normal muscle glycogen levels ($96\,\mathrm{mmol\,kg^{-1}}$), whereas four possessed low muscle glycogen levels ($45\,\mathrm{mmol\,kg^{-1}}$) before the start of the match. The reasons for the low levels of muscle glycogen in the four players was due to the fact that they had a hard training session the previous day and had not consumed enough carbohydrates subsequently. Further muscle biopsy samples were taken at half-time and at the end of the match, and the players'

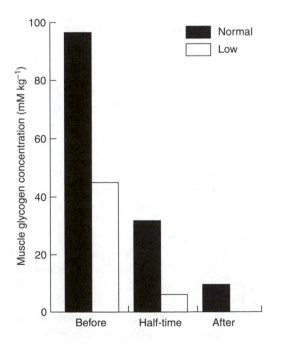

Figure 6.3 Muscle glycogen changes for nine players during a soccer match. (Data from Saltin, 1973.)

movements were analysed using cine-film. A significant glycogen depletion occurred in the muscles during the match (Figure 6.3), and this probably contributed to the comparatively lower distances covered by the four players during the second half, that is, 4100 m as opposed to 5900 m.

The four players with low muscle glycogen spent 50% of the time walking compared to 27% for the other five players, and they only spent 15% of the time sprinting compared with 24% for the other group. Clearly, it is undesirable for soccer players to have reduced muscle glycogen stores prior to a match. Can this happen? Pre-season training camps, evening matches and so on could result in inadequate attention being given to nutritional considerations in aiding recovery of glycogen stores.

Muscle glycogen content has been reported to approach complete depletion after a Swedish First Division match (Karlsson, 1969, cited by Shephard and Leatt, 1987). A 63% reduction was found after a regular match undertaken by Malmo FC players (Jacobs *et al.*, 1982), and a 50% decrease in elite Canadian soccer players during a simulated match (Leatt and Jacobs, 1988).

A recent study using 17 elite football players employed ^{13}C-magnetic resonance spectroscopy, a non-invasive technique, as the method of estimating muscle glycogen content following simulated football activity (Rico-Sanz *et al.*, 1999). Although the test to fatigue lasted only 42 min, muscle glycogen concentration in the gastrocnemius muscles (not the usual vastus lateralis muscle used in biopsy studies) decreased by 36%. One would assume that if this test were to be repeated after a 15-min recovery, thus simulating a true game, a more significant depletion of muscle glycogen stores would result.

The effectiveness of carbohydrate intake on simulated soccer match-play was reported by Balsom *et al.* (1999). They found that 48 h on a diet containing ~65% of the total energy intake as carbohydrate resulted in a 33% greater time spent in high intensity activities than when on a carbohydrate intake where 30% of the energy was from carbohydrates.

6.2.4 *Hypoglycaemia*

In spite of the considerable weight of evidence that muscle glycogen depletion is correlated with fatigue, there is some evidence that hypoglycaemia may be a causative factor in some individuals. Hypoglycaemia may be evident in prolonged exercise, and can contribute to fatigue. This normally occurs as a result of an inadequate restoration of the liver glycogen stores, as it appears that liver gluconeogenesis is unlikely to produce sufficient glucose for maintenance of blood concentrations once glycogen stores are low. The brain and the central nervous system are dependent on glucose for their metabolism. Blood glucose concentrations of about $3.0 \, \text{mmol} \, \text{l}^{-1}$ can cause nervousness and trembling, whilst levels below $3.0 \, \text{mmol} \, \text{l}^{-1}$ can lead to loss of consciousness.

For the skilled performer any slight malfunction of the brain will result in slow or inappropriate decision-making. In soccer, this may result in being caught in possession with the ball, incorrect weighting of a pass, mis-timed tackles or even missed goal scoring opportunities. One may speculate that lapses in concentration could be as a result of hypoglycaemia leading to poor decision-making.

Ekblom (1986) reported lower than normal blood glucose concentrations of $3.8 \, \text{mmol} \, \text{l}^{-1}$ in players at the end of a Swedish First Division match; indeed three players had values of between 3.0 and $3.2 \, \text{mmol} \, \text{l}^{-1}$. These low levels could impair decision-making compared with those players' responses under normal resting blood glucose levels of approximately $5.0 \, \text{mmol} \, \text{l}^{-1}$.

6.3   Importance of fluids

Although reductions in muscle and liver glycogen are understood to be major causative factors in the onset of fatigue during prolonged exercise, the loss of body fluids leading to dehydration may be another important cause. Mild dehydration will impair performance and reduce the capacity for exercise; a decrease in body weight of 5% due to fluid loss (i.e. 3.5 kg for a 70-kg person) can result in a 30% decrease in physical work capacity (Saltin and Costill, 1988).

Fluid loss during exercise is associated with the need to maintain a relatively constant body temperature. Exercise results in an increased production of energy, both metabolic and heat, and it is as a consequence of this that body temperature becomes elevated. Maughan (1991) estimated that the rate of heat production of running a marathon in 2 h 30 min is approximately $80 \, kJ \, min^{-1}$ ($20 \, kcal \, min^{-1}$). Since the major biological mechanism for losing heat during exercise is by evaporation of sweat, this will result in the loss of $2 \, l \, h^{-1}$ if all the 4.8 MJ (1200 kcal) of heat is to be removed by this route. This rate is possible, and would lead to a total loss of 3 l during a soccer match, as its intensity is comparable to the energy expenditure of marathon running (Reilly, 1990). A loss of 3 l represents a 4% loss of body weight for a 70-kg person. It has been reported that a 2% body weight decrease due to dehydration results in impaired performance, and as previously stated a 5% decrease in body weight results in a 30% decrease in work capacity (Saltin and Costill, 1988). Clearly there is a need to ensure that dehydration is minimized.

Are dehydration and fluid balance likely to be of significance for soccer players? Despite the fact that the recent World Cups (from 1982 to 2002 inclusive) were held in countries where high temperatures were evident, comparatively little research has been reported on this topic. Mustafa and Mahmoud (1979) observed that the average evaporative fluid loss from the Sudanese team in World Cup qualifying matches was about 3% body weight. More recently Elias *et al.* (1991) reported that in the 1988 USA youth soccer tournament, 34 players collapsed from heat exhaustion during six days. Training and playing soccer in hot and humid conditions accentuate heat stress-related problems if fluid ingestion is inadequate. Fluid replenishment is manifestly essential, and the most appropriate form must be considered.

6.3.1   Water

Prolonged exercise, such as playing soccer, in a warm environment results in a significant fluid loss. It would seem quite rational therefore to conclude that the best way of replenishing fluid loss would be to ingest water. Certainly evidence was available from the early 1970s that the rate of gastric emptying from the stomach became compromised if glucose was added to water (Costill and Saltin, 1974), and this led to the widespread favouring of water over carbohydrate drinks. Subsequently the gastric emptying characteristics of many different fluids have been studied, and the results confirm the inhibitory effects of adding calories to water (Murray, 1987).

The rate at which fluids are absorbed by the body is a combination of the rate of gastric emptying and the rate of fluid uptake by the small intestine. It is not advisable therefore to draw conclusions about fluid absorption based solely on gastric emptying rates despite the fact that gastric emptying is slower than fluid uptake by the small intestine. It is conceivable that whilst a dilute glucose drink may reduce the rate of gastric emptying in comparison to water, the glucose stimulation of fluid uptake by the small intestine results in a similar overall rate of fluid absorption.

Many studies have been performed to determine the advantages of water or carbohydrate–electrolyte solutions on performance either in the laboratory or in field conditions (for reviews see Maughan, 1991; Shirreffs and Maughan, 2000). Such studies have invariably concentrated on the enhancement of exercise time to exhaustion rather than on improvements in time to complete a set distance. The advantage of some form of fluid over a no-fluid treatment is well established. What is less clear is whether water had any advantages over a carbohydrate-based fluid or vice versa. When dehydration is believed to be the major factor impairing performance, such as in prolonged exercise in the heat, then ingesting as much water as is deemed necessary to offset dehydration is advisable. The consensus view in the UK is that a sports drink which contains an energy source in the form of carbohydrate together with electrolytes is more effective than plain water in maintaining performance (*British Journal of Sports Medicine*, 1993).

6.3.2 Carbohydrate drinks

Many recent studies of the beneficial effects of carbohydrate ingestion on performance have been reported (see Coyle, 1991). Factors such as type of carbohydrate, concentration of carbohydrate, and timing of ingestion have been considered. The primary purpose of carbohydrate ingestion during exercise is to maintain blood glucose concentrations, and if the exercise is prolonged, to maintain carbohydrate oxidation rates in the later stages. This would permit continuation of exercise at an adequate intensity. The purpose of carbohydrate ingestion after exercise is to restore the muscle and liver glycogen stores as quickly as possible.

When carbohydrate supplementation is provided during prolonged exercise, subjects can exercise for longer and produce greater power at the end of such a performance than when given nothing or indeed when provided with water alone (Coggan and Coyle, 1991). So carbohydrate supplementation is recommended whenever the exercise is likely to be severe enough to significantly deplete glycogen stores and so impair performance. This certainly would apply in top-level soccer. But how much carbohydrate needs to be ingested?

In most of the studies that have been reported where carbohydrate ingestion improved performance, subjects were given $30–60\,g\,h^{-1}$ (Coggan and Coyle, 1991). These authors had previously reported that the glucose infusion rate necessary to restore and maintain blood glucose levels late in exercise and thereby delay fatigue by 45 min was $1\,g\,min^{-1}$ or $60\,g\,h^{-1}$ (Coggan and Coyle, 1987). It has been shown that glucose infusion can result in a glucose utilization rate of $1.8\,g\,min^{-1}$ (MacLaren *et al.*, 1994). It appears that the maximum rate at which a carbohydrate can be utilized is approximately $120\,g\,h^{-1}$.

In order to consume $60\,g\,h^{-1}$ of a carbohydrate, it is possible to do so in the following manner:

> 300 ml of a 20% solution
> 600 ml of a 10% solution
> 1200 ml of a 5% solution
> 2400 ml of a 2% solution

The first drink would appear to be too concentrated and may compromise fluid intake whilst the last drink is too large a volume to ingest in 1 h. The two other drinks would appear to be reasonable; although 600 ml may not be sufficient in fluid intake on a hot and humid day, whilst a 10% concentration of glucose (but not maltodextrin) may inhibit gastric emptying.

By drinking relatively large volumes of 5–10% carbohydrate solutions, most athletes can meet their carbohydrate needs and obtain 600–1200 ml h^{-1} of fluid. For the soccer player who uses 200 g of carbohydrate in a match and loses 2–2.5 l of fluid, a regular consumption of a carbohydrate–electrolyte drink is advisable. In what form is this best provided, and when should it be taken?

It would be prudent to drink 200 ml of 5–10% carbohydrate–electrolyte immediately before the match and 1.2 l h^{-1} of the same at intervals throughout the match. However, due to FIFA regulations this is not usually possible at international competitive matches and so a compromise is necessary. In the 1994 World Cup, the regulations were amended to allow players to consume drinks at the side-line; the changes were made in the light of warnings that players might suffer acutely from dehydration. Drinking fluids at half-time is certainly advisable. The concentration and volume of fluid can be changed to suit the external environmental conditions, that is, on a hot and humid day the emphasis should be on a greater volume of fluid with a dilute carbohydrate content (say no more than 5%), whilst on a cold and rainy day the emphasis should be on a smaller volume of fluid intake with a higher carbohydrate content (say 10%). New drinking strategies should never be tried out under match-play conditions; rather they should be used during training first.

No significant differences have been found in the ability of various forms of ingested carbohydrate to maintain blood glucose concentrations, change carbohydrate oxidation rates, or improve performance. These sources have included glucose, sucrose, fructose, maltose and maltodextrins (also known as glucose polymers). Fructose ingestion is problematical because it can cause diarrhoea due to its slower intestinal absorption. Maltodextrins have become popular because they are not as sweet tasting and they are osmotically less active than glucose and sucrose.

6.4   Meals

6.4.1   Pre-competition

There is a significant amount of research to show that a high carbohydrate meal eaten 3–4 h before exercise can improve performance. In one study, athletes were either given no carbohydrate, a meal containing 300 g of carbohydrate 3 h before exercise, no carbohydrate before exercise but an 8% carbohydrate drink during exercise (giving a total of 175 g carbohydrate), or a combination of carbohydrate 3 h before exercise and during exercise. The average time to exhaustion at 70% $\dot{V}O_2$ max was 44% greater than the control (no carbohydrate) when carbohydrate was given before and during exercise, 32% greater than the control when carbohydrate was given during exercise, and 18% greater when carbohydrate was given 3 h before exercise. In two other studies, which purely examined the effects of ingesting carbohydrates 4 h before exercise, performance was improved by between 15 and 22% compared with placebo. The overall conclusion from these and other studies is that a pre-exercise meal high in carbohydrates extends time to fatigue and improves total work done. The timing of the meals given in these studies (i.e. 4 h before exercise) typically coincides with the time of the pre-match meal in football.

Soccer players should leave at least a 3-h interval between a full meal and competition in order to minimize gastrointestinal problems such as nausea and a feeling of fullness. It is recommended that the stomach should be reasonably empty at the time of the match since the digestion and absorption of food will compete with the muscles for a good blood supply. Fatty foods are known to slow down the rate of gastric emptying and therefore should be avoided. The meal should be high in carbohydrates, preferably complex carbohydrates such as bread, cereals, pasta, rice, potatoes, fruits and vegetables. The actual amount of calories consumed will vary between individuals and how much they had eaten previously. Proteins are acceptable to eat as long as they are not fatty proteins, that is, meat high in fats or containing fatty sauces, or fatty cheeses. This meal (and accompanying drinks) should contain about $4\,g\,kg^{-1}$ body weight of carbohydrate; for a 70-kg soccer player this represents an intake of 280 g of carbohydrate. The accompanying drink could be high in carbohydrates, but not in the form of fructose.

Some players feel incapable of eating much (or anything) in the hours before a competitive game. This is quite understandable in the light of nervous tension in the build up to the match. Under such circumstances, the player would be advised to at least drink some high-carbohydrate drinks and even eat a small snack with the rest of the players. In addition, frequent drinking of small volumes of a carbohydrate drink and eating some fruit in the 3–4 h build up should help. It is inadvisable not to eat or drink anything before a match.

The only time that carbohydrates should be avoided is in the 30–60 min immediately before competition or training. For some individuals this timing of feeding may produce a rapid fall in blood glucose levels in the first 20 min or so of exercise and so impair performance (Costill *et al.*, 1977; Hargreaves *et al.*, 1984). Some athletes are sensitive to mild decreases in blood glucose and may display symptoms such as sweating, light-headedness, dizziness and shakiness. Other than experimenting with the timing of this pre-exercise nutritional strategy to ensure that even mild decreases in blood glucose do not occur in these 'sensitive' individuals, it would be desirable to examine the type of carbohydrate ingested. Selection of foods which have a low GI may be appropriate for individuals who are sensitive to small decreases in blood glucose. Foods with a low GI have a low but sustained release of glucose which may benefit the soccer player during a match or during training.

The most important aspects of pre-competition meals are to elevate the body's carbohydrate stores, ensure hydration and yet provide satisfaction for the player. Trying out new foods or significantly altering eating patterns should be discouraged. Experimenting with new pre-competition meals should take place before training or before unimportant games.

6.4.2 Post-competition

The major considerations after completion are to replenish carbohydrate and fluid losses. As already mentioned, it is important to consume carbohydrates as soon as possible after exercise in order to achieve a quick and complete glycogen restoration. The first 2 h post-exercise is the most crucial period for the ingestion of carbohydrates (Ivy *et al.*, 1988), since the glycogen-synthesizing enzymes are very active during this time. A recommendation would be to consume $1.5\,g\,kg^{-1}$ body weight of carbohydrate within the first 30 min after competition or exercise; for a 70-kg player this represents 105 g of carbohydrate. Whether the carbohydrate is in solid or liquid form is immaterial and may be left to the preference of the player. Some athletes do not like to eat after strenuous exercise but are quite willing to drink. A concentrated carbohydrate beverage would prove invaluable to these players.

The problems with muscle glycogen repletion after playing a soccer match or completing intense training is confounded by the fact that this type of activity invariably involves eccentric exercise. Restoration of muscle glycogen following eccentric exercise has been shown to be significantly impeded even if high amounts of carbohydrate are ingested (Costill *et al.*, 1990). Incomplete restoration was observed even after 3 days on a high carbohydrate diet.

Therefore, it is likely that players may not have fully recovered their glycogen stores from a midweek match to a weekend match or vice versa.

In the course of a match or training, muscles are used extensively to generate movement. Until recently, a neglected area for consideration was concerned with protein synthesis and breakdown following exercise. In general, the rate of protein breakdown increases and the rate of protein synthesis decreases during exercise, whereas during recovery from training the rate of protein synthesis increases after a period of time. Increasing attention has been given to nutritional means of promoting faster and more complete restoration of the losses of muscle protein which occur as a consequence of exercise. The infusions of amino acids, ingestion of concoctions of amino acids, ingestion of types of proteins such as whey or soya protein, and combinations of carbohydrate and amino acids mixtures have been examined by Tipton and Wolfe (2001). The overall findings support the view that amino acid ingestion at around 1 and 3 h after exercise stimulates protein synthesis if essential amino acids are given rather than non-essential amino acids. The addition of carbohydrate to the amino acids further stimulates recovery, probably due to the anabolic effect of elevated insulin levels in the blood (in turn stimulated by a carbohydrate intake). A mixture of approximately 40 g of carbohydrate and 6 g of essential amino acids (i.e. isoleucine, leucine, histidine, lysine, methionine, phenylalanine, threonine and valine) ingested between 1 and 3 h following exercise provides a significant means of enhancing protein synthesis (Rasmussen *et al.*, 2000).

Post-exercise rehydration is best served by a pre-determined schedule of fluid intake rather than for the player to rely on sensations of thirst; the thirst mechanism is unreliable in terms of adequate rehydration. Plain water will certainly promote significant rehydration, although increased urine production and a decreased thirst response may lead to failure to replace fully the intracellular water. It has been suggested that the addition of small amounts of sodium (in the form of salt) will increase the rate of recovery of fluid balance by helping to retain water and thereby promote a normalizing of plasma volume (Nadel, 1988). The use of commercial sports drinks can help to increase the voluntary consumption of fluid compared with water, and therefore may be beneficial in achieving rehydration post-exercise (Johnson *et al.*, 1988). The use of carbohydrate–electrolyte drinks may be of special advantage in recovery post-exercise, where the twin aims of rehydration and restoration of glycogen stores are important.

6.5   Ergogenic aids

Ergogenic aids are substances, legal or illegal, which may lead to an improvement in athletic performance. Such substances include alcohol, caffeine, creatine, alkalinizers and vitamins. Anabolic steroids, amphetamines, erythropoietin and blood doping are illegal and are therefore not considered here. The agents discussed are not banned substances and may be taken in small doses.

6.5.1   Alcohol

Alcohol in small doses has been reported to improve hand steadiness and motor control in sports that require accuracy, that is, archery, darts, shooting and snooker (Reilly and Halliday, 1985; Reilly, 1997). Its use elsewhere is unlikely to enhance performance; indeed in moderate to large quantities it is likely to prove disadvantageous (American College of Sports Medicine, 1982). When ingested in large amounts, alcohol reduces exercise ability by impairing psychomotor coordination, increasing diuresis and so possibly leading to dehydration, and compromising the maintenance of body temperature, particularly in cold temperatures.

Two myths which are associated with alcohol concern its energy content and its carbohydrate content. Although alcohol does possess a higher amount of energy than carbohydrate (29 kJ compared with 16 kJ g^{-1}), it is metabolised by the liver like fat, and is energetically not useful to muscle. Secondly, alcoholic drinks do not contain large amounts of carbohydrate, contrary to popular belief. A bottle of beer contains on average 10–15 g of carbohydrate (5–8 g in 'lite' beers). Alcoholic drinks do not provide useful nutrients for enhancing sports performance. The facts that alcohol is a diuretic and results in increased urine water loss, and that alcohol may interfere with glucose metabolism by reducing gluconeogenesis, mean that it is an inappropriate drink for athletes.

There is no evidence that light social drinking will impair performance even on the following day, although moderate to large intakes will adversely affect endurance, strength, speed and reaction times (Williams, 1991).

6.5.2 *Caffeine*

Caffeine is a drug that is found in several drinks (tea, coffee, cola) and foods (chocolate). It is a stimulant in high concentrations, and elicits a number of physiological and psychological responses linked with increasing endurance performance. The effects are mainly due to an elevation in plasma fatty acids and a consequent sparing of muscle glycogen (Giles and MacLaren, 1984). The International Olympic Committee (IOC) has banned caffeine and set a threshold level of 12 µg ml^{-1} for urine samples. This may be achieved by ingesting approximately 800 mg of caffeine, comparable to drinking six cups of coffee or 17.5 cans of cola in a short time. A caffeine intake half this amount, that is, approximately 350 mg, promotes fat utilization (Giles and MacLaren, 1984).

In spite of the evidence that caffeine stimulates the release of fatty acids and possibly spare muscle glycogen stores, the research findings are equivocal (Powers and Dodd, 1984). Consuming carbohydrate with caffeine negates the fatty acid stimulatory effect of caffeine, and so this combination would have no ergogenic influence. It should also be recognized that caffeine is a diuretic and could lead to dehydration.

6.5.3 *Creatine*

Short-duration high-intensity exercise requires the regeneration of ATP primarily from the breakdown of CP and from anaerobic glycolysis. A significant reduction of CP occurs after 6-s of cycle sprinting (Boobis *et al.*, 1987). A 16% decrease in CP after ten 6-s cycle sprints has been reported and this coincided with a reduction in power output (Gaitanos *et al.*, 1993). Depletion is greater in FG than SO fibres (Greenhaff *et al.*, 1992). It seems that the availability of CP is one of the limiting factors for maintaining the high rates of energy necessary for this type of activity. Could supplementing the muscle creatine pool therefore result in enhanced all-out performance?

Harris *et al.* (1992) demonstrated that ingestion of 20–30 g of creatine monohydrate per day for more than two days increased the muscle creatine content by up to 50%. Subsequent studies have shown that creatine supplementation resulted in improved performance during repeated 30-s bouts of maximal isokinetic exercise (Greenhaff *et al.*, 1993b), during repeated bouts of 6-s sprint cycling (Balsom *et al.*, 1993), and during 300 and 1000 m running time (Harris *et al.*, 1993). Furthermore, CP re-synthesis during recovery from intense activity was enhanced following supplementation (Greenhaff *et al.*, 1993a). The general consensus is that creatine supplementation does enhance these types of activity when taken in doses in excess of 20–30 g per day. Whether such an intake is likely to promote soccer performance is uncertain, although a word of caution should be given in that the long-term effects of taking such large doses have not been studied as extensively as short-term studies. Other disadvantages of creatine supplementation may be found in the increases in body mass due to muscle cell water retention. This normally results in an increase in body mass of between 1 and 2 kg, and so may not be particularly beneficial in weight bearing sports. A more extensive review on this topic is presented elsewhere (MacLaren, 2002).

6.5.4   *Alkalinizers*

Sodium bicarbonate and sodium citrate are alkaline salts that possess buffering properties in the human body when ingested. The theory behind their use is relatively simple. If a build-up of lactic acid occurs due to intense exercise, the resultant increase in acidity or reduction in pH is likely to be a contributory factor in the development of fatigue (MacLaren *et al.*, 1989). Alkalinizers increase the normal alkali reserve of the blood and help buffer the acid produced during exercise. Furthermore, the increase in bicarbonate ions in the blood may facilitate the efflux of hydrogen ions from the muscle and thereby maintain a higher pH in the muscle (MacLaren, 1997). Theoretically, alkaline salts should enhance performance in events maximising the use of anaerobic glycolysis (i.e. intense exercise lasting between 30 and 120 s), and also intense intermittent exercise where removal of lactic acid from muscle during recovery is required. Maximal exercise tasks of less than 30 s and prolonged endurance tasks such as 5–10 km runs, which rely primarily on oxidative processes, generally do not benefit from the use of alkalinizers (Gledhill, 1984). Sodium bicarbonate ingestion may improve running time to exhaustion at an exercise intensity corresponding to a lactate concentration of 4 mmol l^{-1}; an increase of 17% in time to exhaustion from approximately 26 to 31 min was observed in runners by George and MacLaren (1988). An alternative source of alkalinizer in the form of sodium citrate has been used with limited success (Parry-Billings and MacLaren, 1986).

Although soccer is a sport in which there are periodic bouts of intense exercise and the blood lactate levels have been reported to rise to 4.0–9.5 mmol l^{-1} (Bangsbo, 1994), it is unlikely that use of alkalinizers would be appropriate. This is due in particular to the prolonged nature of the game, and also the fact that gastrointestinal problems of nausea and diarrhoea can be associated with their ingestion.

6.5.5   *Vitamins and minerals*

Many of the vitamins play key roles in exercise metabolism although they do not possess any energy. The B vitamins are particularly noteworthy for their relevance to energy production in muscle and also in the recovery processes. Table 6.7 highlights the roles played by some of the vitamins. If vitamins are important in energy-producing reactions, then a deficiency of these vitamins will lead to an impaired performance, and an increase in these vitamins may lead to an enhanced performance. A deficiency in at least some of the vitamins will impair exercise performance although there is no evidence that vitamin supplementation improves performance in individuals eating a well-balanced diet (van der Beek, 1985). Physical activity increases the need for extra amounts of vitamins, but these can easily be obtained by consuming a balanced diet. The DHSS has provided guidelines (Table 6.7) for the minimum consumption of the vitamins A, thiamin, riboflavin, nicotinic acid, vitamin C and folic acid (DHSS, 1979).

Table 6.7 Vitamins, their major functions, and their recommended daily intakes (RDI)

Vitamin	RDI	Major role
Vitamin A (retinol)	750 gg	Vision
Vitamin B (thiamin)	1.4 mg	Carbohydrate metabolism
Vitamin B2 (riboflavin)	1.7 mg	Electron transfer
Vitamin B3 (nicotinic acid)	18 mg	ATP synthesis, fat synthesis
Vitamin B6	No RDI	Amino acid, and glycogen synthesis
Folic acid	No RDI	Red blood cell synthesis
Pantothenic acid	No RDI	Fatty acid oxidation
Biotin	No RDI	Fatty acid, and glycogen synthesis
Vitamin B12	No RDI	Red blood cell synthesis
Vitamin C	30 mg	Collagen synthesis
Vitamin D	2.5 µg	Calcium metabolism
Vitamin E	No RDI	Antioxidant
Vitamin K	No RDI	Blood clotting

Dietary surveys of athletic populations have shown that the estimated vitamin intakes usually exceed the recommended dietary intakes (Brotherhood, 1984; van Erp-Bart *et al.*, 1989). It appears that athletes do not usually suffer from an inadequate vitamin intake, and that vitamin supplementation may only be of value if there is likely to be a deficiency.

Similar findings have been reported with regard to mineral supplementation, in that no benefits have been determined. The only mineral which has presented as being problematical in athletes is iron. The so-called sports anaemia has been explained as being caused by a transitory dilutional effect of an expanded plasma volume due to training, and/or due to haemolysis of red cells as the foot strikes the floor in running. Other possible causes could be decreased iron absorption, increased iron losses in sweat, faeces and urine, or poor iron intake (Economos *et al.*, 1993). Female soccer players may be more susceptible to lower iron status as a result of menstrual blood loss. Anaemia will result in impaired aerobic performance and as a consequence may be treated by additional intakes of iron in combination with vitamin C. More in-depth reports can be found in the reviews by Newhouse and Clement (1988) and Haymes and Lamanaca (1989).

Summary

In order to maximize training, it is essential for the soccer player and coach to give serious consideration to correct nutritional practices. It is not possible for a player to give 100% effort in training or during a match if inadequate attention has been devoted to nutrition. The following points may be considered in highlighting such nutritional practices.

- At least 55% of the energy intake should be in the form of carbohydrates – preferably complex carbohydrates. During intense training or game periods this value should be increased to 60–70% of energy intake.
- Carbohydrates should form the basis of the pre- and post-competitive meals. The former should be consumed 2.5–3 h before playing, whilst the latter should be consumed within 2 h after playing. Carbohydrate beverages would be beneficial in these situations either to 'top-up' the meal or because the player does not feel like eating.

- Carbohydrate–electrolyte drinks are beneficial and should be taken before (say up to 60 min), during (every 15–20 min), and after (within, 1–2 h) training or a match. Thirst is a poor indicator of the degree of dehydration.
- A protein intake of 1.4–1.7 g kg^{-1} body weight per day is desirable; this is likely if the player is consuming 15% of energy intake in the form of protein and is meeting the daily total energy intake requirements.
- A varied diet should ensure the players are likely to meet the recommended amino acid, vitamin and mineral requirements. Supplementation of these nutrients is unnecessary unless the diet is inadequate. There is no evidence that extra amounts of amino acids, vitamins or minerals will improve performance.
- Players should be discouraged from trying out new dietary practices prior to a match; these regimes should be tried out first before or during training.
- In spite of the potential benefits of some ergogenic aids, there is an uncertainty regarding the consequences of frequent use of these products.

References

American College of Sports Medicine (1982). Position statement on the use of alcohol in sports. *Medicine and Science in Sports and Exercise*, **14**, ix–xi.

Balsom, P.D., Ekblom, B., Soderlund, K. *et al.* (1993). Creatine supplementation and dynamic high-intensity intermittent exercise. *Scandinavian Journal of Medicine and Science in Sports*, **3**, 143–9.

Balsom, P.D., Wood, K., Olsson, P. and Ekblom, B. (1999). Carbohydrate intake and multiple sprint sports: with special reference to football (soccer). *International Journal of Sports Medicine*, **20**, 48–52.

Bangsbo, J. (1994). The physiology of soccer – with special reference to intense intermittent exercise. *Acta Physiologica Scandinavica*, **151**(suppl.), 619.

van der Beek, E.J. (1985). Vitamins and endurance training: food for running or faddish claims? *Sports Medicine*, **2**, 175–97.

Bergstrom, J., Hermansen, L., Hultman, E. and Saltin, B. (1967). Diet, muscle glycogen and physical performance. *Acta Physiologica Scandinavica*, **71**, 140–50.

Blom, P.C., Hostmark, A.T., Vaage, O. *et al.* (1987). Effect of different post-exercise sugar diets on the rate of muscle glycogen synthesis. *Medicine and Science in Sports and Exercise*, **19**, 491–6.

Boobis, L.H., Williams, C., Cheetham, M.E. and Wooton, S.A. (1987). Metabolic aspects of fatigue during sprinting, in: *Exercise: Benefits, Limits and Adaptations* (eds D. Macleod, R. Maughan, M. Nimmo *et al.*), E & FN Spon, London, pp. 116–40.

British Journal of Sports Medicine (1993). Consensus statement on fluid replacement in sport and exercise. *British Journal of Sports Medicine*, **27**, 34.

Brotherhood, J. (1984). Nutrition and sports performance. *Sports Medicine*, **1**, 50–89.

Christensen, E.H. and Hansen, O. (1939). Hypoglykamie arbeitsfahigkeit und ermuding. *Skandinavisches Archiv fur Physiologie*, **81**, 172–9.

Clark, K. (1994). Nutritional guidance to soccer players for training and competition. *Journal of Sports Sciences*, **12**, S43–50.

Coggan, A.R. and Coyle, E.F. (1987). Reversal of fatigue during prolonged exercise by carbohydrate infusion or ingestion. *Journal of Applied Physiology*, **63**, 2388–95.

Coggan, A.R. and Coyle, E.F. (1991). Carbohydrate ingestion during prolonged exercise. Effects on metabolism and performance. *Exercise and Sports Science Reviews*, **19**, 1–40.

Costill, D.L., Pascoe, D.D., Fink, W.J., Robergs, R.A., Barr, S.I. and Pearson, D. (1990). Impaired muscle glycogen resynthesis after eccentric exercise. *Journal of Applied Physiology*, **69**, 46–50.

Costill, D.L. and Saltin, B. (1974). Factors limiting gastric emptying during rest and exercise. *Journal of Applied Physiology*, **37**, 679–83.

Costill, D.L., Coyle, E.F., Dalsky, G. *et al.* (1977). Effects of elevated plasma FFA and insulin on muscle glycogen usage during exercise. *Journal of Applied Physiology*, **43**, 695–9.

Costill, D.L., Sherman, W.M., Fink, W.J. *et al.* (1981). The role of dietary carbohydrates in muscle glycogen resynthesis after strenuous running. *American Journal of Clinical Nutrition*, **34**, 1831–6.

Coyle, E.F. (1991). Timing and method of increased carbohydrate intake to cope with heavy training, competition and recovery. *Journal of Sports Sciences*, **9**, S29–52.

DHSS (1979). *Recommended Intakes of Nutrients for the United Kingdom*. HMSO, London.

Economos, C., Bortz, S.S. and Nelson, M.E. (1993). Nutritional practices of elite athletes: practical recommendations. *Sports Medicine*, **16**, 381–99.

Ekblom, B. (1986). Applied physiology of soccer. *Sports Medicine*, **3**, 50–60.

Elias, S.R., Roberts, W.O. and Thorson, D.C. (1991). Team sports in hot weather: guidelines for modifying youth soccer. *Physician and Sportsmedicine*, **19**, 67–80.

van Erp-Bart, A.M.J., Saris, W.H.M., Binkhorst, R.A. *et al.* (1989). Nationwide survey on nutritional habits in elite athletes (Part II): mineral and vitamin intake. *International Journal of Sports Medicine*, **10**(suppl 1), S11–16.

Foster-Powell, K. and Brand Miller, J. (1995). International tables of glycemic index. *American Journal of Clinical Nutrition*, **62**(suppl), 871S–893S.

Gaitanos, G.C., Williams, C., Boobis, L.H. and Brooks, S. (1993). Human muscle metabolism during intermittent maximal exercise. *Journal of Applied Physiology*, **75**, 712–19.

George, K.P. and MacLaren, D.P. (1988). The effect of induced alkalosis and acidosis on endurance running at an intensity-corresponding to 4 mM blood lactate. *Ergonomics*, **31**, 1639–45.

Giles, D. and MacLaren, D. (1984). Effects of caffeine and glucose ingestion on metabolic and respiratory functions during prolonged exercise. *Journal of Sports Sciences*, **2**, 35–46.

Gledhill, N. (1984). Bicarbonate ingestion and anaerobic performance. *Sports Medicine*, **1**, 177–80.

Gordon, B., Kohn, L.A., Levine, S.A. *et al.* (1925). Sugar content of the blood in runners following a marathon race. *Journal of the American Medical Association*, **185**, 508–9.

Greenhaff, P.L., Nevill, M.E., Soderlund, K. *et al.* (1992). Energy metabolism in single muscle fibres during maximal sprint exercise in man. *Journal of Physiology*, **446**, 528P.

Greenhaff, P.L., Bodin, K., Harris, R.C. *et al.* (1993a). The influence of oral creatine supplementation on muscle contraction in man. *Journal of Physiology*, **467**, 75P.

Greenhaff, P.L., Casey, A., Short, A.H. *et al.* (1993b). Influence of oral creatine supplementation on muscle torque during repeated bouts of maximal voluntary exercise in man. *Clinical Science*, **84**, 565–71.

Hargreaves, M., Costill, D.L., Coggan, A. and Nishibata, I. (1984). Effect of carbohydrate feeding on muscle glycogen utilization and exercise performance. *Medicine and Science in Sports and Exercise*, **16**, 219–22.

Harris, R.C., Soderlund, K. and Hultman, E. (1992). Elevation of creatine in resting and exercised muscle of normal subjects by creatine supplementation. *Clinical Science*, **83**, 367–74.

Harris, R.C., Viru, M., Greenhaff, P.L. and Hultman, E. (1993). The effect of oral creatine supplementation on running performance during short term exercise in man. *Journal of Physiology*, **467**, 74P.

Haymes, E.M. and Lamanaca, J.J. (1989). Iron loss in runners during exercise: implications and recommendations. *Sports Medicine*, **7**, 1277–85.

Ivy, J.L., Katz, S.L., Cutler, C.L. *et al.* (1988). Muscle glycogen synthesis after exercise: effect of time of carbohydrate in gestion. *Journal of Applied Physiology*, **64**, 148–50.

Jacobs, I., Westlin, N., Karlsson, J. *et al.* (1982). Muscle glycogen and elite soccer players. *European Journal of Applied Physiology*, **48**, 297–302.

Johnson, H.L., Nelson, R.A. and Consolazio, C.F. (1988). Effects of electrolyte and nutrient solutions on performance and metabolic balance. *Medicine and Science in Sports and Exercise*, **20**, 26–33.

Leatt, P.B. and Jacobs, I. (1988). Effect of a liquid glucose supplement on muscle glycogen resynthesis after a soccer match, in *Science and Football*, (eds T. Reilly, A. Lees, K. Davids and W. Murphy), E & FN Spon, London, pp. 42–7.

Lemon, P.W.R. (1994). Protein requirements of soccer. *Journal of Sports Sciences*, **12**, S17–22.

Levine, S.A., Gordon, B. and Derick, C.L. (1924). Some changes in the chemical constituents of the blood following a marathon race. *Journal of the American Medical Association*, **82**, 1778–9.

MacLaren, D.P.M. (1997). Influence of blood acid-base status on performance, in *The Clinical Pharmacology of Sport and Exercise* (eds T. Reilly and M. Orme), Elsevier, Amsterdam, pp. 157–65.

MacLaren, D.P.M. (2002). Creatine, in *Drugs in Sport* (ed. D. Mottram), E & FN Spon, London.

MacLaren, D.P., Parry-Billings, M., Gibson, H. and Edwards, R.H.T. (1989). A review of metabolic and physiological factors in fatigue, in *Exercise and Sport Science Reviews* (ed. K. Pandolf), vol. 17, Williams & Wilkins, Baltimore, MD, pp. 29–66.

MacLaren, D.P., Reilly, T., Campbell, L. *et al.* (1994). The hyperglycaemic glucose clamp technique, glucose utilization, and carbohydrate oxidation during exercise. *Journal of Sports Sciences*, **12**, 143.

Maughan, R.J. (1991). Fluid and electrolyte loss and replacement in exercise. *Journal of Sports Sciences*, **9**, 117–42.

Maughan, R.J. (1997). Energy and macronutrient intakes of professional football (soccer) players. *British Journal of Sports Medicine*, **31**, 45–7.

Murray, R. (1987). The effects of consuming carbohydrate-electrolyte beverages on gastric emptying and fluid absorption during and following exercise. *Sports Medicine*, **4**, 322–51.

Mustafa, K.Y. and Mahmoud, N.E.A. (1979). Evaporative water loss in African soccer players. *Journal of Sports Medicine and Physical Fitness*, **19**, 181–3.

Nadel, E.R. (1988). New ideas for rehydration during and after exercise in hot weather. *Gatorade Sports Science Exchange*, **1**(3).

Newhouse, I.J. and Clement, D.B. (1988). Iron status in athletes: an update. *Sports Medicine*, **5**, 337–52.

Parry-Billings, M. and MacLaren, D.P. (1986). The effect of sodium bicarbonate and sodium citrate ingestion on, aflaerobic power during intermittent exercise. *European Journal of Applied Physiology*, **55**, 524–9.

Powers, S.K. and Dodd, S. (1984). Caffeine and endurance performance. *Sports Medicine*, **2**, 165–74.

Rasmussen, B.B., Tipton, K.D., Miller, S.L., Wolfe, S.E. and Wolfe, R.R. (2000). An oral essential amino acid-carbohydrate supplement enhances muscle protein anabolism after resistance exercise. *Journal of Applied Physiology*, **88**, 386–92.

Reilly, T. (1990). Football, in *Physiology of Sports* (eds T. Reilly, N. Secher, P. Snell and C. Williams), E & FN Spon, London, pp. 371–425.

Reilly, T. (1997). Alcohol: its influence in sport and exercise, in *The Clinical Pharmacology of Sports and Exercise* (eds T. Reilly and M. Orme), Elsevier, Amsterdam, pp. 281–92.

Reilly, T. and Halliday, F. (1985). Influence of alcohol ingestion on tasks related to archery. *Journal of Human Ergology*, **14**, 99–104.

Reilly, T. and Thomas, V. (1979). Estimated daily energy expenditures of professional association footballers. *Ergonomics*, **22**, 541–8.

Rico-Sanz, J., Zehnder, M., Buchli, R., Dambach, M. and Boutellier, U. (1999). Muscle glycogen degradation during simulation of a fatiguing soccer match in elite soccer players examined by ^{13}C-MRS. *Medicine and Science in Sports and Exercise*, **31**, 1587–93.

Saltin, B. (1973). Metabolic fundamentals in exercise. *Medicine and Science in Sports*, **5**, 137–46.

Saltin, B. and Costill, D.L. (1988). Fluid and electrolyte balance during prolonged exercise, in *Exercise, Nutrition, and Metabolism*, (eds E.S. Horton and R.L. Terjung), Macmillan, New York, pp. 150–8.

Shephard, R.J. and Leatt, P. (1987). Carbohydrate and fluid needs of the soccer player. *Sports Medicine*, **4**, 164–76.

Shirreffs, S.M. and Maughan, R.J. (2000). Rehydration and recovery of fluid balance after exercise. *Exercise and Sports Science Reviews*, **28**, 27–32.

Short, S.H. and Short, W.R. (1983). Four year study of university athletes' dietary intake. *Journal of the American Dietetic Association*, **82**, 632–45.

Tipton, K.D. and Wolfe, R.R. (2001). Exercise, protein metabolism, and muscle growth. *International Journal of Sport Nutrition and Exercise Metabolism*, **11**, 109–32.

Vollestad, N.K. and Blom, P.C.S. (1985). Effect of varying exercise intensity on glycogen depletion in human muscle fibres. *Acta Physiologica Scandinavica*, **125**, 395–405.

Vollestad, N.K., Vaage, O. and Hermansen, L. (1984). Muscle glycogen depletion patterns in type 1 and subgroups of type 11 fibres during prolonged severe exercise in man. *Acta Physiologica Scandinavica*, **122**, 433–41.

Williams, M.H. (1991). Alcohol, marijuana and beta blockers, in *Ergogenics: the Enhancement of Sports Performance* (eds D.R. Lamb and M.H. Williams), Benchmark, Indianapolis.

7   Different populations

Thomas Reilly

Introduction

Soccer at the professional level of play is arguably the world's leading sport. The best players receive the adulation of fans and lucrative financial rewards from both employers and sponsors. Yet the popularity of soccer is reflected as equally in the number of active participants in the recreational game as in those who watch the elite play.

Participants in soccer include young boys and girls, those who play the game for purposes of 'keep fit' indoors or outdoors and a growing number of women and veteran recreational players. Indeed the game has been modified for play by visually handicapped people with appropriate audio stimuli for locating the movement of the ball. The officials who regulate the game and control the players are usually neglected when aspects of play are considered.

This chapter is devoted to different populations within soccer. Attention is directed to specific groups including young and veteran players, disabled players and participants in modified forms of the game. Finally, consideration is given to soccer referees, their roles and fitness requirements.

7.1   Women's soccer

Information about women's soccer is integrated within the content of the other chapters as far as possible. Issues not addressed already are considered here, since research into women's soccer has lagged considerably behind the men's game. This is because participation by women in the game has only recently been actively promoted. There was a step-wise increase in women's soccer clubs in the majority of European countries throughout the 1980s; for example, the number of women's clubs registered with the FA increased from 188 to 321 between 1980 and 1991. The women's game gained international credibility in 1991 when the First World Cup for women's teams took place in China. In the winning nation, the USA, the game is now widely played, as it is in Sweden, the host of the Second World Cup Finals in 1995. The USA, host to the 1999 World Cup Finals, now has an estimated 19 million females participating in soccer. Nevertheless, the game still has to be developed in a number of countries world-wide where women's participation in sport is restricted by cultural, domestic and economic circumstances.

It is probably only at the elite or professional level of play that female soccer players engage in strenuous sport-specific training programmes. In Danish soccer clubs, the training of women during the season consists of two to three sessions of 90 min each week. The national team players undergo additional training which consists of running (two to four sessions of 20–30 min per week) and general strength training (30 min once or twice a week). The training frequency is modified to suit individuals according to their domestic or

Table 7.1 Maximal oxygen uptake (ml kg^{-1} min^{-1}) mean values for women soccer players

Team	n	$\dot{V}O_2\,max$	References
England national squad	14	52.2[a]	Davis and Brewer (1992)
Italian elite players	12	49.8	Evangelista *et al.* (1992)
Danish national team	10	57.6[b]	Jensen and Larsson (1993)
Australian elite players	10	47.9	Colquhoun and Chad (1986)
Canadian university team	12	47.1	Rhodes and Mosher (1992)
English university players	10	42.4	Miles *et al.* (1993)
Turkish elite players	22	43.2[a]	Tamer *et al.* (1997)
New Zealand team	20	49.1	Dowson *et al.* (2002)

Notes
a Estimated, after training.
b After training.

occupational constraints (Jensen and Larsson, 1993). The intensity and quality of training are altered as the season progresses, starting with an emphasis on continuous running and culminating in short intermittent exercise later on. The programmes are comparable to those of elite British female soccer players, who train on five occasions each week during the season (Davis and Brewer, 1992).

The relative intensity of exercise at which women play soccer approaches that of male counterparts and averages about 70% of $\dot{V}O_2$ max (see Table 7.1). The overall energy expenditure has been estimated to be about 4600 kJ for a 60-kg player. This compares with an estimate of about 5700 kJ for a 75-kg male player (Ekblom and Williams, 1994). Female players sustain heart rates exceeding 85% of maximal values for two-thirds of the game but overall cover less distance than do male players. Recreational players in small-sided games perform at about 70–75% $\dot{V}O_2$ max but play for shorter periods and can take time off for rests (Miles *et al.*, 1993).

The specific nutritional requirements of women soccer players were considered by Brewer (1994). Elite players were recommended to consume a diet high in carbohydrates. They were encouraged to consume foods high in calcium and iron in order to ensure general health. Those counselling female soccer players on nutrition should be aware of the possible existence of (or potential risk of inducing) eating disorders, particularly with regard to advice on weight loss.

The performance of female soccer players is not necessarily compromised by the phase of the menstrual cycle. In other sports, Olympic medals have been won by athletes during menses and throughout the follicular and luteal phases. Some women experience premenstrual discomfort, others dysmenorrhoea or painful menses. Participation in exercise programmes can attenuate menstrual discomfort and oral contraceptives are used to regulate the cycle by some sportswomen. There is evidence that women are more vulnerable to errors in the pre-menses phase and this has been reflected in the incidence of injury incurred by Swedish players (Moller-Neilsen and Hammar, 1989). Fewer traumatic injuries were noted in women players using oral contraceptive pills to reduce pre-menstrual and menstrual symptoms of discomfort.

Women on high training loads are prone to disruptions to the normal menstrual cycle. This may be indicated by a shortened luteal phase or by absence of menses, known as amenorrhoea. Competitive and personal stress are also implicated in amenorrhoea which is seen as a disturbance of the hypothalamic–pituitary–gonadal axis that regulates the menstrual cycle. A consequence of the low oestrogen levels associated with prolonged training-induced amenorrhoea is a loss of minerals (notably calcium) from bone. The osteoporosis observed in women distance runners is rarely found in women soccer players who tend to have less arduous training programmes and a higher energy intake.

The body composition of women soccer players tends to be closer to other team sports participants than to endurance athletes. Body adiposity tends to be only slightly lower than for women in the normal population of the same age. Percentage body fat values for Danish national players were 20.1 (17.5–25.0)% in mid-season and 22.3 (20.1–28.3)% at the start of systematic training (Jensen and Larsson, 1993). These figures compare with values of 18.3 (±1.7)% in top Turkish players, 24.7 (±2.4)% for top players in the English League, 25.1 (±5.4)% for university players and 26% for age-matched women in the general population (Reilly and Drust, 1997). England national squad members had values of 21.5 and 21.1% as an average at the start and end of a 12-month training programme (Davis and Brewer, 1993).

The muscle strength of women soccer players can be measured as peak torque during iso-kinetic movements. Top women soccer players tend to have greater muscle strength values than university players or non-games players (Reilly and Drust, 1997). The differences are pronounced at fast movement velocities, a characteristic also of male soccer players.

Despite the high aerobic demands imposed by competition on women soccer players, fitness test profiles suggest that the anaerobic system is more developed than the oxygen transport system. This may be due to an imbalance of emphasis in training or to a reluctance to engage in aerobic training on a systematic basis. Nevertheless, the maximal oxygen uptakes of Danish women soccer players have reached values of 57.6 (51.5–63.8) $ml\,kg^{-1}\,min^{-1}$ in mid-season compared with 53.3 (48.0–60.8) $ml\,kg^{-1}\,min^{-1}$ at the start of systematic training (Jensen and Larsson, 1993). Estimated $\dot{V}O_2$ max values of the England squad increased from a mean value of 48.4–52.2 $ml\,kg^{-1}\,min^{-1}$ during a 12-month period of training (Davis and Brewer, 1992). The average value of Australian national squad members was 47.9 $ml\,kg^{-1}\,min^{-1}$ (Colquhoun and Chad, 1986) and elite Italian players had average values of 49.8 $ml\,kg^{-1}\,min^{-1}$ (Evangelista *et al.*, 1992). These are appreciably higher than the mean of 42.4 (±4.3) $ml\,kg^{-1}\,min^{-1}$ for university players (Miles *et al.*, 1993). A spread of values from various studies is shown in Table 7.1.

7.2   Youth soccer

The behavioural building blocks of soccer excellence are laid down during the growth process. For many sports, the physiological determinants of performances are influenced by genetic factors: they must be carefully nurtured during childhood (approximately ages 4–12 years) and adolescence (ages 13–19 years) if exercise and sporting potential are to be realized. A huge problem for soccer scouts is the identification of soccer talent at an early age.

Developing talent poses questions about whether there is a 'golden age' for specialization, if there is an optimal timing for skills acquisition and so on. Such issues relating to talent identification and development are considered in depth in Chapter 21. There are considerations also for the negative consequences of high training loads on the growing skeleton, the possibilities of psychological 'burnout' through specializing too early and the concomitant parental and social pressures. There is some concern about the hours devoted to soccer by those excelling at an early age who are obliged to play for their school, representative teams and participate also in soccer schools for talented players.

Chronological age is not a perfect marker of biological maturity. Consequently, early developers may be at an advantage in sport because of their size. They may orientate towards particular roles (centre-back or centre-forward) due to their advantage. They may be demotivated and drop out of sport later on when their counterparts catch up. Late developers may emerge as potential champions only when growth is finished. Since a near 12-month difference in age can make an enormous difference in performance capability, some boys and girls may be disadvantaged in under-age competitions by virtue of having a birth date later in the competitive year. In contact sports where strength is important, this will lead to risk of injury. The alternative of soccer competitions where children are matched according to biological age is acknowledged to be unrealistic. Growth and development processes are outlined in detail in Chapter 20.

It seems that the most sensitive period for learning new movement patterns is probably between 9 and 12 years of age. The movement and muscle activity patterns of skilled kicking actions are evident in young soccer players by age 11. The practice in some sports is to start specialist training well before this time. Whilst the physiological systems that sustain prolonged training sessions can withstand the strenuous programmes prescribed, the growing skeletons of these elite children may not. The consequences of the stresses imposed on skeletal structures may show up later in the form of overuse injuries.

There are particular difficulties in interpreting fitness data collected on adolescent and youth players. For example, the $\dot{V}O_2$ max increases with age but this is largely due to an increased body size. Anaerobic power and capacity are less well developed in children compared to aerobic power. This is reflected in the relative contribution of aerobic and anaerobic mechanisms in all-out effort. In children, a 6-min run is dependent almost entirely on aerobic metabolism. For a maximal effort of less than 60 s, children derive 60% of their total energy from anaerobic sources compared to 80% in adults. The relatively poor anaerobic capacity in children is confirmed by low levels of lactate production during intense exercise bouts and suggests a low glycolytic rate. There is especially limited potential in the prepubescent child for developing the anaerobic system. Children are less able than adults to effect ATP re-phosphorylation in anaerobic pathways during high-intensity exercise. Anaerobic capacities increase progressively during maturation until reaching adult levels after the teenage years.

The truism that children are not miniature adults is acknowledged by the national governing bodies of soccer; the rules for youth soccer are different in some respects from the game for adults, although the modifications vary from country to country. Competitive matches start at under-8, in the USA for example, where matches are divided into four quarters each of 12 min. At under-10, each half lasts 25 min, at under-12 this becomes 30 min which is progressively lengthened to 45 min at under-19. In tandem with these changes are modifications to the size of the pitch, the number of players and the allowances for substitutions. The regulations are intended to reduce the physiological strain on young players but the game still retains its intermittent high-intensity nature.

The physiological demands of German under-11 and under-12 matches were assessed by Klimt *et al.* (1992). Heart rates were in the range 160–180 beats min^{-1}, values comparable with elite adult players. Blood lactate levels remained in the 3–4 mmol l^{-1} range and reflect the completion of high-intensity efforts by children without major accumulation of lactate.

7.3   Referees

The referee is the key official in regulating behaviour of soccer participants by implementing the rules of play. This role makes demands on mental faculties, visual perception, attention and decision-making. The referee has to be decisive and strict, yet employ discretion where appropriate. The other officials, the two linesmen (assistant referees), offer some advice in cases of controversy but the ultimate decision is charged to the referee.

These decision-making stresses are superimposed on a relatively high level of physiological stress. The referee is expected to keep up with play whatever its tempo in addition to maintaining alertness throughout the game. These demands have implications for fitness required to officiate at a high level.

Referees in the Premier Division of the English Premier League cover approximately 9.5 km during the course of the game. Of this total distance, on average 47% is covered at a jogging pace, 23% walking, 12% sprinting and 18% reverse running (Catterall *et al.*, 1993). Greater distances have been reported for top-class referees in Japan (Asami *et al.*, 1988) and in Italy (D'Ottavio and Castayna, 2002). Mean values of 10.5 km were reported for seven foreign referees at international matches and 11.2 km for 10 referees in the Japan National Soccer League. Mean values of 11.5 km were reported also for 33 top Italian referees. These figures compare favourably with distances covered by professional soccer players and exceed those reported for central defenders (Reilly, 1994).

The physiological strain incurred by soccer referees can be indicated by monitoring heart rate during a match. Short-range telemetry provides a convenient method of doing this as the equipment is lightweight and can be worn by the referee without interfering with him in any way. Measurement on 13 referees during top-class league matches indicated heart rates averaging 165 beats min^{-1} throughout the whole game (Figure 7.1). An estimate of the aerobic loading on referees would be 70–75% $\dot{V}O_2$ max.

A fatigue effect is evident in referees as indicated by a fall-off in work-rate towards the end of play. This happens despite the fact that the heart rate is maintained at about 165 beats min^{-1} (Catterall *et al.*, 1993). The fatigue is linked with diminishing energy stores within the active muscles. The urgency on the part of the losing team to move the ball quickly forward and press for a score means that the work-rate demands on the referee are unrelenting until the game is over.

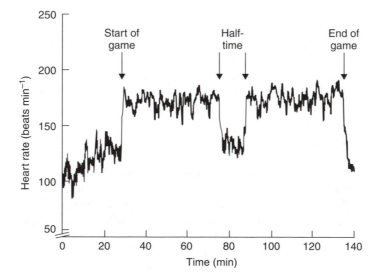

Figure 7.1 Heart rate of a referee throughout a Premier Division match in the English League. (Reproduced with permission from Catterall *et al.*, 1993, with kind permission of Butterworth-Heinemann Journals, Elsevier Science Ltd, Kidlington OX5 1GB, UK.)

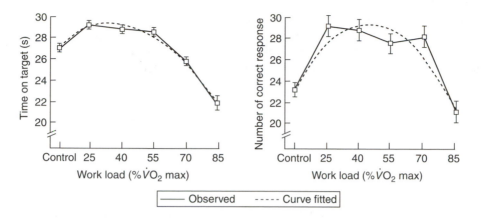

Figure 7.2 Effect of exercise intensity on psychomotor tasks (left) and on cognitive function (right) during exercise. (Data from Reilly and Smith, 1986.)

The activity level may differ with the standard of play being refereed. Harley *et al.* (2002) monitored 14 County League referees. The average distance covered in a game was 7496 m and the mean heart rate was 162 beats min^{-1}. These referees also showed evidence of fatigue, less distance being covered in the second half compared to the first half.

The high level of exercise intensity associated with refereeing has consequences both for mental judgements and for fitness. Decrements in cognitive function are noted once the exercise intensity exceeds about 50% $\dot{V}O_2$ max (Figure 7.2). Decrements in less complex psychomotor functions are observed at only marginally greater exercise intensities (Reilly and Smith, 1986). It is likely, therefore, that decision-making is prone to error at the exercise intensities experienced by referees during professional and international match-play.

The overall pattern of a referee's activity is acyclical and varies in parallel with the players' actions. Players incur additional energy expenditure when directly involved in executing match skills but they have a degree of choice in undertaking high-intensity efforts 'off-the-ball'. In contrast, the referee has to follow the play, irrespective of the intensity of previous movements and sometimes may not be able to keep up with it. The referee must, therefore, be able to anticipate the direction of play in advance. The referee does not display the acute-angled changes of direction which players make, but the overall work-rate profiles and the frequency of discrete evens (a change in category of activity every 6 s) are broadly similar.

It seems that the major physiological demands on referees are imposed on the oxygen transport system. Superimposed on these are short-term high-intensity runs. The work-rate profiles exhibited by referees can be interpreted to mean that aerobic fitness is an important requirement of soccer referees but they also need to be quick 'off-the-mark'. The national governing bodies recognize this and impose strict fitness standards on officials. The need to pass these fitness tests is an ongoing concern of referees since some loss of fitness occurs due to the ageing process. The referees at professional soccer matches are often 15 years older than the average player and so the work-rate must be deemed appreciable. The fact that the top referees no longer have to hold down a job outside football has introduced the concept of the referee as a professional. This brings them little sympathy from soccer spectators when refereeing decisions are not to their liking.

7.4   Veteran players

Professional soccer players in the 1990s have tended to maintain their careers for as long as possible. This contrasts with previous decades where it was the convention to contemplate retirement at about 30. Indeed many players now continue at the highest level well into their thirties, as evidenced by the ages of the oldest players at the World Cup Finals. This is especially true of goalkeepers who have been traditionally able to stay in the game for longer than outfield players.

Another trend in recent years has been the tendency to continue playing at a recreational level following retirement from a professional career. Such participation may be for charitable or recreational reasons or for purposes of health. The emphasis on active sport as a means towards prevention of coronary heart disease has given rise to the growth of veterans' soccer, although this is largely a characteristic of developed nations with 'sport for all' promotional programmes.

Kohno *et al.* (1988) investigated the health status of senior Japanese citizens who had been playing soccer for about 2 h a week for 40 years. These individuals had lower body fat levels than age-matched norms and had $\dot{V}O_2$ max values clearly superior to population values. These observations would support a lifelong commitment to recreational soccer for purposes of maintaining sound cardiovascular health. Nevertheless the following qualifications were made. Matches should consist of 20-min halves; a small ball should be used; play should be on a grass or artificial surface; substitutes should be freely allowed and play should be against opponents of the same age.

The relatively high intensity of recreational soccer calls for attention to physical fitness if it is to be played safely. For this reason, it is important to get fit for the game before making a regular commitment to representing the team. This applies particularly in veterans' soccer in view of the greater health risks that are associated with the ageing process.

7.5   Small-sided games

Modifications to the normal rules are made for young and for veteran players. Professionals may engage in 'conditioned games' during training sessions in order to emphasise particular aspects of skill or tactics. Games for recreational play also have their own local rules: this applies in particular to small-sized matches, with anything from four to nine members on each side.

Small-sided games are especially suited to indoor leagues. Generally, rules prohibit side-tackling, playing the ball above waist (or shoulder) height or scoring from your own half of the pitch. Indoor matches tend to be for short periods, say 10–15 min each half, and are played at a high tempo. This is because the ball is out of play for only a small time, as it remains in play when hit off the side- or back-wall. The high tempo is a feature of the entertainment provided by professional indoor soccer in North America, where free substitution is allowed at all times. This form of play is common also in South America and Australia much more so than in Europe, where its use tends to be for recreational purposes. The extent to which errors and risk of injury are associated with a fast pace of play is unknown.

The physiological load imposed by a 4-a-side soccer incorporates anaerobic and aerobic components. The mean relative exercise intensity exceeds 82% $\dot{V}O_2$ max and mean blood lactate levels of 4.5–4.9 mmol l^{-1} have been reported. Time in possession of the ball is higher than in 11-a-side outdoor matches (MacLaren *et al.*, 1988). These observations lead to the conclusion that small-sided games provide an effective physiological stimulus for preparation for the normal 11-a-side match.

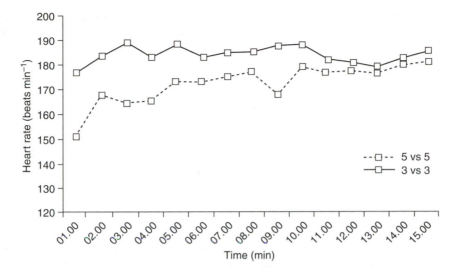

Figure 7.3 Mean heart rates for players during 3 versus 3 and 5 versus 5 matches. (Data from Platt *et al.*, 2001.)

Male and female players may be used in different combinations to make up small-sided teams. Females have virtually the same physiological responses to recreational play in female-only and in mixed (two males and two females) teams (Miles *et al.*, 1993). In the latter condition, their direct participation in play tends to be reduced due to a male dominance in dictating play. This could be overcome by an appropriate matching of skill levels between the sexes.

The size of teams is an important consideration in young players where a major objective is skill acquisition. Platt *et al.* (2001) compared the responses of young players under 12 years of age to 3-a-side and 5-a-side matches which lasted 15 min. In the former, the boys had more touches of the ball, more successful passes and more tackles than in the 5 versus 5 matches. Moreover, they operated at a higher heart rate all through the session. It was concluded that the small-sided game had multiple advantages over the 5 versus 5 contests for skill learning and physiological reasons (Figure 7.3).

7.6   Disability soccer

Traditionally, the game of soccer was modified for play by handicapped individuals. For example, visually impaired players engaging in recreational soccer were assisted by audio stimuli for locating the ball. With the development of the Paralympic Games, there has been a parallel progress in soccer for disabled individuals, culminating in what is known as 'disability soccer'.

In one form of 'disability soccer' the sport is restricted to those who have lost a lower limb. Individuals move on crutches but are disallowed from using the crutch to control the ball, which would be equivalent to a 'hand ball'. The duration of matches is restricted compared with normal regulations but competition is no less intense.

Wilson (2002) described the sports science support system for the England team when participating in the 2001 World Championship in Brazil. The scientific information on heat stress and rehydration was directly transferable from able-bodied athletes to the disabled squad. Moreover, the preparations for the next World Championships were outlined, mirroring the training plans and player profile assessments employed in preparing all competitive squads.

Summary

The popularity of soccer is reflected in the range of participants at a recreational and competitive level. These include under-age and veteran players as well as women and male athletes. The game has been modified for play by visually impaired individuals. It has been modified for conditioned games, for indoor recreational purposes and for competitive tournaments. The match officials are rarely the subject of scientific investigations, despite the physiological and psychological stresses associated in particular with refereeing.

References

Asami, T., Togari, H. and Ohashi, J. (1988). Analysis of movement patterns of referees during soccer matches, in: *Science and Football* (eds T. Reilly, A. Lees, K. Davids and W.J. Murphy), E & FN Spon, London, pp. 341–5.

Brewer, J. (1994). Nutritional aspects of women's soccer. *Journal of Sports Sciences*, **12**, 535–8.

Catterall, C., Reilly, T., Atkinson, G. and Goldwells, A. (1993). Analysis of the work rates and heart rates of association football referees. *British Journal of Sports Medicine*, **27**, 193–6.

Colquhoun, D. and Chad, K.E. (1986). Physiological characteristics of female soccer players after a competitive season. *Australian Journal of Science and Medicine in Sport*, **18**, 9–12.

Davis, J.A. and Brewer, J. (1992). Physiological characteristics of an international female soccer squad. *Journal of Sports Sciences*, **10**, 142–3.

Davis, J.A. and Brewer, J. (1993). Applied physiology of female soccer players. *Sports Medicine*, **16**, 180–9.

D'Ottavio, S. and Castagna, C. (2002). Activity profile of top level soccer referees during competitive matches, in: *Science and Football IV* (eds W. Spinks, T. Reilly and A. Murphy), E & FN Spon, London, pp. 151–6.

Dowson, M.N., Cronin, J.B. and Presland, J.D. (2002). Anthropometric and physiological differences between gender and age groups of New Zealand and national soccer players, in: *Science and Football IV* (eds W. Spinks, T. Reilly and A. Murphy), E & FN Spon, London, pp. 63–71.

Ekblom, B. and Williams, C. (1994). *Foods, Nutrition and Soccer Performance*. E & FN Spon, London.

Evangelista, M., Pandolfi, O., Fanton, F. and Faina, M. (1992). A functional model of female soccer players: analysis of functional characteristics. *Journal of Sports Sciences*, **10**, 165 (abstract).

Harley, R.A., Banks, R. and Doust, J. (2002). The development and evaluation of a task specific fitness test for association football referees, in: *Science and Football IV* (eds W. Spinks, T. Reilly and A. Murphy), E & FN Spon, London, pp. 76–80.

Jensen, K. and Larsson, B. (1993). Variations in physical capacity in a period including supplemental training of the national Danish soccer team for women, in: *Science and Football II* (eds T. Reilly, J. Clarys and A. Stibbe), E & FN Spon, London, pp. 114–17.

Klimt, F., Betz, M. and Seitz, U. (1992). Metabolism and circulation of children playing soccer, in: *Children and Exercise XVI: Paediatric Work Physiology* (eds J. Coudert and E. van Praagh), Masson, Paris, pp. 127–9.

Kohno, T., O'Hata, N., Morita, H. *et al.* (1988). Can senior citizens play soccer safely?, in: *Science and Football* (eds T. Reilly, A. Lees, K. Davids and W.J. Murphy), E & FN Spon, London, pp. 230–6.

MacLaren, D., Davids, K., Isokawa, M. *et al.* (1988). Phsyiological strain in 4-a-side soccer, in: *Science and Football* (eds T. Reilly, A. Lees, K. Davids and W.J. Murphy), E & FN Spon, London, pp. 76–80.

Miles, A., MacLaren, D., Reilly, T. and Yamanaka, K. (1993). An analysis of physiological strain in four-a-side women's soccer, in: *Science and Football II* (eds T. Reilly, J. Clarys and A. Stibbe), E & FN Spon, London, pp. 140–5.

Moller-Nielsen, J. and Hammar, M. (1989). Women's soccer injuries in relation to the menstrual cycle and oral contraceptive use. *Medicine and Science in Sports and Exercise*, **21**, 126–9.

Platt, D., Maxwell, A., Horn, R., Williams, M. and Reilly, T. (2001). Physiological and technical analysis of 3 v 3 and 5 v 5 youth football matches. *Insight: The F. A. Coaches Association Journal*, **4**, 23–4.

Reilly, T. (1994). Motion characteristics, in: *Football (Soccer)* (ed. B. Ekblom), Blackwell Scientific, London, pp. 31–42.

Reilly, T. and Drust, B. (1997). The isokinetic muscle strength of women soccer players. *Coaching and Sport Science Journal*, **2**, 12–17.

Reilly, T. and Smith, D. (1986). Effect of work intensity on performance in a psychomotor task during exercise. *Ergonomics*, **29**, 601–6.

Rhodes, E.C. and Mosher, R.E. (1992). Aerobic and anaerobic characteristics of female university soccer players. *Journal of Sports Sciences*, **10**, 143–4.

Tamer, K., Gunay, M., Tiryaki, G., Cicioolu, I. and Erol, E. (1997). Physiological characteristics of Turkish female soccer players, in: *Science and Football III* (eds T. Reilly, J. Bangsbo and M. Hughes), E & FN Spon, London, pp. 37–9.

Wilson, D. (2002). Sport science support for the England amputee team. *Insight: The F. A. Coaches Association Journal*, **5**(2), 31–3.

Part 2

Biomechanics and soccer medicine

8 Biomechanics applied to soccer skills

Adrian Lees

Introduction

Sports biomechanics offers methods by which the very fast actions which occur in sport can be recorded and analysed in detail. There are various reasons for doing this. One is to understand the general mechanical effectiveness of the movement, another is the detailed description of the skill, yet another is an analysis of the factors underlying successful performance. An important application of sports biomechanics within any sport, and soccer in particular, is the definition and understanding of skills. This can help in the coaching process and as a result enhance the learning and performance of those skills.

There are a wide range of skills which form the foundation of soccer performance. Those which have been the subject of biomechanical analysis are the more technical ones which are concerned directly with scoring. For example in soccer, shooting at goal is an aspect of kicking and is the means by which goals are scored. Similarly, heading the ball and throwing-in can be important elements of attacking play, while goalkeeping skills are important in preventing goals.

Other skills are important in the game but have received much less attention in terms of biomechanical analysis. For example, kicking actions such as passing and trapping the ball, tackling, falling behaviour, jumping, running, sprinting, starting, stopping and changing direction are all important skills in soccer but have received little detailed analysis. The skills in other codes of football have similarly received little attention in terms of biomechanical analysis.

This chapter looks at those skills in which biomechanics has been successfully applied in order to gain an insight into their mechanical characteristics.

8.1 Kicking

Kicking is without doubt the most widely studied skill in soccer. Although there are many variations of this skill due to ball type, ball speed and position, nature and intent of kick, the variant which has been most widely reported in the literature is the maximum velocity instep kick of a stationary ball. Essentially this corresponds to the penalty kick in soccer.

The mature form of the kicking skill has been described by Wickstrom (1975). It is characterized by placement of the supporting leg at the side and slightly behind the stationary ball. The kicking leg is first taken backwards and the leg flexes at the knee. The forward motion is initiated by rotating around the hip of the supporting leg and by bringing the kicking leg thigh forwards. The leg is still flexing at the knee at this stage. Once this initial action has taken place, the thigh begins to decelerate until it is essentially motionless at ball contact. During this deceleration, the shank vigorously extends about the knee to almost full

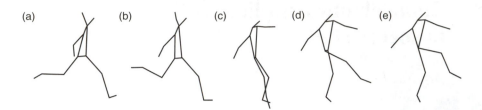

Figure 8.1 Kinetograms of the key positions during a soccer kick showing: (a) maximum hip
retraction; (b) forward movement of the thigh and continued knee flexion; (c) ball contact;
(d) post-impact follow through; and (e) knee flexion as follow through proceeds.

extension at ball contact. The leg remains straight through ball contact and begins to flex
during the long follow through. The foot will often reach above the level of the hip during
the follow through. A series of kinetograms of the kicking action are presented in Figure 8.1.

Kicking, like many of the skills in soccer, is developmental in nature and it has been
shown that it develops from an early age. Bloomfield *et al.* (1979) analysed the kicking
action of young boys from the age of 2–12 years. They looked at various indicators of performance and were able to characterize six levels of development. These ranged from level 1
(average age 3.9 years) where the children often hit the ball with their knee or leg, to level 6
(average age 11.2 years) where the mature kicking pattern as described above had been
achieved by 80% of the children. The intermediate ages for levels 2–5 were 4.11, 4.8, 6.11
and 8.2 years respectively. Although chronological age was not found to be a good predictor
of level of skill development, the age ranges suggest that the skill develops rapidly between
the age of 4 and 6 years. This has implications for skill development for children of a very
young age and is an illustration of the role biomechanics can play in this area.

The definition of the mature skill above suggests that there are four stages to motion. The
first is the retraction of the leg during the backswing; the second is the rotation of both thigh
and shank forwards which occurs as a result of hip rotation and thigh flexion; the third is
when the thigh decelerates and there is corresponding shank acceleration leading to impact
with the ball; and the fourth is the follow through. The two intermediate stages are the most
important from a performance point of view. The interaction between the thigh and shank
can be seen in Figure 8.2 which shows their angular velocities throughout the movement. On
this graph each stage is marked, and it can be seen that during stage 2, both the lower and
upper leg increase in angular velocity. The muscular energy for this must come from the
muscles around the hip and thigh. In stage 3, just before impact, there is an increase in shank
angular velocity, while at the same time a decrease in the thigh angular velocity. There
appears to be an interaction between the two segments. A high angular velocity of the shank
means a high foot velocity, and this is important in the production of a well hit kick. It can
be seen that in order to achieve high foot velocity energy must be built up in the early stage
of the movement. About 50% of the shank angular velocity at impact is built up during stage
2, and the remaining 50% is produced during stage 3. The energy of the thigh is built up in
stages 1 and 2. Therefore the range of movement that the hips and leg move through, and the
muscular strength applied during these stages will determine the maximal speed of the foot
at impact.

It would be expected that there is a relationship between muscle strength and performance, due to the fact that the muscles are directly responsible for increasing foot velocity.

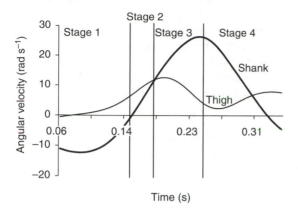

Figure 8.2 Angular velocity of the upper and shank during a soccer kick. The four stages of the kick are marked with impact being at the end of stage 3 just after the shank reaches peak velocity.

Such a relationship has been found by several researchers. Cabri *et al.* (1988) found that there was a high correlation between knee flexor and extensor strength as measured by an isokinetic muscle function dynamometer and kick distance. There was also a significant relationship between hip flexor and extensor strength but this was lower than that for the knee. Similar results have been found by Poulmedis (1988) and Narici *et al.* (1988) who used ball velocity as a measure of performance. If muscle strength is related to performance then it would be expected that training should show positive effects on ball speed or distance. De Proft *et al.* (1988) found that over a season of specific leg muscle strength training, muscle strength increased and so too did kick performance as measured by kick distance. The correlations between leg strength and distance increased from the beginning to the end of the season.

Although the evidence reviewed above suggests that there is a good relationship between muscle strength and performance, there are other factors which contribute to successful kicks. These factors can be appreciated from a consideration of the relationship between foot and ball velocity before and after contact. By considering the mechanics of collision between the foot and ball (following the treatment of Plagenhoef, 1971), the velocity of the ball can be stated as:

$$V_{(ball)} = \frac{V_{(foot)} \cdot [M] \cdot [1 + e]}{[M + m]} \tag{1}$$

where V = velocity of ball and foot, respectively, M = effective striking mass of the leg; m = mass of the ball; and e = coefficient of restitution. The effective striking mass is the mass equivalent of the striking object (in this case the foot and leg) and relates to the rigidity of the limb. A similar expression has recently been reported by Bull-Andersen *et al.* (1999).

The term $M/[M + m]$ gives an indication of the rigidity of impact and relates to the muscles involved in the kick and their strength at impact. Therefore one would expect that the best correlations with performance would be with eccentric muscle strength, and the data from Cabri *et al.* (1988) suggest that this is the case.

The term $[1 + e]$ relates to the firmness of the foot at impact. Because the ball is on the ground, the foot contacts the ball on the dorsal aspect of the phalanges and lower metatarsals. The large force of impact serves to forcefully plantarflex the foot and it will do so until the bones at the ankle joint reach their extreme range of motion. At this stage the foot will deform at the metatarsal–phalangeal joint. There is little to prevent considerable deformation here and this will affect the firmness of impact and the value of '*e*', the coefficient of restitution. If contact between the foot and ball is made closer to the ankle joint then the impact will be firmer. In punts or drop kicks it is possible to achieve this.

The term $M/[M + m]$ would be expected to have a value of about 0.8 based on realistic data for the masses of the foot and the ball. The term $[1 + e]$ would be expected to have a value of about 1.5. Therefore, the product of the two suggests that the ball should travel at about 1.2 times the velocity of the foot, in other words the ball leaves the foot faster than the foot is travelling. The relationship now becomes

$$V_{(ball)} = 1.2V_{(foot)} \tag{2}$$

This ratio of ball to foot velocity is an indicator of a successful kick. Foot velocities for competent soccer players are between about 16 and $22\,\mathrm{m\,s^{-1}}$ and ball velocities are in the range of 24–$30\,\mathrm{m\,s^{-1}}$ (Lees and Nolan, 1998) which always gives a ball/foot velocity ratio greater than 1. For professional players performing a maximal low drive instep kick, Asami and Nolte (1983) reported a ball/foot velocity ratio as low as 1.06 while Plagenhoef (1971) has reported values over 1.5 for side-foot kicks.

For submaximal kicking, there still appears to be a relationship between foot and ball velocity. Zernicke and Roberts (1978) reported a regression equation between the two variables of foot and ball velocity over a ball speed range of 16–$27\,\mathrm{m\,s^{-1}}$:

$$V_{(ball)} = 1.23V_{(foot)} + 2.72 \tag{3}$$

This is reassuringly close to the relationship for maximal kicking (equation 2) suggested above on the basis of theoretical data.

A characteristic of all soccer place kicking and frequently seen in other codes is the angled approach to the ball. Isokawa and Lees (1988) investigated the effect of changing the angle of approach on foot and ball velocities. Six male subjects were required to take one step approach in order to kick a stationary ball from angles of $0°$, $15°$, $30°$, $45°$, $60°$ and $90°$. (The direction of the kick was $0°$.) The foot and ball speed were measured from high speed cine film. Although there were no significant differences between approach directions, the trend in the data suggested that the maximum swing velocity of the leg ball was achieved with an approach angle of $30°$ and the maximum ball velocity with an approach angle of $45°$. Therefore, an approach angle between $30°$ and $45°$ would be considered optimum, and agrees with practical observations. The explanation for this finding is that with an angled approach, the leg also is angled to the ball in the lateral plane, and so can be placed more under the ball and make a better contact. It was noted above that a more solid impact position will produce higher ball velocities.

Other types of kick have also been studied. The side-foot kick is often used to make a pass. In order to make a side-foot kick the foot has to be angled outwards. This prevents the leg from flexing in the same way as it would for an instep kick. Therefore, the foot velocity during a side-foot kick is lower than the instep kick. As contact with the ball is made on the firm bones of the shank and ankle, the foot provides a much better surface for the impact. The resultant ball velocity is much higher than it would be with the same foot velocity for an instep kick. A further advantage is that the flatter side of the foot allows a more accurate placement of the ball. Levanon and Dapena (1998) compared the instep with side-foot kicks and reported that during the side-foot kick for a right footed player, the player orientates the pelvis, the right leg and foot more towards the right. Although the velocity of the foot at impact in the side-foot kick is lower than the instep kick, most of the speed of the foot in both cases is generated by knee extension.

The punt kick and drop kick have been compared by McCrudden and Reilly (1993). They found that the mean range of the best drop kick was 36.1 m while for the punt kick the range was 40.1 m. They did not measure velocities or angles of projection and so the reason for the superior range in the punt may well be due to angle of projection rather than higher ball velocity. Elliott *et al.* (1980) investigated the developmental nature of the punt kick. They found similar results to those reported by Bloomfield *et al.* (1979) in terms of levels and ages of development of the punt kicking skill.

During the kicking motion, the foot rotates about both the medio-lateral and longitudinal axes of the body, and so the true kinematics of the kicking skill can only be fully defined from a three-dimensional (3D) analysis. A small number of 3D studies have been conducted but most have not reported on the movements which occur specifically in the frontal or transverse planes, such as hip and trunk rotations. Where these have been reported, they have provided an insight into the additional variables that may be associated with successful performance. Browder *et al.* (1991) described selected 3D characteristics of the instep kick of female players including data for pelvic rotation and hip adduction. They demonstrated that when comparing a fast to a slow kick, the pelvis showed greater rotational range of motion (ROM) for the faster kick than for the slower kick (18° compared to 13°). This suggests an important role for rotation of the hips in producing high velocity kicks. The authors suggested that pelvic rotation might be a method by which female players enhance the speed of their kick, rather than relying on joint extension at the hip and knee. Tant *et al.* (1991) also claimed that in the performance of the maximal velocity instep kick, males used greater ranges of motion at the hip and knee while females exhibited greater pelvic rotation. Using a speed-accuracy analysis, Lees and Nolan (2002) also showed that increased ball speed was associated with increased range of motion at the hips and increased angular velocity of the knee and hip joints, all of which appeared to be associated with the length of the last stride before the kick was made.

8.2   Soccer throw-in

The throw-in is both a method of restarting the game and a tactical skill. The long-range throw-in can be performed from a stationary position or with a run-up. In a stationary throw-in, the movement is performed with both feet together on the ground. The throw is initiated by bending the knees and taking the ball backwards with both hands behind the head. As the ball is travelling backwards with respect to the body there is an upward extension of the knee joint and a marked pushing of the hips both forward and upward. This serves to prime the upper body for the recoil which will propel the ball forwards. As the upper body starts to come forwards there is a sequential unfolding starting with the hips, then followed by the shoulders, elbows and finally the wrists and hands until ball release.

The sequencing of body segment motions is similar in the run-up throw but the feet placement on the ground is staggered one in front of the other behind as can be seen from the

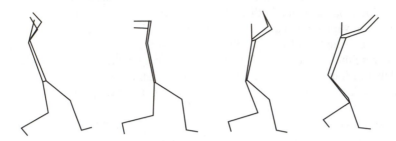

Figure 8.3 A kinetogram of a player performing a run-up throw-in (the ball is omitted from this illustration).

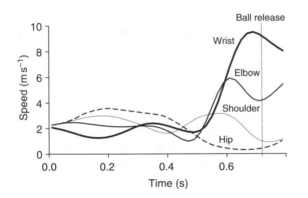

Figure 8.4 The joint velocities for a run-up throw-in as depicted in Figure 8.3. These graphs illustrate the rapid increase in velocity achieved by the hand, wrist and ball during the last phase of the motion.

kinetogram in Figure 8.3. This sequential series of rotations about the medio-lateral axis serves to build up initial rotational velocity using the large muscles of the legs and trunk first, and then to transfer this energy out towards the distal segments in order to gain high end speed velocity as seen in Figure 8.4. This mechanism is identical in principle to that used to attain high foot velocity in the kick. The advantage of the run-up in a running throw-in is that the ball has an initial forward speed. The running action means that one foot will be leading the other into the action. The movement goes from the rear to the front foot, but because both feet need to be on the ground, forward hip movement is restricted. The general segment motions of the upper body and their sequence are identical to those in the standing throw-in.

For the run up throw-in, Levandusky *et al.* (1985) reported an angle, height and speed of release of 29.1°, 2.32 m and 18.3 m s^{-1}, respectively and a range of 23.1 m. Kollath and Schwirtz (1988) investigated the long-range throw-in action of skilled players with and without a run-up. They found that the mean distance achieved using the running throw-in was 24.1 m compared to 20.9 m for the standing throw-in. There were similar angles and heights of release in the two types of throw so the differences in range were attributed to differences in release speed which were 15.3 and 14.2 m s^{-1}, respectively. The running throw-in is clearly superior to the standing throw-in as far as release speed and distance thrown are concerned. Within both types of throw, there were differences in the release parameters chosen to obtain similar ranges. Some players chose to use a low velocity and high trajectory while others used a high velocity and low trajectory. The low-speed high-trajectory throw might be used by players with poor muscular capability, or as a strategy to reach over a defensive wall or to ensure that the ball descends more vertically onto the awaiting players. The high-speed low-trajectory throw will reduce the time the ball is in the air and might be used as a strategy to reduce the time for the opposition to regroup for the defence.

Other forms of the throw-in skill have appeared in order to take advantage of its attacking capability. One such variant is the 'handspring' throw-in which first appeared in US collegiate football during the 1980s (Messier and Brody, 1986). The player runs up with the ball in both hands, the ball is placed on the ground and the player rotates over it as in a handspring. After the feet hit the ground the body is rotating forwards with the arms and ball over the head. During the recovery from the handspring the ball is released. The rules state that for a throw-in to be legal, the player must face the field of play, have both feet in contact with the ground, be on or outside the touchline at release and deliver the ball from behind and over the head using both hands equally. The handspring throw-in is a novel approach to the throw-in which does not contravene the rule. The advantage of this type is that it is thought to have a greater velocity of release and hence a greater range.

Messier and Brody (1986) described the mechanics of the conventional running throw-in compared to the handspring throw-in. They studied 13 university-level players performing the conventional throw-in and four performing the handspring throw-in. Data for the two groups at release are shown in Table 8.1.

These data show that the handspring throw-in achieves considerably greater distance with a lower angle of projection than the conventional throw-in. A model for the conventional running throw-in is one which is characterized by the body as a whole moving forwards and upwards and rotating forwards. On to this is added the action of shoulder and elbow extension. There is a sequencing of segmental actions going from the large to small segments distally, that is, from trunk to upper arm to lower arm. The model for the handspring throw-in is one which is characterized by a body which is moving forwards at higher speed but dropping. It is also rotating forwards at a higher speed. The end point velocity is therefore enhanced by the rotation of the whole body. There is sequential extension of the arms but this is less important. As a result the handspring throw-in is faster (23 compared to $18\,\mathrm{m\,s^{-1}}$) with a lower angle of projection (23° compared to 28°). As a consequence of these release parameters the ball goes further. It would therefore be more suitable for playing strategies requiring a fast long-range ball.

Table 8.1 Mechanical data for the conventional and handspring throw-ins at release

	Conventional	*Handspring*
Ball velocity (m s^{-1})	18.1	23.0
Angle of release (deg)	28.0	23.0
Distance (m)	29.3	44.0
Centre of mass velocities		
Horizontal (m s^{-1})	1.8	3.2
Vertical (m s^{-1})	0.9	−0.8
Angular (rad s^{-1})	−4.5	−5.5

8.3   Goalkeeping

Goalkeeping skills are important in preventing opponents scoring. The goalkeeper has to anticipate attacks on goal and be positioned accordingly. There are various movement skills that the goalkeeper needs to master, but few of them have been subjected to biomechanical analysis.

With regard to this positioning, Figure 8.5 shows the position the goalkeeper adopts to reduce the angle of shot from the opponent.

The diving motion made by goalkeepers in saving a set (penalty) shot is reported by Suzuki *et al.* (1988). They analysed two skilled and two less skilled goalkeepers in terms of their ability to dive and save. They found that the more skilled keepers dived faster ($4\,\mathrm{m\,s^{-1}}$ as opposed to $3\,\mathrm{m\,s^{-1}}$) and more directly at the ball. In this case, the skilled keeper was able to perform a counter-movement jump and launch himself into the air and then turn to meet the ball. The less skilled keeper failed to perform a counter-movement, thereby restricting his take-off velocity. He also failed to turn his body effectively to meet the ball. In this analysis both quantitative and qualitative methods were used to clarify the differences in performance between the two groups. No other goal-keeping actions have been studied in this way.

In an analysis of goalkeeping dive actions, Graham-Smith and Lees (1999) conducted an investigation whereby a ball was projected into several areas of the goalmouth (A–L in Figure 8.6). Seven basic goalkeeping actions were identified as indicated in the shaded areas

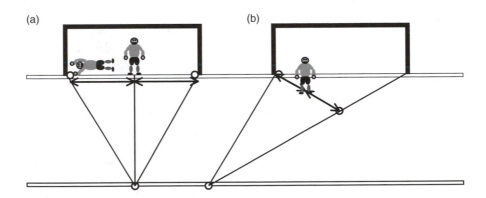

Figure 8.5 Diagram to show how the goalkeeper's diving range changes with ball position.

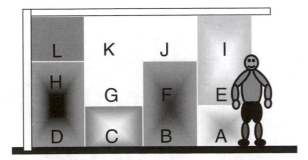

Figure 8.6 Diagram to show how the 12 goalkeeping areas and their relationship to the seven diving actions.

Table 8.2 Goalkeeping action types used to save the ball in each of the 12 goal areas of Figure 8.6

Goalkeeping action	Cross-over step	Side step
Type 1 area A Collapse both legs and drop to ground	0	0
Type 2 areas E and I Single or two legged jump to get in line with ball	0	0
Type 3 areas B and F Right leg comes under body and dive off left leg	0	1
Type 4 area C Small step to right followed by low dive driving off left leg	0	1
Type 5 areas G, J and K Small step followed by drive upwards off right leg	0	1
Type 6 areas D and H Cross-over step before diving off left leg	1	0
Type 7 area L Two cross-over steps and dive off right leg	2	0

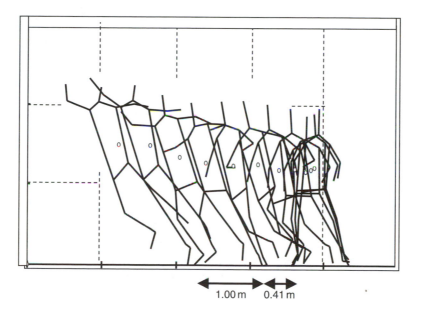

1.00 m   0.41 m

Figure 8.7 An example of diving action – Type 7 with a two cross-over steps followed by a dive from the right leg.

of Figure 8.6 and tabulated in Table 8.2. These actions were characterized by the use of a cross-over and side-steps when reaching for the ball (Figure 8.7).

Summary

This overview has illustrated the way in which biomechanics can be applied to gain an insight into the performance of soccer skills. Many skills in the games of soccer are amenable to biomechanical analysis, but relatively few of them have been analysed in depth. There are still many opportunities for biomechanists to apply their analytical methods to soccer skills and to contribute to the development of science and soccer.

References

Asami, T. and Nolte, V. (1983). Analysis of powerful ball kicking, in *Biomechanics VIII-B* (eds H. Matsui and K. Kobayashi), Human Kinetics, Champaign, IL, pp. 695–700.

Bloomfield, J., Elliott, B.C. and Davies, C.M. (1979). Development of the punt kick: a cinematographical analysis. *Journal of Human Movement Studies*, **6**, 142–50.

Browder, K.D., Tant, C.L. and Wilkerson, J.D. (1991). A three dimensional kinematic analysis of three kicking techniques in female players, in *Biomechanics in Sport IX* (eds C. L. Tant, P. E. Patterson and S. L. York), Ames, IA, ISU Press, pp. 95–100.

Bull-Andersen, T., Dorge, H.C. and Thomsen, F.I. (1999). Collisions in soccer kicking. *Sports Engineering*, **2**, 121–6.

Cabri, J., De Proft, E., Dufour, W. and Clarys, J.P. (1988). The relation between muscular strength and kick performance, in *Science and Football* (eds T. Reilly, A. Lees, K. Davids and W.J. Murphy), E & FN Spon, London, pp. 186–93.

De Proft, E., Cabri, J., Dufour, W. and Clarys, J.P. (1988). Strength training and kick performance in soccer, in *Science and Football* (eds T. Reilly, A. Lees, K. Davids and W.J. Murphy), E & FN Spon, London, pp. 108–13.

Elliott, B.C., Bloomfield, J. and Davies, C.M. (1980). Development of the punt kick: a cinematographical analysis. *Journal of Human Movement Studies*, **6**, 142–50.

Graham-Smith, P. and Lees, A. (1999). Analysis of technique of goalkeepers during the penalty kick. *Journal of Sports Sciences*, **19**, 916.

Isokawa, M. and Lees, A. (1988). A biomechanical analysis of the instep kick motion in soccer, in *Science and Football* (eds T. Reilly, A. Lees, K. Davids and W.J. Murphy), E & FN Spon, London, pp. 449–55.

Kollath, E. and Schwirtz A. (1988). Biomechanical of the soccer throw-in, in *Science and Football* (eds T. Reilly, A. Lees, K. Davids and W.J. Murphy), E & FN Spon, London, pp. 460–7.

Lees, A. and Nolan, L. (1998). Biomechanics of soccer – a review. *Journal of Sports Sciences*, **16**, 211–34.

Lees, A. and Nolan, L. (2002). 3D kinematic analysis of the instep kick under speed and accuracy conditions, in *Science and Football IV* (eds W. Spinks, T. Reilly and A. Murphy), E & FN Spon, London, pp. 16–21.

Levanon, J. and Dapena, J. (1998). Comparison of the linematics of the full-instep and pass kicks in soccer. *Medicine and Science in Sports and Exercise*, **30**, 917–27.

Levendusky, T.A., Clinger, C.D., Miller, R.E. and Armstrong, C.W. (1985). Soccer throw-in kinematics, in *Biomechanics in Sports II* (eds J. Terauds and J.N. Barham), Academic Publishers, Del Mar, CA, pp. 258–68.

McCrudden, M. and Reilly, T. (1993). A comparison of the punt and the drop-kick, in *Science and Football II* (eds T. Reilly, J. Clarys and A. Stibbe), E & FN Spon, London, pp. 362–8.

Messier, S.P. and Brody, M.A. (1986). Mechanics of translation and rotation during conventional and handspring soccer throw-ins. *International Journal of Sport Biomechanics*, **2**, 301–15.

Narici, M.V., Sirtori, M.D. and Morgan, P. (1988). Maximum ball velocity and peak torques of hip flexor and knee extensor muscles, in *Science and Football* (eds T. Reilly, A. Lees, K. Davids and W.J. Murphy), E & FN Spon, London, pp. 429–33.

Plagenhoef, S. (1971). *The Patterns of Human Motion*. Prentice-Hall, Englewood Cliffs, NJ.

Poulmedis, P. (1988). Muscular imbalance and strains in soccer. *Proceedings, Council of Europe Meeting: 'Sports Injuries and their Prevention'*, Papandal, The Netherlands, 53–7.

Suzuki, S., Togari, H., Isokawa, M., Ohashi, J. and Ohgushi T. (1988). Analysis of the goalkeeper's diving motion, in *Science and Football* (eds T. Reilly, A. Lees, K. Davids and W.J. Murphy), E & FN Spon, London, pp. 468–75.

Tant, C.L., Browder, K.D. and Wilkerson, J.D. (1991). A three dimensional kinematic comparison of kicking techniques between male and female soccer players, in *Biomechanics in Sport IX* (eds C.L. Tant, P.E. Patterson and S.L. York), ISU press, Ames, IA, pp. 101–5.

Wickstrom, R.L. (1975). Developmental kinesiology, in *Exercise and Sports Science Reviews* (ed. J. Wilmore), vol. 3, 163–92.

Zernicke, R. and Roberts, E.M. (1978). Lower extremity forces and torques during systematic variation of non-weight bearing motion. *Medicine and Science in Sports*, **10**, 21–6.

9 The biomechanics of soccer surfaces and equipment

Adrian Lees and Mark Lake

Introduction

Sports biomechanics is concerned not only with the mechanical functioning of the human body in sport but also with the equipment and apparatus used. In all codes of football, the equipment used has a major effect on the way the game is played. The ball itself is of a certain size, construction, weight and pressure, all of which affect the way it responds in play. The ground on which the game is played also affects the nature of the game. Surfaces for football, particularly soccer, have evolved in response to both performance and economic requirements. The controversy aroused by the introduction of the synthetic surface for soccer has led to biomechanical investigations into the performance and protection characteristics of all types of surfaces. The boot is an important piece of players' equipment and well-fitting boots can aid not only comfort but also provision of a positive interaction between player and surface to create traction for stopping, starting and turning. It can also aid player–ball interaction for passing and shooting. Boots must also protect the player and have a resistance to the stresses imposed on them during the game. Boots are not the only piece of equipment providing a protective function. Shin guards are essential in the modern game of soccer for protecting against kicks and blows, but have until recently been neglected in biomechanical investigations.

In this chapter an attempt is made to look at the major items of equipment used within soccer and detail the results of biomechanical investigations on such equipment. Some anticipated future developments in this research area are also discussed.

9.1 Playing surfaces

There has been much controversy in Britain concerning the use of artificial turf for playing soccer. It has been more readily accepted in North America, Scandinavia and the Middle East where there has not been the same long standing traditions associated with the game, or where environmental considerations are important. In other parts of Europe particularly, there is little contemplation of anything other than a natural turf surface. Artificial turf is an issue in soccer not least because of the economic benefits that may accrue from its use, but also the pressure for its use in many parts of the world.

There is much conjecture concerning the merits of artificial surfaces, but only a little scientific evidence. Much of this was collected in England by a commission headed by Winterbottom (1985) and supported by the Football Association and the Sports Council. The first artificial pitch was installed in the UK in 1971, and the first Football League artificial pitch was installed at Queens Park Rangers (then in the Second Division of the Football League) in 1980. The opinions of players, managers and club chairmen were that soccer could be played to a high standard on artificial pitches but necessitated a modification to the playing of the game which often did not suit the 'British' game. Therefore, a 3-year moratorium was placed on the installation of artificial pitches for League football until Winterbottom's report had been considered fully. The report attempted to obtain scientific data on the comparative performance of both natural and artificial pitches. The author took several examples of each class of pitch from various levels of play and conducted a series of tests of performance characteristics. The tests were concerned with aspects of ball–surface interaction, player movement and player–surface interaction.

Two tests were used to establish the interaction between ball and surface. These were rebound resilience and rolling resistance. The former test established how a ball reacts after hitting the surface and relates to ball bounce, while the later test established how quickly a ball slows down when rolling over the surface. Rebound resilience was found to be 3–6% higher on artificial rather than real turf surfaces, although variations in ball type (all nominally at the same pressure) were between 3 and 7%. Variation in rebound resilience due to variations in ball pressure accounted for 4–5%. Other factors which affected the results were the spatial location on the pitch where there was less variation for artificial surfaces, and wetness which tended to reduce the rebound resilience. Various non-systematic effects were produced from the grass species and the type of material used in the construction of artificial surfaces. In addition, as turf surfaces became older they became more compact and harder, thereby increasing the rebound resilience. The test for rolling resistance showed that this was about 20% less for artificial surfaces. This proportion was reduced when the surface was wet. On turf surfaces spatial location could affect the results. Where the turf was more lush the rolling resistance was greater, and where worn it was lower. In both of these tests there would appear to be little difference between the two types of surface under optimum conditions. The turf surface would appear to be the more variable.

The tests used for assessing player movement were tests for torsional traction and sliding resistance between the boot and surface. The torsional traction was measured using a studded plate which was loaded onto the surface and rotated, simulating a player pressing the boot down onto the ground and twisting. The torque produced when slipping occurred was recorded. A torsional traction coefficient was calculated, and it was found that there was great variation in this for all types of surfaces. It ranged from 1.1 to 2.2 for turf surfaces, and from 1.0 to 2.8 for artificial surfaces. It tended to be lower for sand-filled artificial surfaces compared to open weave surfaces. A major factor affecting these results was the type of stud and stud pattern, and the results presented on this topic are some of the few in any research which has considered the effect of these variables. Different stud types and patterns yielded a range of torsional coefficients from 1.7 to 2.5. It was concluded that the surfaces per se are not clearly different with respect to torsional traction. The test for sliding friction produced essentially the same results. There was found to be no difference between surface type, but factors such as surface pile, moisture, stud pattern, and whether the front or rear of the shoe was used, are all of some importance. Therefore, other factors are more important with regard to sliding friction than the surfaces themselves.

Two tests were used to quantify player–surface interaction. These were based on impact characteristics of both the foot and the head. The former test used a specially constructed mechanical device known as the 'Stuttgart Artificial Athlete' which simulated the impact force produced during running. Again it was found that there was little difference between surface types, although artificial surfaces showed a little more deformation. Other important factors were whether the surface was sand filled or open (the former absorbing energy better) and the degree of compaction of the turf. In both cases the effect of stud penetration had not been taken into account. The final test was of impact severity and was applicable to head and face impacts. A large spherical mass approximating to the head was dropped directly onto the surface from a small height. The impact deceleration was used to form an impact severity index. It was found that this was much higher on artificial surfaces, particularly the open weave type. It was thought that the reason for this was that the surface was 'bottoming out'. In other words, the surface had compressed to its fullest extent and had begun to interact with the hard sub-surface layer. The addition of sand, moisture and a more shock absorbing sub-layer are construction techniques which can be used to reduce the impact severity of these surfaces.

The general conclusion of Winterbottom's Committee was that in many respects there were little differences between the two types of surface, but in some important respects there were. In these, artificial surfaces could be designed to make their performance within that acceptable for natural surfaces, and the performance of pitches already laid could be controlled by the use of water to deaden a lively pitch and to provide a better energy absorbing surface. Although artificial surfaces could be tailored to suit playing requirements, one feature of their performance does not readily match that of real turf, and that is its variability. Generally, a natural surface is more varied both between surfaces and within an area of a pitch, and this is thought to be a crucial element in the game of soccer. In 1989, the Football League published its final report after the period of moratorium and concluded that artificial surfaces were not suitable for the playing of the game at a high level in the English League. They were deemed suitable for lower level play where economic advantages of the artificial surface were also of importance. This finding was endorsed by the international authorities of the game (FIFA), and all competitive international matches are played on real turf. It should be noted that the conclusion reached by the Football League regarding the suitability of artificial surfaces was as much to do with the subjective judgements of how the game should be played as to their performance, economic or injury characteristics.

More recent studies on natural turf surfaces for soccer have begun to examine the complex traction properties of footwear–soil interaction. The particulate nature of soil provides a non-linear, irreversible load deformation behaviour at each ground contact and obviously the classical laws of friction do not apply to boots sliding on such surfaces (Barry and Milburn, 1999a). Barry and Milburn (1999b) used a specially designed shear test machine to examine the mechanical properties of boot–turf interaction during simulated soccer-like ground contacts. A weighted (400 N) vertical shaft holding a boot was precisely slid or rotated in a turf sample attached to a force platform using stepping motors. Traction force was applied for 0.4 s to mimic the stance time during braking or turning moves during soccer play. Traction was found to be modified by factors such as moisture content, root density and slip planes within the surface. Importantly, boots from different manufacturers developed dissimilar traction forces on a given turf surface. Grip was influenced by the orientation of the boot with respect to the boot's motion, with lower values at 45° and 135°. Boots with blades developed larger traction at 45° compared to other boots tested. It is anticipated that further work in this area of terramechanics applied to soccer may lead to better guidelines on the characteristics of the surface required for adequate traction and safety during dynamic manoeuvres.

9.2   The soccer ball

The full-size ball is required to have an outer casing or cover which should be of leather or another approved material which does not prove dangerous to players. It should have a mass between 14 and 16 oz (0.396 and 0.453 kg), a circumference between 27 and 28 in. (0.685 and 0.711 m), and an internal pressure of 0.6–1.1 atm (60 600–111 100 Pa). The ball may be constructed in two main ways. The sewn ball is constructed of panels of leather or similar artificial material, or a moulded ball made from rubber with cover panels bonded to the surface or from plastic with the cover panels painted on. The sewn ball may be treated to prevent the ingress of moisture, but some moisture will inevitably seep into it and increase its weight. All of these factors affect the way the ball will play. While its size only varies for junior play, its material and method of construction, and internal pressure may all vary. The effect of these variations on both performance and potential injury can be established.

The performance characteristics of the ball describe the way it flies through the air, bounces on the ground, and reacts to being kicked or headed. The three main performance determinants of a soccer ball are its mass, its surface roughness and its internal pressure. The mass of the ball is restricted by the rules of the game. Small variations in mass can occur due to the ingress of water through the seams of the ball, or by absorption through the material. Both of these are less important nowadays due to the developments in ball materials and construction methods. Nevertheless, a heavier ball will have a lower velocity when kicked, although it will retain more of this velocity during flight. Of more importance are the aerodynamic forces acting on the ball. The aerodynamic forces are drag (air resistance) and lift. An important characteristic determining how air resistance will affect the ball's flight is the critical Reynold's number. This number is a function of the diameter of the ball and its speed. If the ball has exceeded the critical Reynold's number then there is a reduced air resistance on the ball and it will fly further. For moderate to hard hit kicks, this will be true and the ball will have a greater range than if the critical Reynold's number had not been exceeded. A lift force acts if spin is put onto the ball. The lift force may not always act vertically to lift the ball but may act horizontally, depending on the direction of spin, to cause the ball to swing away from its intended direction of flight. This spin swing effect is known as the 'Magnus effect', and is used tactically during set plays in soccer. The 'inswinger' and 'outswinger' are frequently used kicks from corner positions and the direction and amount of swing are determined by the direction and amount of spin put on the ball. In addition, a curved ball flight is often used to get around the defensive wall from free kicks. This is exploited to full effect by such players as Roberto Carlos (Brazil) who intuitively knows how to curve the ball by hitting it at a particular velocity and with a particular spin. The physics of this motion of the ball has been described by Asai *et al.* (1998b). They predicted that over a 30-m trajectory, the lift force from ball spin could make the ball deviate by as much as 4 m from its normal straight-line course. A high-speed camera operating at 4500 Hz was used to examine ball velocity and spin during curve-ball kicking. They measured boot–ball contacts of around 9 ms, ball spins of 8–10 revolutions per second and ball speeds of approximately 23 m s^{-1}. Asai *et al.* (1998a), using finite element modelling of the boot–ball impact, found that off-centre foot impacts generated more spin up to a threshold but that ball speed was always compromised. If the offset distance was increased, then the foot was found to touch the ball for a shorter time and over a smaller area, which caused both the spin and velocity of the ball to decrease. There appeared to be an optimum place to hit the ball for maximum spin: if the ball was hit too close or too far from the centre of gravity then little spin was acquired. Future developments in this area of research will likely focus on specific characteristics of the ball–boot interaction during impact and the influence of boot design and upper materials on the generation of ball velocity.

Considering the short contact duration, a ball construction which allows good grip between foot and ball is important. The sewn method of construction is not only a practical method for manufacturing the ball but gives the opportunity for additional grip to be gained between the foot and ball to apply spin. In Rugby and American football, the longitudinal shape of the ball enables it to be given a spin about the longitudinal axis by sweeping the foot across the ball during the punt kick. This spin causes the ball to drift towards the touchline on its downward descent. This type of kick in rugby is often referred to as the 'torpedo' kick. Its more detailed mechanical description is given by Daish (1972).

The way in which the ball responds when bouncing from the ground depends on its internal pressure, ground characteristics and surface–ball frictional properties. The higher the internal pressure the better will be the bounce of the ball. On soft turf surfaces, the condition of the surface is important. If the surface is too soft the surface dominates the behaviour of the ball. Only when the surface is reasonably hard will the true effects of the ball pressure be seen. When a ball bounces on the ground, the ball tends to skid, slightly reducing its forward velocity. How much forward velocity is lost depends on the frictional interaction between the ball and surface. If the surface is well lubricated by water the ball will skid more easily and lose less forward velocity. As the ball also loses vertical velocity when bouncing, the effect of a ball bounce is to lose considerable energy. It will slow down and bounce less high. The interaction of the ball surface and surface condition serves to alter this effect in any one case.

The force imparted to the player during contact with the ball can lead to injury. Although repeated kicking of the ball will lead to overuse injuries of the toes and ankle (Masson and Hess, 1989), the main concern is usually in heading the ball. The possible injurious effect of heading the ball has been the subject of recent biomechanical investigations as a result of potential legal cases over the misuse of equipment for young players. This has furnished some useful data on ball characteristics.

Levendusky *et al.* (1988) investigated the impact characteristics of a stitched and moulded soccer ball and measured the force of impact using a force platform. They found that for velocities of impact of about $18 \, \text{m s}^{-1}$ the force of impact was about 6% higher in the stitched compared to the moulded ball. This finding has implications for the risk of injury of the players when heading a ball. Armstrong *et al.* (1988) continued this investigation by considering the effect of ball pressure and wetness on the impact force. They found that if a ball was wet it could increase the impact force by about 5% due to the extra weight as a result of water retention, and if a ball had a pressure increase from 6 to 12 psi (1 psi = 6975 Pa) there would be an increase in impact force of about 8%. These results clearly show the effect that poor combinations of conditions could have on the impact load sustained by the head during heading.

Levendusky *et al.* (1988) gave examples from the literature of where heading the ball can cause damage due to: (i) surface deformation of the head leading to a broken nose, eye damage and lacerations; (ii) damage due to direct impact causing compression waves travelling through the brain creating high internal pressures; and (iii) rotational accelerations causing shearing between the brain and the skull. The levels of impact which are likely to cause injury are about 80 g for loss of consciousness and 200 g for fatalities. For rotational accelerations, values greater than $5500 \, \text{rad s}^{-2}$ are likely to lead to a loss of consciousness. Burslem and Lees (1988) investigated the acceleration on the head during a modest speed header (ball velocity about $7 \, \text{m s}^{-1}$) and found that accelerations were about 60 g, and rotational accelerations about $200 \, \text{rad s}^{-2}$. Clearly, there is more danger from the direct impact. In a mathematical simulation of impact Townend (1988) estimated that the average acceleration of impact was about 25 g, but increased with the reduction of mass of the player and the increase in mass of the ball, supporting the results of Armstrong *et al.* (1988). The conclusion which can be drawn here is that although heading is below the injury threshold, it is sufficiently close to it for care to be shown, particularly in dealing with young children in the development of the skill. The skill of heading can lead to greater head and neck rigidity thereby reducing the effect of the impact. This skill must be taught properly and carefully, and a reduced ball mass should be used for children.

9.3   The boot

The typical football boot is one which is still based on a leather construction, generally cut below the ankles, and with a hard outsole to which studs are attached. Over recent years minor adjustments in football boot design have gradually been added to take account of the requirements of players and the trends within the game. Previously, relatively large changes in soccer boot construction have been introduced to improve playing performance. In the mid-1950s, a German shoemaker named Adi Dassler developed boots for the first time with screw-in studs that could be changed to suit pitch conditions (Vetten, 1998). This was a big advantage for the German team members who used the boot but that was not the only aspect of the boot he improved. In distinct contrast to the rigid, heavy, high-topped boots of that era, with his new 'Dassler' boot in 1954 he introduced boots with cut-away ankles and soft leather with a 50% weight reduction. The German Dassler boots were over twice as expensive (approximately £5 versus £2 for the average British boot) but because they were so much better, the future of the 'Adidas' athletic shoe company within soccer was assured.

The thinness of the outsole provides the boot with its flexibility, while its hardness provides a firm surface for the attachment of studs. The studs may be either moulded as a part of the boot or detachable, and great variety is seen in sole stud patterns. Boots have a firm heel cup but do not usually include a heel counter as found in running shoes. Some boots have a raised heel to provide both heel lift and a midsole for shock absorption. Most boots will have a foam insock to aid in the provision of comfort and fit.

Although manufacturers take a systematic approach to boot design, there have been virtually no reported scientific investigations of soccer boot performance which have then been fed back into design. Notable exceptions are the work reported by Valiant (1988) and Rodano *et al.* (1988). Both of these studies have presented data on the vertical and horizontal forces acting on the boot. Essentially, the vertical force serves to press the studs into the ground and to compress the sole of the boot, whereas the horizontal forces serve to provide traction and to deform the boot by the action of the foot on the boot leading to deformation of the heel cup, stretching or even splitting of the boot material.

There are some general principles governing the function of footwear which can be applied to the soccer boot. The boot in soccer, as indeed any form of footwear, provides an ergonomic function. It must be comfortable to wear and not be an encumbrance to the player or the play required of an individual. It must relate to: (a) the demands of the game; (b) provide protection for the foot; and (c) enable the foot to perform the functions demanded of it. These aspects can be considered in turn.

9.3.1   The demands of the game

The demands of the game on the boot can be established by notation analysis techniques. While these have been conducted in soccer and rugby in order to investigate the physiological demands and strategic development (Reilly and Thomas, 1976; Treadwell, 1988), there is little of these data that can be used for an ergonomic assessment of the requirements of the boot. Therefore, the functions that the boot is required to perform are based on anecdotal evidence and the experience of players.

Studies such as those by Lees and Kewley (1993) provide a suitable model for assessing the demands of the game from the perspective of footwear. They looked at the physical demand which is placed on the boot during soccer playing and training. They did this by identifying the major categories of playing movements made during a game of soccer, and recording their frequency of occurrence during both training and playing. The actions made by a player in soccer are many and varied and each is likely to put a unique demand on the strength of the boot. In order to obtain an indication of the role the boot has in the game it is necessary to investigate the types and numbers of actions which are made during the game. In an early study, Reilly and Thomas (1976) performed a motion analysis of the different positional roles in professional soccer. Although differences were found between positions of play, the authors averaged the distances covered among the different positions to give the data shown in Table 9.1. Also in this table are further comparative data from Withers *et al.* (1982). Both sets of data show differences, but the general trends in distances covered, and their respective percentages, for each activity can be clearly seen.

In addition to this perspective, both studies reported on the frequency of occurrence of some other more specific actions. Reilly and Thomas (1976) reported that the average number of jumps per individual per game was 15.5 and shots was 1.4. Withers *et al.* (1982) reported that the average number of tackles per individual per game was 13.1, jumps 9.4, turns 49.9 and contacts with the foot 26.1. Despite the differences between the two studies, general trends are evident and clearly indicate that the general locomotor use of the boot is by far the most frequent. It might be expected that the more forceful actions (i.e. jumping, turning, sprinting) would be the actions putting the greater strain on the boot. Although small in number, the intensity of these actions could well be a critical factor which determines the life of the boot.

An estimate of the demand put on the boot was obtained by measuring the horizontal force during each of the categories of movements. The data from these two approaches were then combined to give an overall estimate of the demand on the boot, and related to the problems experienced by the players. The horizontal data were presented as a vector plot together with a 'stress-clock' (Figure 9.1). The stress clock was produced by adding up the magnitudes of the force during foot contact which appeared in each of twelve 30° or 'hourly' segments. The 'hour', total stress and number of counts were given on the graphical output, together with their total and the sample rate. The accumulated force was converted to a 'severity index' for each of the playing actions. The direction of force was related to the occurrence of

Table 9.1 Distances (m) of major movement categories

Activity	Reilly and Thomas (1976)		Withers et al. (1982)	
	m	%	m	%
Walking	2150	24.7	3026	27.0
Jogging	3187	36.8	5139	45.8
Cruising	1810	20.8	1506	13.4
Sprinting	974	11.2	666	5.9
Backing	559	6.5	874	7.9
Total distance	8680		11 211	

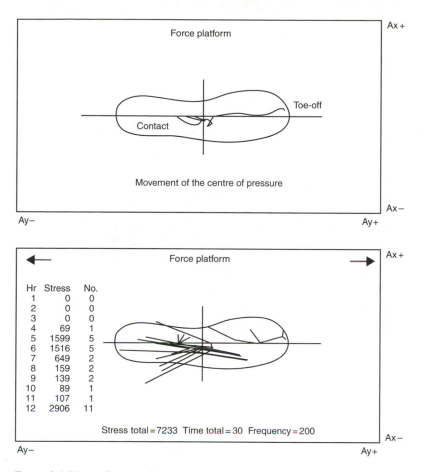

Figure 9.1 Upper diagram showing the movement of the centre of pressure across the boot during running from midsole contact to toe-off. Lower diagram showing the horizontal friction forces acting on the shoe and a stress clock indicating the 'hourly' directions where most stress is located.

splits in the forefront and outside regions correspond very well with the main directions of the stress on the boot. It was estimated that over a period of 90-min playing or training the stress on the boot was three times greater in training than in playing. This has consequences for the type of boot that is used for both types of play. It was concluded that this approach to the assessment of the demand on the boot is useful, being one which had not previously been described in the literature. A subjective evaluation of the direction of stress had also provided a useful insight into the nature of the directions of stress on the shoe.

The occurrence of splits in the forefront and outside regions correspond very well with the main directions of the stress on the boot. Some of the playing actions identified were not amenable to the measurement of force (e.g. dribble, trap and tackles) and so these were omitted from the analysis. Others were more relevant and were included (e.g. locomotor actions). The average stress level on the boot which serves as a severity index can be integrated with the number of actions of each type occurring during both playing and training.

If this is done the accumulated severity for 90 min of activity is:

Professional training   161 kN
Professional match-play   58 kN
Amateur match-play   50 kN

   While these data must be interpreted with care as they cover only a selection and not all of the actions occurring in the game, results illustrate that the demand put on the boot is likely to be considerably (i.e. three times) more severe in training than in match-play.

9.3.2   Protection

The boot must be comfortable, be a good fit to the foot, but at the same time protect from external forces, spread the pressures over the sole of the boot, and control foot movement particularly rear foot movement. Comfort is a difficult term to define objectively. It often relies on subjective experiences of players. It is something that can change with time, or with conditions of play (e.g. wetness, foot microclimate, properties of the boot material). The fit of a boot is related to the type of last (the foot shape used for boot construction) and the materials used for its construction. There are substantial ethnic differences in foot shape, and it is unlikely that a boot designed for an American, Italian or oriental foot will fit a British foot well. Manufacturers either use standard lasts, or lasts which have been developed for other types of footwear (e.g. running shoes). Certain types of leather (e.g. kangaroo leather) have the properties of yielding to accommodate different foot shapes, while still providing a strong material, resistant to splitting. This material helps to improve fit and enhance comfort.

   The boot should be constructed so as to protect the foot from external forces which may arise from the ground, other players or by contact with the ball. When the foot contacts the ground the typical ground reaction force exceeds 2.5 times body weight (Figure 9.2). This force can increase as a result of running speed or type of landing action used. The force will also be higher on hard as opposed to soft grounds. The boot should have built into it materials designed to reduce the effect of these forces, but they often do not.

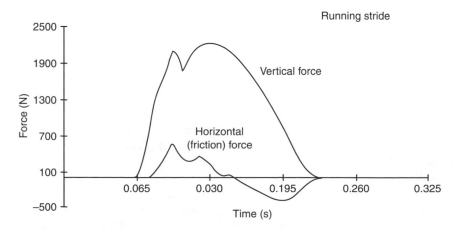

Figure 9.2 A typical vertical ground reaction force in running.

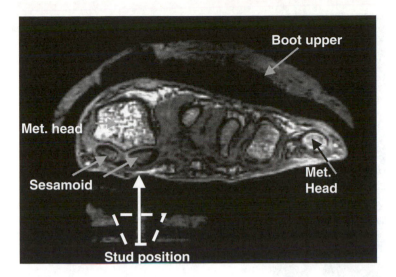

Figure 9.3 Magnetic resonance image of a foot inside a soccer boot while under load. This slice through the centre of the first metatarsal head is also the exact location of a stud on the boot (identified by an oily paste).

The boot should also be able to distribute the force so that it is not concentrated in certain areas, such as for example under the heel, or more particularly under the head of the first metatarsal. The positioning of studs is particularly critical in this regard, as well as the method of attachment of stud to the boot. Little research has examined the distribution of loading under the foot in studded footwear even though they are typically poorly cushioned and players run, on average, 10 km during the course of a match. Incidences of foot pain have been found to coincide, in many cases, with the positioning of studs on the boot outsole (Lake, 2000), particularly in the regions of the medial and lateral metatarsal–phalangeal joints. One problem area is the first metatarsal head and imaging techniques can be used to demonstrate the location of these bone directly over a stud on the fore part of the boot (Figure 9.3).

Although no definite relationship has been established, stud configuration, penetration and sole plate stiffness have been identified as factors that elevate localized foot pressures inside the boot during soccer. The traditional six-studded boot has been found to increase the peak pressures under the foot during running in comparison to other boots with more studs and different configurations (Coyles and Lake, 1999). The in-boot pressures are illustrated in Figure 9.4 where dark areas of high pressure can be seen that coincide with stud locations particularly under the first metatarsal head. In-boot plantar pressures have recently been measured during other soccer manouevres such as cutting, jumping and sprinting and the peak forces are typically experienced under the first metatarsal head or hallux (Eils *et al.*, 2001; Asai *et al.*, 2002). This has implications for shoe design with protection of the first metatarsal head becoming important. Boot manufacturers are attempting to solve this problem by either moving studs slightly away from sensitive foot structures in the forefoot or designing the studs or sole plate to redistribute the forefoot loading. Eils *et al.* (2001) also indicated that rapid wear and compression of the sockliner inside the boot can increase localised loading of the foot by over 10%. To protect the foot, they recommended the use of either compression resistant shockliners or replacing them after a short period of use.

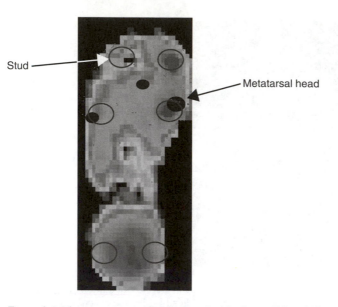

Stud

Metatarsal head

Figure 9.4 Plantar pressure distribution inside the traditional six-studded football boot during running. Notice the higher (darker) pressure areas in the forefoot correspond to locations where studs are positioned.

The foot is susceptible to knocking and treading by the feet of other players, and so the material of the boot should be able to provide some form of protection. The use of sound or padded leather is necessary. When the ball is kicked, there is a contact force in excess of 1000 N (1.5 times body weight). This force will deform the foot but could also lead to bruising on the dorsal aspect of the foot. The force can also be a function of ball wetness, ball inflation pressure, and ball construction as noted above (Levendusky *et al.*, 1988).

In running shoe design, great attention is paid to the reduction of the shock force associated with heel strike. This force is characterised by a sharp force peak whose magnitude can reach about three times body weight in sprinting (Figure 9.2). This force can also be assessed by the use of acceleration measures on the lower tibia (Lafortune, 1991). Peak tibial deceleration in running can be up to $10\,g$ $(100\,m\,s^{-2})$. While these techniques have not been applied to study field games, the types of boot construction with thin soles suggest that this is an unimportant factor for soccer players. While it may be less important due to the generally softer surfaces that soccer competitors play upon, nevertheless it is still a feature of any heel contact locomotor action. This becomes progressively more important as the ground becomes harder. It is also likely to be more important as players become more used to the softness of everyday shoes and gradually lose the ability to withstand repeated hard heel impacts. The use of shock absorbent materials placed in the heel of the boot is a standard method for reducing the severity of impact, but this is rarely provided in the boot. In addition, such protective materials would serve to raise the heel, which would put less stress on the calf muscle–tendon complex, as a result of functional shortening from habitual wearing of raised heel footwear in everyday life.

During the game the most frequently used action of the boot is a normal running stride. This can be seen from the data in Table 9.1. In such a stride the foot typically contacts the ground on the lateral border of the heel. The foot then rolls over, and goes into a position of pronation. The amount of movement of the rear foot can be affected by the boot. In running shoes the shoe may actually increase this range of pronation, and special construction methods are required to control it. The boot generally does not have these anti-pronation devices, and players with excessive rear foot movement may benefit from some rearfoot control.

Of major interest in this context is the role of the boot for protecting against ankle inversion or eversion sprains. The boot was traditionally made with a high ankle support. The advent of a faster running game has led to a preference for the low-cut soccer-type boot. This boot allows greater movement of the subtalar joint, and may as a consequence lead to more frequent and more severe ankle injuries.

The ankle joint is one of the most vulnerable joints for a soccer player, and the boot is often relied upon to protect this joint from an inversion/eversion sprain or more serious damage. The role of the boot in protecting the ankle joint was investigated by Johnson *et al.* (1976). They examined the torsional stiffness of different designs of boot uppers. They modelled the lower leg by a mass–spring–dashpot system which gave the joint its load response characteristics. The boot added another resistive layer to the outside of the ankle allowing the natural stiffness of the joint to be supplemented by the properties of the boot. The low-cut boot protected the subtalar joint, while the higher-cut boot protected both this and the ankle joint. In a simulation of the effect of using materials with differing stiffness, they found that if a low cut boot was used it should be made of low stiffness material. This was because the subtalar joint had a certain amount of mobility, and if the ankle was turned a low-cut boot would allow the subtalar joint to accommodate most of the movement. If the low-cut boot was of stiff construction, then the boot would transfer some of the load away from the subtalar joint to the ankle joint. As this does not have any degree of flexibility in the inversion/eversion direction the additional load would be taken up by the collateral ligaments, leading to a greater likelihood of ligamentous damage. On the other hand, the high-cut boot should be made with stiff material because it already has a protective function with regard to the ankle joint and collateral ligaments. The stiffer the material, the more the load is taken by the boot material rather than the ligaments themselves. It should be noted, however, that the high-cut boot with stiff material is only about twice the stiffness of the low-cut, low-stiffness material boot, and that for a severe inversion movement even the high-cut boot would be insufficient to prevent damage occurring.

9.3.3 Performance

During performance, the boot must allow the player to perform without encumbrance and if possible to enhance the playing of the game. The boot must allow the player to run easily, and so lightness is a major consideration. However, there is some incompatibility between the lightness of materials and their ability to protect the foot. The boot must not inhibit the normal joint movement in many phases of the game, particularly running. The low-cut boot has shown itself to allow normal joint function both in terms of plantar and dorsi-flexion, and in supination and pronation. The lack of protection against excessive ranges of motion has been indicated above.

The studs are important for providing traction on a variety of surfaces. The grip provided is a function of the depth of penetration of the stud and the firmness of the turf. Very wet turf, possibly having a high surface water content, will mean that short studs fail to penetrate into the firmer ground underneath. On the other hand, very hard turf will not allow good penetration, and lead to pressure areas on the foot at the heel or forefoot of the boot. Studs of varying length help to overcome some of these problems, but studs are also a source of injury to other players. In games such as rugby the traction requirements of the game differ according to playing position. The traction needed to push in a scrum is over five times that required in running. As all players must be able to apply force in a scrummage, the traction provided by the boot is overspecified for many playing actions.

Surprisingly, the role of the boot in providing traction with the ground is one area which has received little attention. The amount of grip provided by a surface is an important component of playing quality. If there is too little grip the players will slip and fall, while if there is too much there is a danger that players will suffer knee and ankle injuries as their feet become locked during turns and manoeuvres. Recent comparative evaluations of soccer boot designs during soccer turning moves have begun to incorporate subjective evaluations of lower limb discomfort and traction performance (Coyles *et al.*, 1998). Often discomfort of the knee and ankle can be associated with a rating of too much grip during a dynamic manoeuvre.

In a report for the Football Association, Winterbottom (1985) initiated an investigation into the effects of stud configuration which was later extended by the Football League (Football League, 1989). They found that the relationship of traction between boot and surface was a very complicated one, and identified two categories of movement important to players. These were sliding and turning movements. The researchers found that the sliding resistance was affected by turf wetness as well as stud configuration. Differences between extreme conditions were as much as 300%. The torsional traction coefficients for different boot sole types ranged from 2.5 at the highest to 1.0 at the lowest. Therefore, the type of sole and the stud configuration can lead to a 250% change in the degree of traction offered. The degree of torsional traction should be matched with the pitch conditions, but is only ever done so subjectively, and is an area which deserves greater research attention.

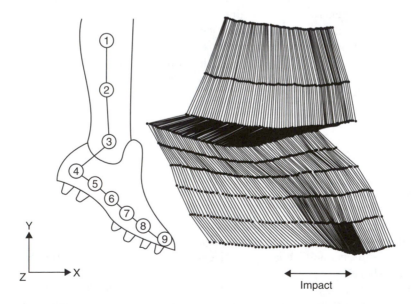

Figure 9.5 Deformation of the kicking foot during impact with the ball (using high-speed video sampling at 4500 Hz). (Taken from Asai, T. *et al.*, 1998a. Computer simulation of curve-ball kicking in soccer, in *The Engineering of Sport* (ed. S.J. Haake), with permission of Blackwell Publishers.)

A point which is little considered is the ability of the boot to provide a sound dorsal surface for kicking. The foot makes contact with the ball on its dorsal surface (Asami and Nolte, 1983) and is very deformable under the large force applied during the kick. The deformation of the foot during kicking was revealed using high-speed filming by Asai *et al.* (1998a). Substantial backward deformations of the foot during impact were recorded with boot–ball contact lasting only 8 ms (Figure 9.5). This deformation leads to reduced kicking effectiveness. The boot could strengthen the foot during this action, but would require a high flexion stiffness of the sole. However, this high stiffness would inhibit the normal flexion of the foot during the more common locomotor actions. The sole of the boot then could be designed with a hinge locking mechanism, and behave in a similar fashion to the elbow joint, for example, providing flexion in one direction but strength in the other.

9.4   Shin guards

The shin guard is used to protect the lower leg from impact injuries. These injuries can range from the severe (such as direct contact between the opponent's boot and the leg as in a poorly executed tackle) to the minor (such as bruises and scratches from glancing blows). The shin guard offers protection from some of these injuries.

The need for shin protection was evident in the early days of soccer when players used to put rolled up newspapers down the front of their socks. The shin pad evolved which was also inserted between the sock and the leg and consisted of a foam-backed plastic shield which was reinforced by wooden strips placed longitudinally down the guard. Recent trends in equipment design have spawned a wide range of different shin guard types, all following similar principles but more sophisticated in construction. A typical contemporary shin guard is integrated within an "over sock" which also incorporates shells to cover the ankle joint. The protective guard is constructed with a hard outer casing and a softer inner layer. The material used for the outer casing is usually thermoplastic moulded to the curvature of the leg, with a shock absorbing inner material made of ethylene vinyl acetate (EVA) or other foam type material.

There are no performance standards for shin guard protection, although a European Standard is in the process of being introduced. Consequently, there are no accepted means whereby the protective quality of the shin guard can be quantified. Manufacturers base their design on the logic of material behaviour, the efficacy of construction methods and on the opinions of players.

An attempt to quantify shin guard performance was performed by Lees and Cooper (1995). He tested the ability of five types of contemporary shin guards to the reduce the impact from a direct blow. The methodology used followed the methods used for the testing of cricket pads (for which there is a British Standard: BS 6183, part 1 1981, British Standards Institution, London). This method involved dropping a weighted mass directly onto the pad from a set height and monitoring its deceleration value. The pad was placed over an aluminium leg form which was rigidly held and hit by a 5-kg mass which was dropped from a height of 40 cm. The face of the striking mass was hemispherical of diameter 7.3 cm. The deceleration was measured by an accelerometer attached to the striking mass. These conditions were designed to simulate the energy of impact delivered by a cricket ball. A similar series of conditions were used by Lees and Cooper except a wooden leg form was used for convenience, and the diameter of the striking mass was smaller to represent the characteristics of the striking boot. The shin guards tested showed a reduction in deceleration ranging from 28 to 56% relative to the impact deceleration obtained from impacts on the wooden leg form without a shin guard in place. There is clearly a large difference between shin guard types. The poorer guard was constructed of a thermoplastic outer casing with a foam inner layer, while the better guard was of a similar thermoplastic outer shell but with an EVA inner layer. The outer layer serves to spread the load reducing the local pressure, while the shock absorbent inner layer serves to reduce the effect of the impact load.

The shin guard also absorbs energy, but its capacity for doing this is restricted by the quantity of material within the shin guard construction, which is generally small. Therefore, the shin guard can reduce the effect of bruising, glancing blows and scraping by the ground or an opponent's studs. It is unlikely to be effective against high-energy direct blows which may lead to fracture. Nevertheless, the shin guard provides an important protective function and its design and materials used in construction make it an important piece of equipment for the player.

Summary

This overview of the biomechanics of soccer equipment has shown that there are many factors which interact to affect the role of the equipment within the game. The equipment itself has mechanical characteristics which are subject to variation, but which can be reasonably well quantified. The interaction between the player and the equipment is also a source of variation. This variation is more difficult to quantify and to predict its effect on both performance and the protection. Nevertheless, a good understanding of the general principles can be gleaned from the above examples. This should help with the application of sports biomechanics techniques to a wider range of equipment and football codes in the future. New measurement techniques such as high-speed video analysis and in-shoe pressure distribution measurements should allow the protective function of soccer equipment to be further enhanced in the future.

References

Armstrong, C.W., Levendusky, T.A., Spryropoulous, P. and Kugler, L. (1988). Influence of inflation pressure and ball wetness on the impact characteristics of two types of soccer balls, in *Science and Football* (eds T. Reilly, A. Lees, K. Davids and W.J. Murphy), E & FN Spon, London, pp. 394–8.

Asai, T., Akatsuka, T., Nasako, M. and Murakami, O. (1998a). Computer simulation of curve-ball kicking in soccer, in *The Engineering of Sport* (ed. S.J. Haake), Blackwell Science, London, pp. 433–40.

Asai, T., Akatsuka, T. and Haake, S.J. (1998b). The physics of football. *Physics World*, **11**, 25–7.

Asai, T., Igarashi, A., Murakami, O., Carre, M. and Haake, S. (2002). Foot loading analysis of primary movements in football, in *Proceedings of the 4th International Conference on Sports Engineering*, Kyoto, Japan.

Asami, T. and Nolte, V. (1983). Analysis of powerful ball kicking, in *Biomechanics VIII-B* (eds H. Matsui and K. Kobayashi), Human Kinetics, Champaign, IL, pp. 695–700.

Barry, E.B. and Milburn, P.D. (1999a). A mechanism explaining traction of footwear on natural surfaces, in *Proceedings of Fourth Symposium on Footwear Biomechanics* (eds E.M. Hennig and D.J. Stefanyshyn) University of Calgary, Canada, pp. 22–3.

Barry, E.B. and Milburn, P.D. (1999b). The traction of football boots, in *Science and Football IV* (eds W. Spinks, T. Reilly and A. Murphy), Routledge, London, pp. 3–7.

Burslem, I. and Lees, A. (1988). Quantification of impact accelerations of the head during the heading of a football, in *Science and Football* (eds T. Reilly, A. Lees, K. Davids and W.J. Murphy), E & FN Spon, London, pp. 243–8.

Coyles, V.R., Lake, M.J. and Patritti, B.L. (1998). Comparitive evaluation of soccer boot traction during cutting manoeuvres – methodological considerations for field testing, in *The Engineering of Sport* (ed. S.J. Haake), Blackwell Science, London, pp. 183–90.

Coyles, V.R. and Lake, M.J. (1999). Forefoot plantar pressure distribution inside the soccer boot during running, in *Proceedings of Fourth Symposium on Footwear Biomechanics* (eds E.M. Hennig and D.J. Stefanyshyn), University of Calgary, Canada, pp. 30–1.

Daish, C.B. (1972). *The Physics of Ball Games*. EUP, London.

Eils, E., Streyl, M., Linnenbecker, S., Thorwesten, L., Volker, K. and Rosenbaum, D. (2001). Plantar pressure measurements in a soccer shoe: characterization of soccer specific movements and effects after six weeks of aging, in *Proceedings of the 5th Symposium on Footwear Biomechanics* (eds. E. Hennig and A. Stacoff), Zurich, pp. 32–3.

Football League (1989). *Commission of Enquiry into Playing Surfaces – Final Report*. The Football League, Lytham, St Annes.

Johnson, G., Dowson, D. and Wright, V. (1976). A biomechanical approach to the design of football boots. *Journal of Biomechanics*, **9**, 581–5.

Lafortune, M.A. (1991). Three dimensional acceleration of the tibia during walking and running. *Journal of Biomechanics*, **24**, 877–86.

Lake, M.J. (2000). Determining the protective function of sports footwear. *Ergonomics*, **43**, 1610–21.

Lees, A. and Cooper, S. (1995). The shock attenuation characteristics of soccer shin guards, in Spree, Leisure and Ergonomics (eds G. Atkinson and T. Reilly), E & FN Spon, London, pp. 130–5.

Lees, A. and Kewley, P. (1993). The demands on the boot, in *Science and Football II* (eds T. Reilly, J. Clarys and A. Stibbe), E & FN Spon, London, pp. 335–40.

Levendusky, T.A., Armstrong, C.W., Eck, J.S., Spryropoulous, P., Jeziorowski, J. and Kugler, L. (1988). Impact characteristics of two types of soccer balls, in *Science and Football* (eds T. Reilly, A. Lees, K. Davids and W.J. Murphy), E & FN Spon, London, pp. 385–93.

Masson, M. and Hess, H. (1989). Typical soccer injuries – their effects on the design of the athletics shoe, in *The Shoe in Sport* (eds B. Segesser and W. Pforringer), Wolfe Publishing, London.

Reilly, T. and Thomas, V. (1976). A motion analysis of work rate in different positional roles in professional football match play. *Journal of Human Movement Studies*, **2**, 87–97.

Rodano, R., Cova, P. and Vigano, R. (1988). Design of a football boot: a theoretical and experimental approach, in *Science and Football* (eds T. Reilly, A. Lees, K. Davids and W.J. Murphy), E & FN Spon, London, pp. 416–25.

Townend, M.S. (1988). Is heading the ball a dangerous activity?, in *Science and Football* (eds T. Reilly, A. Lees, K. Davids and W.J. Murphy), E & FN Spon, London, pp. 237–42.

Treadwell, P.J. (1988). Computer aided match analysis of selected ball games (soccer and Rugby union), in *Science and Football* (eds T. Reilly, A. Lees, K. Davids and W.J. Murphy), E & FN Spon, London, pp. 282–7.

Valiant, G.A. (1988). Ground reaction forces developed on artificial turf, in *Science and Football* (eds T. Reilly, A. Lees, K. Davids and W.J. Murphy), E & FN Spon, London, pp. 406–15.

Vetten, D. (1998). *Making a difference*. Adidas-Salomon AG, Herzogenaurach: Hoffmann und Campe Publishers, Germany.

Winterbottom, Sir W. (1985). *Artificial Grass Surfaces for Association Football – Report and Recommendations*. The Sports Council, London.

Withers, R.T., Maricic, Z., Wasilewski, S. and Kelly, L. (1982). Match analyses of Australian professional soccer players. *Journal of Human Movement Studies*, **8**, 159–76.

10 Injury prevention and rehabilitation

Thomas Reilly, Tracey Howe
and Nigel Hanchard

Introduction

Soccer entails physical contact in the course of tackling or contesting possession of the ball with opponents and this inevitably leads to injury of varying severity. A majority of injuries are unintentional, resulting from an error on the part of the player concerned or by another player. The error may lead to an accident (or unplanned event) and some of these accidents lead to injury. Inflicting injury intentionally on another player is severely punished both by the laws of the game and, where the evidence is clear-cut, by civil law also.

There are many extrinsic factors which may cause injury, besides the behaviour of players. These include the state of the pitch, the weather conditions, inappropriate choice of footwear and inattention to warm-up. There are also intrinsic factors which embrace the mental state of the player, the level of fitness and the existence of predisposing factors such as muscle weakness or a previous injury.

Detailed considerations of injuries in soccer are provided in texts such as Ekblom (1994) and Lillegard and Rucker (1993). The aim of this chapter is to outline the most common injuries that occur to players, consider some predisposing factors and preventive measures. The main methods of treating injuries are delineated. First, it is important to define an injury and examine the incidence of injury in soccer.

10.1 Factors affecting injury occurrence

There is little in the way of standardization in the presentation of injury statistics. There is no common definition that has been generally accepted by those studying the epidemiology of sports injury. In consequence, it is difficult to make comparisons between analyses carried out in earlier decades with statistics from the game as currently played. Unless methodologies are similar, it is also impossible to make inferences about differences between countries.

Generally, analyses of injuries tend to be retrospective studies of the records held at the professional clubs or compiled by the medical team. Records usually detail the type of injury and the timing of the occurrence along with concise descriptive detail. The period of treatment is recorded and the diary is maintained until the player is fully recovered to play in competition. Players may feign injury as an excuse for poor performance but persistence in doing so may compromise his/her selection for the next game. Consequently, the operational definition of an injury might be one that prevents the player from training for two consecutive sessions. An injury that renders the player from competing for a sustained period would constitute a severe injury.

For comparison with other sports in terms of risk, information in addition to the frequency of occurrence is required. The incidence of injury refers to the occurrence of new injuries in a particular time frame; the prevalence of injuries refers to the overall number of sufferers at a particular time. Exposure to injury risk may be expressed from statistics referring to injuries per 1000 h of play or injuries per player exposure. Each game provides a reference in that it can be considered as exposing 22 players to injury over 90 min.

Soccer provides a different profile to running in that a majority of injuries occur during match-play or competition (Figure 10.1). In contrast, three out of four injuries to runners are attributable to training error. Training practices in soccer can be devised whereby collisions and full tackling are discouraged. Besides, players may reduce the intensity of their efforts in practices if they know their place in the team is not in jeopardy. Runners may be reluctant to do this, feeling their own performances in racing may be adversely affected by a disturbed training programme.

It seems that the level of competition is influential in the incidence of injury. Player exposure risks in an English Premier Division side (Table 10.1) show that the risk is higher in the First Team, next highest in the A and Youth's team, and lowest in the Reserves' matches. Injury was defined as being unable to train for two sessions on successive days.

In a recent large, prospective study, Peterson *et al.* (2000) investigated the occurrence of injuries over a wider range of skill levels, including players from adult top-level, third

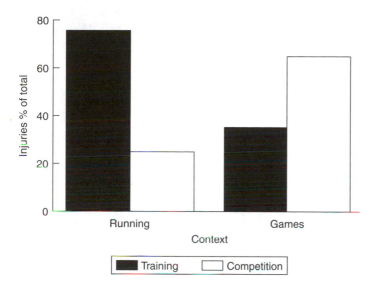

Figure 10.1 Injuries attributable to training and competition in runners and in games players (soccer, rugby, hockey and handball). (Data are based on survey of Reilly and Stirling, 1993.)

Table 10.1 Injuries and player exposure risks for three categories of representative competition in a professional Premier Division English Club over one full season

	Number of games	Player exposures	Injuries	Injury per player exposure
First team	48	528	45	1/11.7
Reserves	45	495	19	1/26.1
A and Youths team	44	116	7	1/16.6
Aggregate	137	1139	71	1/16

league, amateur and local teams, as well as high- and low-level youths in each of the 14–16 and the 16–18 categories. In this case, injury was defined as any tissue damage due to football, irrespective of whether this necessitated time off training or match play. It was found that adult players for local teams sustained three to four times more injuries than the other adult groups and high-level youths. In addition, injury rates in the low-level youth teams were twice as high as those among their high-level counterparts.

Each sport has its own characteristic profile of injuries. The majority of injuries to soccer players are soft-tissue (tendon and muscle) and joint trauma. These occur predominantly in the lower limbs; the joints most frequently affected are the knee and ankle. The muscle injuries are a combination of locomotor functions of running off the ball during play or bruising due to physical contact with other players. Joint trauma can result from specific gross injury or, less dramatically, the cumulative effects of excessive loading. In either event the long-term consequence, in elite players especially, is some predisposition towards degenerative arthritis of the lower limb joints (Dvorak and Junge, 2000).

Soccer injuries are not exclusively confined to the lower limb, however. Upper limb injuries may occur, especially in goalkeepers. Back injury may be more disabling than lower limb muscle injury (Figure 10.2) and facial damage can occur due to opponent's elbows or clashes of heads in contesting possession of the ball in the air. Concussion is a risk in the latter case and also, perhaps less obviously, as a result of heading the ball. There is now some evidence that the frequency of such events may correlate with cognitive impairments in both professional and amateur players (Dvorak and Junge, 2000).

Some injuries promote changes in the rules of play or in their implementation by the referee. Examples were the use of elbows in the English League during the 1994 season and the tackle from behind, outlawed in the 1994 World Cup.

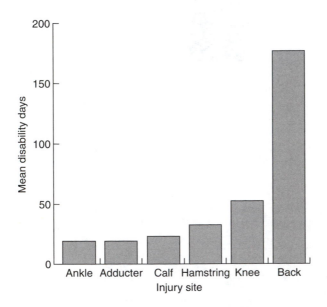

Figure 10.2 Mean disability time for severe injuries to professional soccer players. The back and knee joints have the longest period off training.

10.2 Aetiology of injuries

As soccer is a contact sport, players frequently sustain direct blows to the body, for example, during a block tackle or from a collision with another player. These blows result in contusion injuries, a disruption of the blood vessels within the soft tissues leading to haematoma formation or bone fractures.

Indirect injuries result from forces generated within the musculoskeletal structures during an activity. Damage may be sustained to muscles, tendons, ligaments, joint structures and bone. These types of injury often occur during the early or late stages of a game due to inadequate warm-up, poor flexibility or fatigue as discussed previously. Playing actions implying increased risk of injury are more evident in the first 15 min and the last 15 min than at other periods of the game (Rahnama *et al.*, 2002).

Overuse injuries are caused by continued or repetitive actions or as a result of exposure of a structure to high loads. These types of injuries occur as a result of training errors, biomechanical abnormalities, inadequate or inappropriate footwear and terrain. Training errors include inadequate warm-up, excessive training regimes, sudden increases in the duration, frequency or intensity of training, and inadequate rehabilitation from injury. Biomechanical abnormalities include, leg length discrepancies, soft-tissue inflexibility, incorrect biomechanical alignment and joint stiffness. Problems with footwear include, poor shock absorption qualities, poor grip and poor fit. Terrain may be responsible for injuries, for example, if too much running is done on cambered road surfaces or hills or too lengthy a session on sandhills. Stress fractures, microfractures in cortical bone, resulting from excessive tensile loading may be classified in this category of injury (Corrigan and Maitland, 1994). Such injuries are common in the metatarsals, fibula and tibia.

10.3 Soft-tissue injuries

Overstretching of a ligament results in a sprain and overstretching of a muscle results in a strain. There are three categories of ligament sprain and four categories of muscle strain. In Grade I sprains, only a few fibres of the ligament are damaged. In Grade II sprains, more fibres are damaged and the ligament is partially ruptured. In Grade III sprains, the ligament is totally ruptured, the integrity of the joint is compromised and surgical repair is required. Grade I strains are minor injuries where only a few muscle fibres are damaged and the muscle sheath is intact. In Grade II strains, the muscle sheath is still intact but there is considerable damage to blood vessels. In Grade II strains, the muscle sheath is partially torn, bleeding is diffuse and a large area of muscle is involved. A Grade IV strain is a complete rupture of the muscle belly a palpable gap is present and surgical repair is necessary. Sprains and strains may be chronic or acute injuries.

Whilst muscle damage may entail rupture of muscle fibres which is severe enough to prevent training for some days or weeks, microtrauma to muscle may not give rise to immediate pain. Soreness may be delayed to peak 48–72 h after completing training regimens that require active muscle groups to work eccentrically (Newham *et al.*, 1983). Examples of such regimens include repetitive bounding drills or routines referred to as plyometrics. This describes repeated stretch–shortening cycles of active muscle groups. Damage to muscle is indicated by increased release of creatine kinase which leaks through the muscle membrane into the bloodstream and by disruption of the myofibrils within the muscle (Newham *et al.*, 1983). It seems that no permanent damage is done by such drills and players habituate to it with repeated sessions.

Muscle cramp is a condition where the muscle goes into spasm and fails to relax. It is seen mostly towards the end of a game or during extra time. The muscles mostly affected are the calf muscle (*gastrocnemius*) and the quadriceps. The condition is relieved by stretching the affected muscle. It is associated with reduced energy stores and hydration status within the muscle (Edwards, 1988).

Tendinitis is an inflammation of a tendon. Tenosynovitis is inflammation of the synovial lining of a tendon's sheath. Both conditions are accompanied by tenderness and swelling and, when the tendon is put on tension or moved within its sheath by contraction of the associated muscle, pain. A common site is the Achilles tendon often caused by ill-fitting boots or poor flexibility. Achilles tendinitis and plantar fasciitis (insertion tendinitis of the plantar aponeurosis at the calcaneus) are injuries which the soccer player shares in common with runners, and which are associated with transmission of large forces.

Shin splints is an overuse injury, often affecting soccer players who train on hard surfaces or with inappropriate shoe studs. Whilst 'shin splints' is a blanket term describing pain in the anterior lower leg, it mainly refers to a musculotendinous inflammation of the medial margin of the tibia (Lennox, 1996).

Bursae are sacs of synovial fluid which occur at points of friction, for example, where the patella tendon rubs over the tibia. Bursitis is an inflammation of such a structure. A bursa may become inflamed as a result of a direct blow, for example, the prepatellar bursa during a fall on the knee. At some other sites, for example, around the hip, bursitis is more likely to occur for underlying biomechanical reasons, precipitated by overuse. Bursitis is often a painful condition which is slow to resolve.

10.3.1   *Injuries of the thigh*

Nearly a quarter of soccer-related injuries involve the thigh, and a further 10% the groin (Hawkins *et al.*, 2001). Hamstring strains are commonly reported in soccer players. Generally, damage is at the musculotendinous junction although injury may occur within the belly of the muscle. Mostly they result from a quick forceful stretch of the muscle in an attempt to accelerate, decelerate or stretch for the ball. Injuries to the quadriceps may be a result of imperfect kicking or a blocked tackle. The muscle group may suffer contusions from direct blows. Occasionally a large haematoma within the quadriceps will require surgical drainage. Myositis ossificans, in which part of the muscle tissue becomes ossified, is a complication of a thigh muscle injury, particularly following a direct blow to the muscle (Lennox, 1996).

Injuries to the adductors of the thigh are also a feature of soccer. These muscles work strongly when the kicking leg crosses the non-kicking leg: injuries may occur on either side and are associated with tightness in this muscle group. *Rectus femoris, iliopsoas* and *rectus abdominis* as well as *adductor longus* may be associated with groin pain. The condition referred to as 'Gilmour's groin' implicates the inguinal lining in groin pain.

The iliotibial band refers to a thick sheet of fascia that runs down the side of the leg between the iliac crest of the hip crossing the knee joint to insert on the lateral condyle of the tibia. Running may cause an irritative condition known as iliotibial band friction syndrome (Corrigan and Maitland, 1994) at the site where the band slides backwards and forwards over the lateral femoral epicondyle during flexion and extension of the knee joint.

10.3.2   *Jumper's knee*

Jumper's knee is another overuse injury sometimes incurred by soccer players. The syndrome gets its name due to its frequency in jumpers and basketball players. It includes tendinitis, degeneration and sometimes partial rupture of the patellar tendon. It manifests in anterior knee pain and tenderness around the patella and is aggravated by contracting the knee extensors.

10.4 Joint injuries

Specific injuries to the joints may result from direct blows but more commonly occur as a result of forces generated within the musculoskeletal structures during an activity.

10.4.1 Knee joint

The knee joint is particularly vulnerable in a sport such as soccer, being a hinge joint between two long levers, and is the site of 17% of soccer-related injuries (Hawkins *et al.*, 2001). Besides the articulation between femur and tibia, the knee also includes the articulation between the patella (knee cap) and the anterior surface of the distal end of the femur. There are ligaments on medial and lateral sides of the knee and, within the joint, the cruciate ligaments which cross each other in the shape of an 'X'. The anterior cruciate ligament runs upwards, posteriorly and laterally from the anterior surface of the intercondylar area of the tibia to the medial surface of the posterior aspect of the lateral femoral condyle. Its function is to withstand forces displacing the femur backwards with respect to the tibia, or vice versa. The posterior cruciate ligament runs downwards, anteriorly and medially from the posterior surface of the intercondylar area of the tibia to the lateral surface of the anterior aspect of the medial femoral condyle. Its function is to withstand forces displacing the femur forwards with respect to the tibia, or vice versa. Thus, the integrity of the cruciate ligaments is important in securing the stability of the joint.

Injury to the anterior cruciate ligament (ACL) is a major threat to the career of the professional soccer player. Contemporary methods of ACL reconstruction after rupture, using the middle portion of the patella tendon or carbon fibre grafts, have meant that players may return to match play despite a rupture of the ligament. A prolonged period of rehabilitation for restoring the strength of the muscles around the joint and proprioception is needed after surgery. Particular emphasis should be placed on the hamstrings as these muscles have a similar function to the ACL, preventing forward displacement of the tibia with respect to the femur (Ekstrand, 1995). Unfortunately, the performance of repaired or reconstructed ACLs is not as good as intact ACLs due to the diminished proprioceptive feedback which may impair function and does not protect the joint sufficiently from re-injury.

It has been suggested that female soccer players are more prone to injuring the ACL than their male counterparts. The identification of oestrogen receptors on the ACL and the link between oestradiol and a reduction in tensile force lend some support to this theory. The injury is likely to occur on riding a tackle, landing with an extended knee which induces a huge rise in anterior shear force. The greater liability to ACL injury in female soccer players may be more due to their motion patterns and a tendency to land on a straightened knee when subject to a perturbation.

The medial and lateral collateral ligaments prevent valgus and varus subluxation of the knee joint respectively. A blow to the outside of the knee, forcing the joint to open medially (valgus subluxation), will damage the medial collateral ligament. Conversely, a blow to the medial side of the knee, forcing the joint to open laterally (varus subluxation) will damage the lateral collateral ligament. Such injuries commonly occur during block or high tackles and result in instability of the knee joint.

Damage to the medial meniscus, historically referred to as 'cartilage' damage in soccer players is another hazard of match-play. It is usually associated with weight-bearing when the foot is fixed to the ground and the knee joint is rotated medially in relation to the foot. This can happen when a player is hit from the side in a tackle and the knee joint is rotated whilst the foot is still on the ground. A swivel-boot was designed to allow the foot also to rotate in such an event in American Football but the design was never taken seriously for soccer. Advances in the surgical management of such injuries, employing arthroscopic techniques, in conjunction with thorough rehabilitation programmes, have allowed players an early return to competition and, in stable knees, give good long-term success rates. These methods also appear to avoid the tendency to secondary degenerative arthritis which was previously associated with meniscal injuries (DeHaven and Arnockzy, 1994).

A most damaging injury to soccer players is where the medial meniscus, medial collateral ligament and anterior cruciate are injured together, a combination known as O'Donoghue's triad. The stability of the joint will be severely affected as a result. Medial collateral ligament (MCL) strains – even those of Grade III severity – can usually be successfully managed conservatively. On the other hand, if function is to be preserved, reconstruction of the ACL is mandatory. This reconstruction, and, if appropriate, repair of any concomitant meniscal damage, may be deferred until such time as the medial ligament has healed, pain and swelling have resolved and full range of movement has been restored to the joint (Shelbourne and Patel, 1995). In the interim, a period of 4–8 weeks or more, the chartered physiotherapist has an important role in achieving these ends, as well as in strengthening the joint's supporting musculature and educating the player in the importance of post-operative rehabilitation. Needless to say, intensive rehabilitation also follows such surgery.

A fractured patella is an uncommon injury in soccer; however, a direct blow to the medial side of the patella, for example, in a high tackle, can cause it to dislocate laterally.

10.4.2   Ankle joint

Ankle injuries constitute 17% of all injuries to soccer players (Hawkins *et al.*, 2001). Most injuries affect the lateral ligaments of the joint, due to inversion and plantarflexion of the ankle. Injuries to the deltoid ligament on the medial side also occur but these are less common, being associated with pronation and outward rotation of the foot.

'Footballer's ankle' is a condition where bony growths, exostoses, develop on the anterior and posterior margins of the tibia and talus. These are thought to be due to repeated trauma to the joint (O'Neill, 1981). During powerful kicking the ankle joint is in a position of full plantarflexion. This results in the apposition of the posterior aspect of the lower border of the tibia and the posterior aspect of the talus. Conversely during the push-off phase of running the anterior borders of the tibia and talus hit each other. The force of this is increased during acceleration (Corrigan and Maitland, 1994). The exostoses can be dislodged within the joint leading to persistent pain in the ankle. Surgical removal is often necessary (Biedert, 1993).

10.4.3   Foot injuries

Injuries to the feet are inevitable in soccer players due to direct blows from opponents when shots are blocked or to physical contact in tackling or contesting possession. They may also be due to faulty footwear or interactions with the playing surface. 'Turf toe syndrome', for example, refers to a sprain of the plantar capsule of the metatarsophalangeal joint of the big toe. Its cause is forceful dorsiflexion of this toe because of increased friction between the shoe and a hard or artificial playing pitch. The joint is especially painful during the push-off in a fast run.

Dorsiflexion to 90° at the metatarsophalangeal joint of the big toe is needed at push-off. Dorsiflexion is limited at the big toe of soccer players by a chronic condition known as hallux rigidus (Ekblom, 1994). This is a chronic injury due to repeated minor injury to this joint. Use of a stiffer sole may reduce pain but surgery is sometimes carried out to treat the problem.

The feet are subject to a host of minor niggling injuries following match-play. Sub-ungual haematoma underneath the toe nail can produce acute pain. Blisters on the soles of the feet and on the toes can be extremely discomforting. They are especially seen when players start pre-season training or use new boots. Tendons and ligaments on the feet are also subject to strain. One danger of carrying these conditions into strenuous training or match-play is that they may cause other musculotendinous injuries due to asymmetry as the player favours the use of the most comfortable limb.

10.4.4   The shoulder joint

Dislocation of the shoulder joint results from a fall on the outstretched arm. Often this involves tearing of the rotator cuff muscles which give the shoulder joint its stability and control of movement. This injury may delay or prevent a player from returning to training or competition as the upper limbs play an important role in the maintenance of balance during running, kicking and tackling.

10.5   Bony injuries

Direct and indirect blows to the body may result in fracture. Fractures of the ribs may occur when one player lands on another following a collision or a tackle. This injury can be debilitating and may require a long period out of the game until the fracture has healed. Fractures of the clavicle and bones of the upper extremity occur from direct blows or falls on the outstretched arm. Blows to the head during contesting the ball in the air often result in fractures of the facial bones and skull. These injuries may have serious consequences and often necessitate surgery.

Fractures of the lower limb, especially the tibia and fibula, may occur as a result of a block tackle. Such fractures require long periods of rehabilitation. Players frequently 'fall over the ball'. In this instance a player's time to react is dramatically reduced due to the body being close to the ground and, in rare cases, fractures of the lower portion of the tibia and fibula may result.

10.6   Treatment and rehabilitation

The recovery from injury depends on accurate diagnosis in the first instance, proper first-aid and secondary treatment, a planned period of rehabilitation and a graded progress towards return to competition. These are topics on which a great deal has been written (e.g. Reilly, 1981; Harries *et al.*, 1998) and are outside the scope of this text.

Knowledge of first-aid is essential for both training and paramedical staff in attendance at practice sessions and matches. It is important to have basic first-aid facilities available on-site. There must also be a quick route of access to specialist medical facilities through the club's network of consultants or local hospitals. Modern medical imaging techniques, notably nuclear magnetic resonance imaging, have enhanced diagnostic facilities for soft-tissue and joint injuries. Arthroscopy has enabled exploratory surgery and visualization of intact and damaged structures within joints prior to decisions about the wisdom of open surgical interventions.

The body has its own mechanisms of repairing damaged tissue, whether this is bone, tendon, ligament or muscle. For bone to reunite in correct biomechanical alignment, the separated portions must be repositioned correctly. This frequently involves surgical insertion of plates, screws and wires to hold bone ends and fragments together. This procedure accelerates the repair process and rehabilitation. Ligaments and tendons have much poorer blood supply than skeletal muscle and so their recovery takes longer. Mature skeletal muscle has a great capacity for regeneration and this process starts very soon after the damage to its cells occurs. The repair processes include formation of non-contractile collagenous fibres as well as the regeneration of new muscle cells.

The primary aim of the immediate treatment of soft tissue injuries is to control haematoma formation and avoid further damage to the soft tissues. The body part should be rested (R) for 24–48 h and ice (I) applied intermittently with compression (C) and elevation (E) of the body part to reduce blood flow and subsequent oedema (RICE).

Although immobilization is essential immediately post-injury, even short periods of immobilization (48 h) may have undesirable effects. Muscle tissue begins to atrophy, the biomechanical properties of ligaments alter and changes in the histological properties of articular cartilage occur. To prevent these changes it is essential that a carefully controlled active exercise regimen is commenced as soon as possible. This should be prescribed and monitored by a chartered physiotherapist and should be within the limits of pain tolerance of the injured player. Such a regimen will improve local circulation, facilitate the reabsorption of the haematoma and tissue exudate, maintain or restore muscle strength, fatigue resistance and flexibility of the injured muscle, proprioception of the injured joint and coordination of the limb. Progress may be monitored regularly using standard performance tests or more complex performance measures such as peak muscular torques during isokinetic movements (Perrin, 1993).

Electrotherapeutic modalities may also be employed by the chartered physiotherapist; these include, ultrasound, transcutaneous electrical nerve stimulation (TENS), interferential therapy (IFT), pulsed short-wave diathermy (PSWD), laser and percutaneous neuromuscular stimulation. These modalities may be used for their analgesic effects, to reduce inflammation, to accelerate healing and maintain or improve muscle strength and size. The chartered physiotherapist may also use manual techniques that include mobilization and manipulation techniques, massage and proprioceptive neuromuscular facilitation (PNF) (Corrigan and Maitland, 1994).

Drugs such as non-steroidal anti-inflammatories (NSAIDs) have analgesic, anti-pyretic and anti-inflammatory properties and thus have a role in the treatment of soft tissue injuries (Stankus, 1993). Topical creams may be applied to facilitate absorption of superficial haematomas.

Orthotic devices or supports may be required to enable the player to bear weight on an injured lower limb or return to training and protect the injured body part against re-injury. Such devices may include crutches, walking sticks, knee braces, ankle supports, strapping and orthotic inserts inside shoes.

10.7 Preventive measures

It is axiomatic to state that prevention is easier to cure. Identification of injury predisposition is a first step towards prevention, although this is often neglected even at the highest

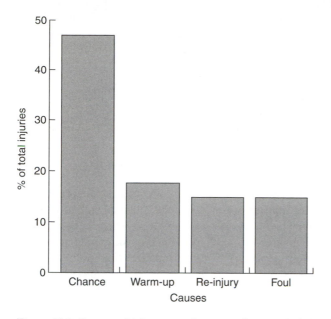

Figure 10.3 Causes which games players attribute to their injury. (Data were collected in the course of study by Reilly and Stirling, 1993, the games players including rugby, hockey and handball as well as soccer players.)

level of soccer play. Besides, games players are reluctant to recognize intrinsic factors that are responsible for incurring injuries and attribute about 50% of their injuries to chance (Figure 10.3).

Players who carry muscle weaknesses into competition are likely to experience situations where the muscle fails. Such weaknesses can be identified if players have a regular profiling of their muscle strength capabilities. This facility is available for teams with a systematized 'sports science support' programme. Muscle strength profiling should also show up asymmetries between left and right limb, the weaker of which is the side most likely to be affected in locomotor sports. Asymmetry may also be reflected in improper hamstrings to quadriceps ratios. Soccer players may acquire strong quadriceps but must also balance this by training the hamstrings. It seems that attention should be paid to eccentric as well as concentric contractions in training, in view of the eccentric role of the hamstrings in actions such as kicking a ball. The peak torque of the knee flexors in eccentric actions may be compared to that of the knee extensors in concentric mode in computing a 'dynamic control ratio'.

Strength angle profiles can be determined using isokinetic force data at a selection of angles throughout the range of movement at a particular joint (Perrin, 1993). This is especially relevant in avoiding re-injury, since reduction of strength may be evident only in a restricted range of motion. This deficiency could be corrected by recommending isometric exercises for the range of motion where muscle strength had been reduced.

In a study of Swedish soccer players, personal factors such as joint stability, muscle tightness, inadequate rehabilitation and lack of training were deemed responsible for 42% of all injuries observed (Ekstrand and Gillqvist, 1982). In an extension of this research, Ekstrand (1982) reported that 67% of soccer players had tight muscles and such players were vulnerable to injury. Tightness was pronounced in the hamstring and hip adductor muscle groups.

As a programme of flexibility training among Swedish professional soccer players over a complete season was found to reduce the incidence of injury, there is little doubt that flexibility is an important protective factor against injury.

Stretching muscles prior to training and match-play affects flexibility short-term. Flexibility routines can be incorporated into the warm-up. As flexibility is particular to each joint rather than representing a whole-body characteristic, it is important that the stretching routine is appropriate for footballers (Reilly and Stirling, 1993). The incidence of injuries over a season was less in games players who paid attention to jogging, technique work (to rehearse game skills) and lower-body flexibility exercises than those who warmed up for the same duration but who had a more general type of warm-up regimen. Warm-up is especially important in cold conditions in order to raise body temperature for the more strenuous training drills to follow.

Between 17 and 30% of injuries may be attributed to an incompletely recovered previous injury at the same site. Secondary injuries of this type tend to be severe (Hawkins *et al.*, 2001), emphasizing the importance of appropriate and complete rehabilitation. Such rehabilitative intervention should also be timely. Complaints of pre-existing joint or muscle pain and strapping of acute problems during play have been identified as important correlates of injury (Dvorak *et al.*, 2000).

Summary

The emphasis placed in this chapter has been on the occurrence of the major injuries in soccer, the main methods of treatment and the importance of preventive strategies. Whilst some injuries can be prevented and the risks of re-injury reduced, damage due to reckless play (especially by opponents) cannot be anticipated. Nevertheless, training practices can include collision-avoidance drills and routines that improve the ability to ride tackles safely. This demonstrates that training and coaching staff can contribute towards injury prevention and that the medical care of the player is a team effort. This team includes coaching personnel and management as well as chartered physiotherapist, paramedical and medical personnel.

References

Biedert, R. (1993). Anterior ankle pain in football, in *Science and Football II* (eds T. Reilly, J. Clarys and A. Stibbe), E & FN Spon, London, pp. 396–401.

Corrigan, B. and Maitland, G.D. (1994). *Musculoskeletal and Sports Injuries*. Butterworth-Heinemann, Oxford.

DeHaven, K.E. and Arnockzy, S.P. (1994). Basic science, indications for repair and open repair [instructional course lectures, The American Academy of Orthopaedic Surgeons: meniscal repair part I]. *Journal of Joint and Bone Surgery (Am.)*, **76-A**, 140–52.

Dvorak, J. and Junge, A. (2000). Football injuries and physical symptoms: a review of the literature. *American Journal of Sports Medicine*, **28**(5), S3–9.

Dvorak, J., Junge, A., Chomiak, J., Graf-Baumann, T., Peterson, L., Rösch, D. and Hodgson, R. (2000). Risk factor analysis for injuries in football players: possibilities for a prevention programme. *American Journal of Sports Medicine*, **28**(5), S69–74.

Edwards, R.H.T. (1988). Hypotheses of peripheral and central mechanisms underlying occupational muscle pain and injury. *European Journal of Applied Physiology*, **57**, 275–81.

Ekblom, B. (1994). *Football (Soccer)*. Blackwell Scientific, Oxford.

Ekstrand, J. (1982). Soccer injuries and their prevention. Medical Dissertation No. 130, Linköping University.

Ekstrand, J. (1995). Knee ligament injuries in soccer players, in *Science and Football III* (eds T. Reilly, J. Bangsbo and M. Hughes), E & FN Spon, London, pp. 150–5.

Ekstrand, J. and Gillqvist, J. (1982). The frequency of muscle tightness and injuries in soccer players. *American Journal of Sports Medicine*, **10**, 75–5.

Harries, M., Williams, C., Stanish, W.D. and Micheli, L.J. (1998). *Oxford Textbook of Sports Medicine*, 2nd Edition, Oxford University Press, New York.

Hawkins, R.D., Hulse, M.A., Wilkinson, C., Hodson, A. and Gibson, M. (2001). The association football medical research programme: an audit of injuries in professional football. *British Journal of Sports Medicine*, **35**, 43–7.

Lennox, C.M.E. (1996). Muscle injuries, in *The Soft Tissues. Trauma and Sports Injuries* (eds G.R. McLatchie and C.M.E. Lennox), Butterworth-Heinemann, Oxford.

Lillegard, W.A. and Rucker, K.S. (1993). *Handbook of Sports Medicine*. Andover Medical Publishers, London.

Newham, D.J., Mills, K.R., Quigley, B.M. and Edwards, R.H.T. (1983). Pain and fatigue after concentric and eccentric contractions. *Clinical Science*, **64**, 55–62.

O'Neill, T. (1981). Soccer injuries, in *Sports Fitness and Sports Injuries* (ed. T. Reilly), Faber and Faber, London, pp. 127–32.

Perrin, D.H. (1993). *Isokinetic Exercise and Assessment*. Human Kinetics, Champaign, IL.

Peterson, L., Junge, A., Chomiak, J., Graf-Baurmann, T. and Dvorak, J. (2000). Incidence of football injuries and complaints in different age groups and skill-level groups. *American Journal of Sports Medicine*, **28**(5), S51–7.

Rahnama, N., Reilly, T. and Lees, A. (2002). Injury risk associated with playing actions during competitive soccer. *British Journal of Sports Medicine*, **36**, 354–9.

Reilly, T. (1981). *Sports Fitness and Sports Injuries*. Faber and Faber, London.

Reilly, T. and Stirling, A. (1993). Flexibility, warm-up and injuries in mature games players, in *Kinanthropometry IV* (eds W. Duquet and J.A.P. Day), E & FN Spon, London, pp. 119–23.

Shelbourne, D.K. and Patel, D.V. (1995). Instructional course lectures, The American Academy of Orthopaedic Surgeons: management of combined injuries of the anterior cruciate and medial collateral ligaments. *Journal of Bone and Joint Surgery (Am.)*, **77-A**(5), 800–6.

Stankus, S.J. (1993). Inflammation and the role of anti-inflammatory medications, in *Handbook of Sports Medicine* (eds W.A. Lillegard and K.S. Rucker), Butterworth-Heinemann, London.

11 Psychology and injury in soccer

Frank Sanderson

Introduction

Injury blights the lives of many soccer players and affects the fortunes of many teams. When an elite player is injured, there are not only personal costs to the player but also potentially major repercussions for the teams he/she represents. For example, David Beckham's foot injury, sustained at a critical stage in Manchester United's season and a few weeks before the 2002 World Cup, triggered intense media interest – not surprising, given his pivotal role at club level and in the national team.

Sports participation rates and training intensities have both increased dramatically in recent years (Ninedek and Kolt, 2000) and this has inevitably led to sharp increases in the incidence of sports injuries (Crossman, 1997, 2001). This trend has stimulated increasing investment in the prevention and treatment of injuries, with growing numbers of sports scientists and medical practitioners specializing in sports medicine.

But how much attention is actually being paid to psychological aspects of injury? Are there psychological dispositions which make injuries more likely to occur? What about the psychological implications of injury? Are there lessons we can learn from psychology which will help in the prevention and treatment of injury?

The aim of this chapter is to explore answers to these questions, and in the process, promote an awareness that there is an important psychological dimension to injury. Initially, the many links between psychological factors and injury occurrence are outlined. In the second part of the chapter, the psychology of rehabilitation is considered, drawing from literature that has focused on emotional reactions to injury, social support and psychological interventions.

11.1 Psychological factors and injury occurrence

11.1.1 Stress and injury

Andersen and Williams (1988) linked athletic injury with stress levels, the latter triggered by such psychological factors as trait anxiety, daily hassles and general personality disposition. Other researchers have focused particularly on 'life stress', arguing that it is cumulative in its effects, enhancing the likelihood of injury by affecting the player's concentration on the task in hand. For example, Bramwell *et al.* (1975) noted that those players with the greatest accumulated life change (as caused by stressful life events, such as 'death of a spouse', 'getting divorced' or even 'getting married') were more likely to experience injury due to attentional narrowing. More recently, it has been found that the frequency of injury is related to negative rather than total or positive life stress (Passer and Seese, 1983; R.E. Smith *et al.*, 1990). The latter also reported that only those athletes low in coping skills and in social support demonstrated a significant relationship between life stress and injury. Hanson *et al.* (1992) found that coping resources, negative life stress, social support and competitive anxiety were predictive of injury severity. May *et al.* (1985) suggested that athletes' experience of psychological stress has negative effects on their concentration and general emotional balance, with the implication that they are more injury-prone as a consequence.

Some of these issues are considered in greater detail in the following section.

11.1.2   *Personality and injury-proneness*

There are conceptual connections between an individual's personality and susceptibility not only to illness but also injury. For example, neurotic individuals tend to be cautious, timid, indecisive and easily stressed. These characteristics are likely to predispose the individual to injury in that the nervous player might, for example, be less than fully committed in a 50–50 tackle and so sustain an injury as a consequence. Despite the intuitive conceptual link, research evidence in support of this hypothesis is limited. Reilly (1975) found a relationship between apprehensiveness amongst professional soccer players and the number of joint injuries sustained in a season. Jackson *et al.* (1978) found that 'tender-minded' American Football players were more likely to be injured and that the more 'reserved' players tended to have the most severe injury. In contrast, others have failed to find a relationship between personality factors and injury (e.g. Kraus and Gullen, 1969; Brown, 1971).

Nideffer (1989) has promoted the concept of 'attentional style'. He argued that individuals have a preferred attentional style, such as a broad external focus or a narrow internal focus, and that, under stress, they become more heavily dependent on this style, irrespective of its appropriateness in the situation. Anderson (2001) noted that an individual who perceives a situation as stressful may suffer from a narrowing of attentional focus, a kind of 'tunnel vision', and thereby become more injury-prone.

Research suggests that the intervening variable between personality and injury is stress in that an individual's personality may predispose him or her to experience stress in a wide variety of situations. Support for this link between stress and injury was provided by Davis (1991) who found that swimmers and American Football players experienced a reduction in the incidence of injury after they had commenced a regular progressive relaxation programme.

11.1.3   *Anxiety reactions and injury*

Despite a lack of tangible evidence, there is interest in the way psychoanalytical perspectives can increase our understanding of the athletic experience in general and sports injury in particular (Strean and Strean, 1998).

Sanderson (1981a) outlined possible links between injury-proneness and the individual's subconscious attempts to cope with anxiety reactions. Anxiety can be caused, for example, by the player's ambivalence about competition and aggression. Positive feelings about participation are encouraged because there are rewards, such as victory and prestige, to be had. However, the game also offers discomforting aggression and possible injury. The anxiety and tension which are associated with such conflict can seriously affect performance and increase the likelihood of real or imagined injury, as an unconscious means of reducing tension.

11.1.3.1   Injury resulting from counterphobia

A player who finds the game very anxiety-inducing may attempt to counteract the anxiety by meeting it head-on, by being overtly aggressive and fearless – the physical version of 'whistling in the dark'. Soccer will have some individuals of this kind, although Moore (1967) maintained that they tend to be attracted to high-risk sports such as down-hill skiing, boxing, rugby or motor-racing.

Closely related is the situation where an anxious player uses injury as a badge of courage, an overt sign of masculinity. He lacks real confidence, needing the visible scars of battle to confirm his manhood.

11.1.3.2   Injury as a weapon

Conversely, there is another kind of athlete who uses injury as a means of punishing another or others in an indirect way. Ogilvie and Tutko (1971) gave the example of the reluctant player forced to play because of an athletically frustrated father. By being injured, he can accomplish several objectives: he can make his father feel guilty for pressurizing him; he can frustrate his father's displaced aspirations; and he can avoid the undesired competition. A soccer player in dispute with the club could well use injury in this kind of way to frustrate the management. He/she may wish to cause difficulties for the team and/or the coach because of real or imagined grievances – a particularly effective technique when the player is valuable. He/she lacks courage for a confrontation, and so reacts in this indirect but effective way.

11.1.3.3   Injury as an escape

There are players who fear competition so much that injury provides an ideal way of reducing anxiety. With injury, whether real or imagined, the feared competition can be avoided, perhaps even without squad membership being jeopardized. The ego can also be kept intact: had there not been so much injury, the individual can believe that he/she would have been an (even more) outstanding player. A player's disability can be used by the team-mates as a rationalization for any shortcomings in the team's performance.

11.1.3.4   Psychosomatic injury

Unconscious and powerful psychological forces can sometimes precipitate psychosomatic injury. The player frequently complains of injury and yet no physical reasons can be obtained to substantiate the claim. The player does not respond to conventional treatment. If he/she is a key member of the squad, it is extremely frustrating to all concerned and may lead to a build-up of resentment in the team, thereby exacerbating the individual's psychological problems. Once the underlying emotional difficulties are resolved, the physical problems disappear.

In each of the above cases, the fundamental cause of the injury is psychological, with a recurring theme being that the individual, often unconsciously, is attempting to cope with anxiety reactions. However, there is a need for caution in categorizing injured players in this way:

- It is difficult to gain conclusive proof that a player's injury-proneness is psychological in origin.
- Erroneous assumptions about causation of injury could lead to inappropriate treatment.

Nevertheless, it is advantageous that all those concerned with ensuring the player's complete recovery should have as much information about the individual as possible. Communication amongst players, coaches, trainers, physiotherapists and physicians is essential. It is possible for vulnerability profiles to be established on the basis of comprehensive information, thus allowing positive injury prevention.

11.1.4   Symptoms of injury-proneness

Lysens *et al.* (1989) examined the development of accident- and overuse-prone profiles of young players. Whilst acknowledging the importance of physical traits in predisposing a player to injury, they stressed that psychological factors need to be considered, even in relation to profiling the overuse-prone player.

Various symptoms associated with the psychologically vulnerable player have been suggested, and they are listed in Table 11.1.

11.1.5   Exercise addiction

Some players become obsessed with exercise to the extent that they risk a variety of health problems, including multiple overuse injuries. Wichman and Martin (1992) pointed out that such players are difficult to treat, often requiring a catastrophe (e.g. consequential marriage break-up) for them to appreciate that they have a problem.

Table 11.1 Symptoms associated with the psychologically vulnerable player

Symptoms	Comments
Discrepancy between ability and aggressiveness	A player with modest ability who is overly aggressive is vulnerable
Success phobia	Fear of failure is a common phenomenon and well understood, but the incidence of fear of success should not be underestimated
Unihibited aggressiveness	The player presents a danger to himself and others
Feelings of invulnerability	Associated with reckless behaviour
Excessive fear of injury	The apprehension causes over-cautious play, paradoxically making injury more likely in, say, 50–50 tackles
Extensive history of injuries	Repeated injury may indicate physical and/or psychological vulnerability
Concealment or exaggeration of injuries	This indicates probable underlying psychological problems
Marked anxiety proneness	The over-nervous player's performance is detrimentally affected and injury is more likely

11.2   Psychological implications of injury

11.2.1   Introduction

Reaction to injury

Mentally, I'm a bit fragile about the injury. If something negative happens in the rehabilitation, I get depressed about it. My initial reaction when I saw my foot facing the wrong way was, "That's my career gone". For the first couple of days in hospital, I thought I had lost everything I had worked for over the years.

Graeme Le Saux on the dislocation and fracture of his ankle bone whilst playing for Blackburn Rovers during the 1995–96 season.

Many players sustain injuries which trigger significant negative psychological reactions. Weiss and Troxel (1986) in a survey of injured collegiate/elite players found disbelief, fear, rage, depression, tension and fatigue to be common emotional responses. Common somatic complaints were upset stomach, insomnia and loss of appetite. Many found great difficulty in coping with the enforced inactivity, the powerlessness and the long rehabilitation. Such negative reactions can lead to recurrent injury problems, an inadequate recovery period and extended rehabilitation. Brewer (2001) descibed the period following injury as emotional 'duress'. He notes that in some cases athletes have attempted suicide after falling into a post-injury depression, commenting that, generally, feelings of depression and frustration can occur at all phases of a rehabilitation cycle.

The implications of injury were profound for Graham Tutt, the Charlton Athletic goalkeeper who was forced to retire after a kick in the face at the age of 20:

It's impossible to erase from my memory the moment of impact and pain when the boot of … made contact with my face at full force. The physical and mental scars of that accident will be with me for the rest of my life … I reached the ball and at the split-second that … was poised to strike it. His boot whacked me in the face … I couldn't hear anything. Everything was hazy and strange.

"He's kicked my eye out" was the immediate thought that ran through my mind. There was also a great deal of blood spurting from my nose and more blood coming from my cheek. My eyelid was split as well. But the most frightening thought was that I had lost an eye.

(Harris and Varney, 1977, p. 78)

In a long and detailed account of the post-trauma events, Tutt did not indicate that his psychological 'scars' were appreciated by those involved in the treatment. In fact, when Tutt reached the dressing room, the manager's reaction was hardly sensitive:

By this time both my eyes were closed and I was coughing blood, as he said to me, "Can you go on?"

(Harris and Varney, 1977)

His severe long-term problems with his 'nerves' seem to have gone totally untreated.

11.2.2   Factors affecting psychological reaction to injury

11.2.2.1   The individual's history of injury

If the injury background is extensive then psychologically negative reactions are likely to be more intense. Frustration, anger, resignation and despair may be intensely felt, creating an apathetic attitude during the recovery phase. In this kind of psychological state, the recovery phase is likely to be seriously extended and the chances of re-injury enhanced (Sanderson, 1981b).

11.2.2.2   The cause of the injury

It can be hypothesized that a player who gets injured in a 'fair' 50–50 tackle will have less psychological repercusssions than if he believes that the tackler meant to hurt him, or if he believes himself to have been very unlucky – in the case of Graeme le Saux, journalist Steven Seaton commented in October 1996 that 'It is almost worse that such a sickening injury resulted from nothing more sinister than catching his foot in the ground.'

11.2.2.3   The nature of the injury

Other things being equal, the psychological trauma will tend to increase as a function of the severity of the injury, but only to the extent that the individual is aware of the severity. As Stein (1962) has noted, it is often the case that the graver the injury, the fewer are the emotional complaints. This is partly because of immediate post-trauma shock which can leave a player amnesic and anaesthetized against feelings of pain. When full awareness returns, the process of rationalization has already begun.

11.2.2.4   The nature of the sport

Few participants in sport fully expect to be injured but it is clear that the likelihood of injury varies markedly across sports, with soccer being one of the more dangerous sports. Injury in golf is relatively rare, whereas the risk of disabling injury in soccer is thousands of times higher than in underground mining (Mongillo, 1968). It seems reasonable to hypothesize that the psychological trauma associated with a particular injury is a decreasing function of the general level of risk entailed in the sport. All else being equal, a particular sports injury will generate more emotional trauma in a low-injury risk golfer than in the high-injury risk soccer player.

11.2.2.5   The situation favourableness

For a professional soccer player, the timing of the injury would be a factor influencing the psychological reaction to injury, and hence the chances of a successful and speedy return to competition. For example, a season-ending injury in the last game of the season is likely to be less traumatic than the same injury mid-season (Brewer, 2001). This sense of injury as an off-time crisis is recognized within the life development intervention literature (Danish *et al.*, 1992; Petitpas *et al.*, 1999). A player can also be fortunate in terms of health status, and having a body which heals well (Gould *et al.*, 1997).

11.2.2.6   The nature of the injury interacting with the nature of the sport

The interaction is important in that, for example, the psychological effect of a cut eye would be greater for a boxer than, say, for a free-style wrestler. Although both are contact sports in which injury risk is relatively high, the nature of boxing ensures that a cut eye is a particularly devastating injury, encouraging greater psychological repercussions.

11.2.2.7   *The level of competition in which the injury has occurred*

The casual soccer player who plays infrequently and who sprains his wrist is unlikely to find the experience as traumatic as the player who earns his living from the game. The implication of this is that the psychological effects of injury are only worthy of serious study in relation to what might be termed 'serious' sport. The recreational skier who fractures his leg may suffer considerable personal trauma but it is of no general significance. Physical fitness is also important; the player who is physically fit having become adapted to fairly severe physical stresses before injury, can more easily adapt to the demands of a physical rehabilitation programme – he is less likely to 'acquiesce' to the disability (Bender *et al.*, 1971).

Where the injury is so severe as to end the career of the player, there may be long-term psychological implications, particularly for those at the professional level. Kleiber and Brock (1992) found lower self-esteem and life satisfaction amongst former professional athletes 5–10 years after they had sustained a career-ending injury.

11.2.2.8   *The player's personality*

The key importance of personality in reaction to injury can be demonstrated by reference to the theories of Hans Eysenck who has studied the structure of personality. Eysenck (1957) identified the two major independent dimensions of extroversion and neuroticism, along which the personality of individuals can vary. From Eysenck's theory, it can be deduced that personality might affect the player's reaction to injury as shown in Table 11.2.

Crossman and Jamieson (1985) found that players who overestimated the seriousness of an injury displayed more anxiety and greater feelings of inadequacy, anger, loneliness and apathy. Overestimation was more common amongst those players competing at lower levels. They suggested that such players might benefit from counselling concerning coping strategies.

The neurotic extrovert, being reckless, impatient, unreliable and optimistic, tends to under-react to injury and may undergo inadequate rehabilitation. Neurotic extroverts have been found to be particularly vulnerable to neurotic breakdown when sustaining injury towards the end of their careers as their over-valued physical abilities go into decline (Little, 1969). Little concluded:

> Like exclusive and extensive emotional dependence on work, on key family relationship bonds, intellectual pursuits, physical beauty, sexual prowess or any other over-valued attribute or activity, athleticism can place the subject in a vulnerable pre-neurotic state leading to manifest neurotic illness in the event of an appropriate threat, or actual enforced deprivation, especially if abrupt and unexpected.

Table 11.2 Effect of personality on reaction to injury

Personality type	Characteristics	Reaction to injury
Stable extrovert	Sociable, carefree	Straightforward
Neurotic extrovert	Impulsive, assertive, reckless, changeable, pain tolerant	Tendency to under-react to injury. Inadequate rehabilitation and resumption of activity too early
Stable introvert	Reflective, cautious	Straightforward
Neurotic introvert	Nervous, unconfident, pessimistic	Tendency to over-react to injury, leading to lengthy rehabilitation

By contrast, the very stable extrovert/introvert benefits from emotional stability, thereby ensuring relatively straightforward rehabilitation.

11.2.3   Psychological hardiness

Kobasa (1979) described a kind of individual who has abundant 'coping resources' in response to stressful events. Psychological hardiness is characterized as 'a constellation of characteristics such as curiosity, willingness to commit, seeing change as a challenge and stimulus to development, and having a sense of control over one's life'. It could be argued that Eysenck's self-assured stable extrovert matches this definition most closely.

Tunks and Bellissimo (1988) stated that, 'some individuals seem able to transform calamities into opportunities for growth while others transform everyday hassles into overwhelming adversities'. Whilst it is undoubtedly true that some players will be more psychologically hardy than others, it is important to recognize that coping skills can be learned even by those with a tendency to 'catastrophize'. This latter group, who are more likely to experience problems with the rehabilitation process, represent the greatest challenge to the care-givers in that they will need active and sympathetic support during rehabilitation.

The psychologically hardy players will tend to take injury in their stride. For example, when Alan Shearer of Newcastle United and England ruptured his cruciate ligament in the mid-1990s, he was initially 'bewildered' by the diagnosis, 'It was unbelievable. I was expecting it to be something … but nothing that serious. It's one of the worst injuries of the game.' But he quickly adjusted, 'It took me a few days to come to terms with what was happening, but it wasn't too bad after that'. He had the benefit of strong support from family and the club, as well as a phone call from Paul Gascoigne to say that he had recovered from a similar injury and that there was no reason why he shouldn't. He was clearly well-briefed about the rehabilitation process. 'There were dark times sure. Times when the knee never seemed to be getting any better. But I expected them and I was ready for them. I kept in touch with the lads. I went into training every day …'

11.2.4   Attribution and injury

Causal attribution is concerned with the individual's perceptions of the causes of outcomes such as success, failure, illness or injury. Studies of accident victims and those recovering from major illness suggest that causal attributions can play a significant role in psychological and physical recovery. By the same token, it is reasonable to expect that attributions would have a potentially important mediating role in recovery from athletic injuries. For example, if a player focuses on blaming an opponent for the injury, the resultant frustration, anger and sense of injustice could well hinder the recovery process.

The related construct of 'locus of control', the extent to which individuals perceive that they have power over what happens to them, is relevant here. Those with an internal locus of control perceive themselves to be responsible for their own lives, whereas those with an external locus perceive themselves to be at the mercy of outside forces such as fate, luck and powerful others. The practical implications of this idea in relation to injury are immediately apparent. An internal locus of control would generally be more functional, whereas an external locus focusing on blame would be counter-productive.

Achievement motivation is higher in those who believe they are responsible for their own destiny. Those acting in support of athletes would generally wish to encourage the internal disposition, thereby enhancing the likelihood that the injured player will be optimally committed to rehabilitation. Those players with an internal locus of control who are undergoing treatment would tend to be more aware of the importance of their own efforts in achieving full recovery.

Dweck (1986) described internal locus individuals as 'mastery oriented', as opposed to those exhibiting 'learned helplessness'. Learned helplessness exists when the individual, having experienced repeated failure (or repeated injury) together with inappropriate attributions, concludes that 'nothing I do matters' and that failure is unavoidable.

Relevant research has concentrated on *health* locus of control and health behaviour, health status, and exercise adherence (O'Connell and Price, 1982; Dishman, 1986). Results have generally been equivocal although Slenker *et al.* (1985) found that joggers were significantly more internal than non-exercisers.

11.2.5   Individual differences in reaction to pain

Historically, a stimulus–response (S–R) model of pain was accepted, that is, the intensity of the pain is a reliable guide to the severity of the injury. It is now more widely recognized that many variables other than the severity of injury determine the reaction to pain. Pain perception can be influenced by: health, age, personality, suggestion (the doctor who warns the patient that 'this is going to be painful' encourages perception of pain), expectancy, experience, attentional focus (in a competitive soccer match, a gashed shin may go unnoticed) and cultural factors.

Taylor and Taylor (1998) provide a useful review of pain management in the rehabilitation from sports injury.

Having considered psychological factors which will influence the individual's reactions to injury, what are the threats as perceived by the player?

11.2.6   The player's psychological reactions to injury

The vulnerable player

> Once hurt, they begin to question their invulnerability, and many return to competition before they are psychologically ready. This is seen in athletes who become much more tentative in their play or protective of the injured area. They often lose the spontaneity and assertiveness that allowed them to excel. Their cautious play transforms into performance decrements, which can further erode their confidence and lead to more stress and frustration. The end result of this process can be reinjury, injury to another area of the body, temporary or permanent performance problems, or emotional upsets that further drain motivation and the desire to compete.
>
> Petitpas and Danish (1995)

Following injury, the player is faced with a range of threats which have the potential to delay or prevent the process of healing. They are threats to:

- *Body image*: players typically have a positive body image and attach great importance to their physical appearance/integrity. An injury, particularly a disfiguring injury, threatens the sense of being a whole and undamaged person [what Eldridge (1983) has called narcissistic disfigurement].

- *Self-esteem*: closely linked with body image, self-esteem is also affected by, for example, the loss of autonomy and control, the need to depend on others and the loss of feelings of personal invulnerability.
- *Life-style*: loss of important roles, separation from friends, thwarting of plans, uncertainty about the future and, perhaps, loss of income.

There is no doubt that some players suffer extreme mental torment as a consequence of disabling injury. Kubler-Ross (1969) noted a five-stage grief process which Rotella and Heyman (1986) argued has relevance to the player who experiences traumatic injury (Table 11.3).

Recognizing that the reaction to injury can be traumatic, Rotella and Heyman (1986) emphasized the need for positive 'crisis intervention' to help the athlete to be focused on immediate practical and manageable concerns rather than seeing the injury as an 'overwhelming, engulfing catastrophe'.

The 'grief' reaction to injury

Life is absurd. Just when I begin to pull it all together, I pull this muscle. I'm so depressed. Why me? Why now? … The stress is unbearable, to say nothing of the physical pain itself. It's just not fair. I feel like dying. A terrible loss.
>   An elite runner following a severe groin injury during Olympic Trials

It is important to acknowledge, as Kubler-Ross did in relation to the grief response, that these stages are not absolute in that not everyone goes through every stage, in a predictable sequence and within a predictable time-frame. Stages may be omitted, the sequence may be different and there are likely to be regressions (e.g. Alan Shearer's 'dark times'). However, an awareness of possible stages which might be experienced by an injured player will be of assistance to those involved in the rehabilitation process.

Brewer (2001) noted that stage models have been replaced by cognitive appraisal models of injury rehabilitation. The core notion here is that emotional reactions are understood to be directly linked to the appraisal processes undertaken by individuals, that is, emotions are thought to be dynamically related to the situational antecedents experienced by the injured athlete at any moment rather than to any 'stage'.

Table 11.3 Five-stage grief process applied to sport

Stage	Description
1 Denial	A defence mechanism whereby the player downplays or ignores the reality of injury. 'There's no problem … it'll be okay'
2 Anger	Sometimes indiscriminate but often directed at the perceived cause, for example, a particular opponent, the coach, inwardly to self
3 Bargaining	The player attempts to rationalize away the injury, indicating that the reality has not been fully accepted
4 Depression	Reality begins to emerge and the player finds it difficult to imagine making a full recovery
5 Acceptance	A necessary stage to be reached for the rehabilitation process to be effective. 'I'm injured but life goes on'

11.2.7　*The psychosocial process*

A four-phase model of the psychosocial processes associated with injury has been developed by Rose and Jevne (1993); Figure 11.1.

- Phase 1 – Getting injured: determined by the pre-injury context.
- Phase 2 – Acknowledging the injury: this in itself encapsulated the 'grief' stages, from denial to acceptance. For the less traumatized player, the acceptence may be virtually immediate.
- Phase 3 – Dealing with the impact: this involves acting on the lessons and dealing with the psychological implications, as well as obtaining appropriate treatment.
- Phase 4 – Achieving a physical and psychosocial outcome: this may involve re-evaluating the importance of participation in activities and acknowledging physical limitations.

It is argued that successful negotiation of these phases, with the lessons being learned and acted upon, will lead to a lower risk of injury in future.

11.3　The treatment process

11.3.1　*Intervention strategies*

The physician and the physiotherapist are not treating an injury. They are treating an injured player. There may be complex psychological antecedents to the injury and there are certainly psychological consequences. The specialist in sports medicine should be sensitive to such factors in order that the treatment process is most effective.

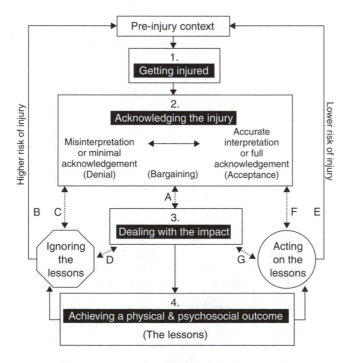

Figure 11.1 Psychosocial processes.

A.M. Smith *et al.* (1990) detailed various psychological intervention strategies in relation to the most frequently reported emotional responses.

- *Coping with frustration*: clear explanation of treatment is necessary, allowing the player to make informed decisions. Well-defined and achievable goals should be provided and revised in the light of progress.
- *Coping with depression*: within the constraints imposed by the injury, here should be a resumption of activity, which serves to encourage the perception of progress and reduce helplessness. In this context, it is important that relationships are maintained with the team and the coach.
- *Coping with anger*: discussion with the player about the source of the anger may reduce misunderstandings and irrational beliefs. The latter should be replaced with positive, realistic and rational thoughts.
- *Coping with tension*: first, there should be identification of the causes of the tension (e.g. fear of re-injury), followed by the use of mental imagery or relaxation procedures.

A.M. Smith *et al.* (1990), on the basis of their research into the emotional responses of athletes to injury, stressed the importance of the prompt recognition of emotional disturbance in facilitating the rehabilitation process and the safe return to competition. Davis (1991) provided evidence of the beneficial effects of progressive relaxation during team work-outs on reducing the incidence of injury.

11.3.2 Guidelines for support staff

There follows a list of broad guidelines to help the medical team and other support staff to be most effective in treating the injured player.

- Provide the player with 'quality time': make him/her feel that you really want to help by being supportive and reassuring and committed (Silva and Hardy, 1991). Implicit here is the importance of good access to all the appropriate medical resources (Gould *et al.*, 1997).
- Take every opportunity to educate the patient. Be good at communication: reduce anxiety and uncertainty by providing accurate and clear information about the diagnosis and the prescribed treatment, including a recovery timetable (Ford and Gordon, 1993). Kahanov and Fairchild (1994) found evidence of miscommunication between athletes and trainers in that 52% of a sample of injured athletes did not understand how the rehabilitation process related to their injury. Webborn *et al.* (1997) surveyed athletes attending sports injury clinics in the southeast of England. They observed 150 consultations but found only 22 athletes were prescribed exercises, and 17 of this group had misunderstood some aspect of the prescription. Written instructions, although rarely provided, increased the understanding of what was expected.
- Listen carefully to the player: you are likely to gain information which will improve the treatment. The player may even suggest rehabilitation techniques which are more effective than the official ones.
- Negotiate a treatment plan: help the player set challenging, attainable and measurable goals. A simple approach to goal-setting involves answering three questions: who? will do what? by when? (Wiese and Weiss, 1987). The treatment plan should take account of the fact that the injury is likely to have resulted in a significant increase in unstructured time.

- Show commitment.
- Liaise with other support staff.
- Be patient about the player's return to competition and about his or her behaviour. The rehabilitation process must take its proper course, with no premature return to competition. Coaches or team members can sometimes put undue pressure on a key player who is undergoing treatment for an injury – the question, 'When are you going to be back?' can make the injured player feel that he/she is letting the team down. The highly competitive extrovert player, impatient to return to competition, and used to being the centre of attention, can be exasperating to treat (Silva and Hardy, 1991).
- Recognize the motivational role of relaxation, positive self-talk, imagery and social support (Rotella and Heyman, 1986; Grove and Gordon, 1991; Green, 1992; Gould *et al.*, 1997; Bianco and Eklund, 2001). The value of social support is evident from this comment from Graeme Le Saux, 'As soon as my family and friends started to visit me in the hospital, I came to terms with the injury. I had to look further than my profession and realise there are other things in life.' However, problems can arise if the wrong kind of support is provided, for example, the player's family being over-protective or transmitting their own uncertainties and anxieties to the injured player.
- Know the sport: this is not a problem at professional clubs where the club doctor and paramedical staff will be very familiar with soccer and the particular club culture.
- Peer modelling – the pairing of an injured player with someone who has successfully recovered from a similar injury – and injury support groups whereby players find comfort from sharing their concerns with other injured players (Wiese and Weiss, 1987).
- Get patients actively involved early with the aim of making the player self-directed. The more the player is actively involved, the more he/she will feel ownership of the treatment process and attain the appropriate 'mind-set' for recovery (Green, 1992).
- Follow up early and frequently.

11.3.3   Exercise compliance

There has been a large amount of research in recent years into compliance or adherence to rehabilitation programmes – so-called secondary prevention. Detailed examination of adherence research is not possible here. However, those with a particular interest in the area should read Dishman (1986) for a useful review of research on exercise compliance.

Examination of the available research suggests the following conclusions:

- Blue-collar workers, smokers and overweight individuals are less likely to adopt and maintain a fitness regimen (Dishman, 1986).
- High self-motivation is associated with compliance/adherence (Fisher and Hoisington, 1993), as is having a positive attitude towards the injury and to life in general (Larson *et al.*, 1996).
- For some individuals, perceived inconvenience and lack of time are associated with dropping out (Fisher and Hoisington, 1993).
- Support from significant others, for example, health professionals, coaches and family members, is particularly important for adherence to rehabilitation (Rotella and Heyman, 1986; Fisher and Hoisington, 1993).
- Achievement goals and feelings of well-being are relatively important for adherence to the programme (Dishman, 1986).

In the context of adherence in the rehabilitation of sports injuries the following research findings may be useful.

Eichenhofer *et al.* (1986) found that players high on somatic (physical) anxiety have difficulty adhering to prescribed treatment.

Fisher *et al.* (1993) reported that adherers tended to perceive greater social support, be higher in self-motivation (intrinsic motivation), and believe they work harder.

Duda *et al.* (1989) noted that adherers amongst a sample of intercollegiate players believed in the effectiveness of the treatment, perceived greater social support, were more self-motivated and were more focused on personal mastery and improvement in sport. They suggested that injured players should be assessed for degree of self-motivation and that special (goal-setting) support be given to those who are low in this variable. They also argued that the presence of social support networks such as family, friends and fellow players is likely to facilitate the treatment process.

Weiss and Troxell (1986) maintained that players high in self-esteem are more likely to persist in rehabilitation programmes. Lampton *et al.* (1993) found that injured athletes low in 'self-esteem certainty' and with high ego involvement in tasks tended to miss the most rehabilitation appointments. Given that injury itself can negatively affect self-esteem, it is imperative that those concerned with the rehabilitation process seek to promote the player's functional self-esteem and confidence.

Worrel (1992) explained how behavioural techniques, involving the setting of short-term functional goals, and cognitive techniques aimed at promoting positive thoughts and actions, can be used to enhance the likelihood of compliance to rehabilitation. Ievleva and Orlick (1991) examined slow and fast healers from knee and ankle injuries and found that those athletes using cognitive techniques such as goal-setting, positive self-talk and the use of healing imagery, tended to be the fastest healers. There is evidence of the benefits of goal-setting (Gilbourne *et al.*, 1996; Gilbourne and Taylor, 1998) and imagery (Green, 1992) in the rehabilitation of injured athletes.

Rapport with the injured player is also important: as Danish (1986, p. 347) observed, 'when patients do not feel understood, do not find the health care professional warm and friendly, and are intimidated by the technical terminology, non-compliance is likely'.

Summary

The successful treatment of an injured player requires that support staff are sensitive not only to the physical, but also to the psychological antecedents and consequences of injury. Personality, exposure to life stress and psychodynamic factors have been linked with injury-proneness. The psychological implications of injury can be profound, and are affected, for example, by the severity of the injury, the player's history of injury and the player's personality.

It is important that the support staff acknowledge the psychological dimension in the treatment process. Giving the player full attention, emphasizing good communication and using motivation techniques to maximize the likelihood of compliance with the treatment process are necessary.

References

Anderson, M.B. (2001). Returning to action and the prevention of future injury, in *Coping with Sports Injuries* (ed. J. Crossman), Oxford University Press, Oxford, pp. 162–73.

Andersen, M.B. and Williams, J.M. (1988). A model of stress and athletic injury: prediction and prevention. *Journal of Sport and Exercise Psychology*, **10**, 294–306.

Bender, J.A., Renzaglia, G.A. and Kaplan, H.M. (1971). Reaction to injury, in *Encyclopedia of Sports Sciences and Medicine* (ed. L.A. Larson), Macmillan, New York, pp. 1001–3.

Bianco, T. and Eklund, R.C. (2001). Conceptual considerations for social support research in sport and exercise settings: the case of sport injury. *The Journal of Sport and Exercise Psychology*, **23**, 85–107.

Bramwell, S.T., Matsuda, M., Wagner, N.D. and Holmes, T.H. (1975). Psychosomatic factors in athletic injuries: development and application of the Social and Athletic Readjustment Rating Scale (SARRS). *Journal of Human Stress*, **1**, 6–20.

Brewer, B.W. (2001). Emotional adjustment to sports injury, in *Coping with Sports Injuries* (ed. J. Crossman), Oxford University Press, New York, pp. 1–19.

Brown, R.B. (1971). Personality characteristics related to injuries in football. *Research Quarterly*, **42**, 133–8.

Crossman, J. (1997). Psychological rehabilitation from sports injuries. *Sports Medicine*, **23**, 333–9.

Crossman, J. (2001). *Coping with Sports Injuries*. Oxford University Press, New York.

Crossman, J. and Jamieson, J. (1985). Differences in perceptions of seriousness and disrupting effects of athletic injury as viewed by players and their trainer. *Perceptual and Motor Skills*, **61**, 1131–4.

Danish, S.J. (1986). Psychological aspects in the care and treatment of athletic injuries, in *Sports Injuries: The Unthwarted Epidemic* (eds P.F. Vinger and E.F. Hoener), PSG Publishing, Boston, MA, p. 345.

Danish, S.J., Petitpas, A.J. and Hale, B.D. (1992). A developmental–educational intervention model of sport psychology. *The Sport Psychologist*, **6**, 403–15.

Davis, J.O. (1991). Sports injuries and stress management: an opportunity for research. *The Sport Psychologist*, **5**, 175–82.

Dishman, R.K. (1986). Exercise compliance: a new view for public health. *Physician and Sportsmedicine*, **14**, 127–45.

Duda, J.L., Smart, A.E. and Tappe, M.K. (1989). Predictions of adherence in the rehabilitation of athletic injuries: an application of personal investment theory. *Journal of Sport and Exercise Psychology*, **11**, 367–81.

Dweck, C.S. (1986). Motivational processes affecting learning. *American Psychologist*, **41**, 1040–8.

Eichenhofer, R. Wittig, A.F., Balogh, D.W. and Pisano, M.D. (1986). Personality indicants of adherence to rehabilitation treatment by injured athletes. *Midwestern Psychological Association Conference*, Chicago, May.

Eldridge, W.E. (1983). The importance of psychotherapy for athletic-related orthopedic injuries among adults. *International Journal of Sports Psychology*, **14**, 203–11.

Eysenck, H.J. (1957). *The Dynamics of Anxiety and Hysteria*. Routledge and Kegan Paul, London.

Fisher, A.C. and Hoisington, L.L. (1993). Injured athletes' attitudes and judgements toward rehabilitation adherence. *Journal of Athletic Training*, **28**, 48–50; 52–54.

Fisher, A.C., Scriber, K.C., Matheny, M.L., Alderman, M.H. and Bitting, L.A. (1993). Enhancing athletic injury rehabilitation adherence. *Journal of Athletic Training*, **28**, 312–18.

Ford, I. and Gordon, S. (1993). Social support and athletic injury: the perspective of sport physiotherapists. *Australian Journal of Science and Medicine in Sport*, **25**, 17–25.

Gilbourne, D., Taylor, A.H., Downie, G. and Newton, P. (1996). Goal-setting during sports injury: a presentation of underlying theory, administrative procedure, and an athlete case-study. *Sport Exercise and Injury*, **2**, 192–201.

Gilbourne, D. and Taylor, A.H. (1998). From theory to practice: the integration of goal-perspective theory and life development approaches within an injury specific goal-setting programme. *Journal of Applied Sport Psychology*, **10**, 124–39.

Gould, D., Udry, E., Bridges, D. and Beck, L. (1997). Coping with season-ending injuries. *The Sport Psychologist*, **11**, 379–99.

Green, L.B. (1992). The use of imagery in the rehabilitation of injured athletes. *The Sports Psychologist*, **6**, 416–28.

Grove, J.R. and Gordon, A.M.D. (1991). The psychological aspects of injury in sport, in *Textbook of Science and Medicine in Sport* (eds J. Bloomfield, P.A., Fricker and K.D. Fitch), Blackwell, London, pp. 176–86.

Hanson, S.J., McCullagh, P. and Tonymon, P. (1992). The relationship of personality characteristics, life stress and coping resources to athlete injury. *Journal of Sport and Exercise Psychology*, **14**, 262–72.

Harris, H. and Varney, M. (1977) *The Treatment of Football Injuries*. MacDonald James, London.

Ievleva, L. and Orlick, T. (1991). Mental links to enhanced healing: an exploratory study. *The Sports Psychologist*, **5**, 25–40.

Jackson, D.W., Jarrett, H., Bailey, D., Kausek, J., Swanson, J. and Powell, J.W. (1978). Injury prediction in the young athlete: a preliminary report. *American Journal of Sports Medicine*, **6**, 6–14.

Kahanov, L. and Fairchild, P.C. (1994). Discrepancies in perceptions held by injured athletes and athletic trainers during the initial injury evaluation. *Journal of Athletic Training*, **29**, 70–5.

Kleiber, D.A. and Brock, S.C. (1992). The effect of career-ending injuries on the subsequent well-being of elite college athletes. *Sociology of Sport Journal*, **9**, 70–5.

Kobasa, S.C. (1979). Stressful life events, personality, and health: an inquiry into hardiness. *Journal of Personality and Social Psychology*, **37**, 1–11.

Kraus, J.F. and Gullen, W.H. (1969). An epidemiologic investigation of predictor variables associated with intramural touch football injuries. *American Journal of Public Health*, **59**, 2144–56.

Kubler-Ross, E. (1969). *On Death and Dying*. MacMillan, London.

Lampton, C.C., Lambert, M.E. and Yost, R. (1993). The effects of psychological factors in sports medicine rehabilitation adherence. *Journal of Sports Medicine and Physical Fitness*, **33**, 292–9.

Larson, G.A., Starkey, C. and Zaichowsky, L.D. (1996). Psychological aspects of athletic injuries as perceived by athletic trainers. *The Sport Psychologist*, **10**, 37–47.

Little, J.C. (1969). The player's neurosis – deprivation crisis. *Acta Psychiatrica Scandinavica*, **45**, 187–97.

Lysens, R.L., Ostyn, M.S., Vanden Auweele, Y., Lefevre, J., Vuylsteke, M. and Renson, L. (1989). The accident-prone and overuse-prone profiles of the young player. *American Journal of Sports Medicine*, **17**, 612–19.

May, J.R., Veach, T.L., Reed, M.W. and Griffey, M.S. (1985). A psychological study of health, injury and performance in players on the US Alpine Ski Team. *Physician and Sportsmedicine*, **13**, 111–15.

Mongillo, B.B. (1968). Psychological aspects in sports and psychosomatic problems in the athlete. *Rhode Island Medical Journal*, **51**, 339–43.

Moore, R.A. (1967). Injury in athletics, in *Motivation in Play, Games and Sports* (eds R. Slovenko and J.A. Knight), Charles C. Thomas, Springfield, IL.

Nideffer, R.M. (1989). Anxiety, attention and performance in sports: theoretical and practical considerations, in *Anxiety in Sports: An International Perspective* (eds D. Hackfort and C.D. Spielberger), Hemisphere, New York, pp. 117–36.

Ninedek, A., and Kolt, G.S. (2000). Sport physiotherapists' perceptions of psychological strategies in sport injury rehabilitation. *Journal of Sport Rehabilitation*, **9**, 191–206.

O'Connell, J.K. and Price, J.H. (1982). Health locus of control of physical fitness-program participants. *Perceptual and Motor Skills*, **25**, 925–6.

Ogilvie, B. and Tutko, T.A. (1971). *Problem Athletes and How to Handle Them*. Pelham Books, London.

Passer, M.W. and Seese, M.D. (1983) Life stress and athletic injury: examination of positive versus negative events and three moderator variables. *Journal of Human Stress*, **9**, 11–16.

Petitpas, A.J. and Danish, S.J. (1995). Caring for injured athletes, in *Sport Psychology Interventions*, (ed. S.M. Murphy), Human Kinetics, Champaign, IL.

Petitpas, A.J., Giges, B. and Danish, S.J. (1999). The sport psychologist athlete relationship: implications for training. *The Sport Psychologist*, **13**, 344–57.

Reilly, T. (1975). An ergonomic evaluation of occupational stress in professional football. Unpublished PhD thesis, Liverpool Polytechnic.

Rose, J. and Jevne, R.F.J. (1993). Psychosocial processes associated with athletic injury. *The Sport Psychologist*, **7**, 309–28.

Rotella, R.J. and Heyman, S.R. (1986). Stress, injury and the psychological rehabilitation of players, in *Applied Sports Psychology: Personal Growth to Peak Performance* (ed. J.M. Williams), Mayfield, Palo Alto, CA.

Sanderson, F.H. (1981a). The psychology of the injury-prone player, in *Fitness and Sports Injuries*, (ed. T. Reilly), Faber and Faber, London, pp. 31–6.

Sanderson, F.H. (1981b). The psychological implications of injury, in *Fitness and Sports Injuries* (ed. T. Reilly), Faber and Faber, London, pp. 37–41.

Silva, J.M. and Hardy, C.J. (1991). The sport psychologist, in *Prevention of Athletic Injuries: The Role of the Sports Medicine Team* (eds F.O. Mueller and A.J. Ryan), F.A. Davis Co, Philadelphia, PA, pp. 114–32.

Slenker, S.E., Price, J.H. and O'Connell, J.K. (1985). Health locus of control of joggers and non-exercisers. *Perceptual and Motor Skills*, **61**, 323–8.

Smith, A.M., Scott, S.G. and Wiese, D.M. (1990). The psychological effects of sports injuries: coping. *Sports Medicine*, **9**, 352–69.

Smith, R.E., Smoll, F.L. and Ptacek, J.T. (1990). Conjunctive moderator variables in vulnerability and resiliency research: life stress, social support and coping skills, and adolescent sport injuries. *Journal of Personality and Social Psychology*, **58**, 360–70.

Stein, C. (1962). Psychological implications of personal injuries. *Medical Trial Technique Quarterly*, 17–28.

Strean, W.B. and Strean, H.S. (1998). Applying psychodynamic concepts to sport psychology practice. *The Sport Psychologist*, **12**, 208–22.

Taylor, J. and Taylor, S. (1998). Pain education and management in the rehabilitation from sports injury. *The Sport Psychologist*, **12**, 68–88.

Tunks, T. & Belissimo, A. (1988) Coping with the coping concept: a brief comment. *Pain,* **34**, 171–4.

Webborn, A.D.J., Carbon, R.J. and Millar, B.P. (1997). Injury rehabilitation programs: 'what are we talking about?', *Journal of Sport Rehabilitation*, **6**, 54–61.

Weiss, M.R. and Troxell, R.K. (1986). Psychology of the injured player. *Athletic Training*, **21**, 104–9.

Wichman, S. and Martin, D.R. (1992) Exercise excess: treatment patients addicted to fitness. *Physician and Sportsmedicine*, **20**, 193–200.

Wiese, D.M. and Weiss, M.R. (1987). Psychological rehabilitation and physical injury: implications for the sports medicine team. *The Sport Psychologist*, **1**, 318–30.

Worrel, T.W. (1992). The use of behavioral and cognitive techniques to facilitate achievement of rehabilitation goals. *Journal of Sport Rehabilitation*, **1**, 69–75.

12 Environmental stress

Thomas Reilly

Introduction

Soccer is played worldwide and in highly varied environmental circumstances. In some instances the climatic conditions are too hostile or are temporarily unsuitable for playing and there is a lull in the competitive programme. This applies in northern climates in winter and in tropical countries during the rainy season. In the former it becomes impossible to maintain playing pitches and the weather is too cold to play in comfort. At another extreme is the stress imposed by a hot environment and the difficulty of coping with high heat and humidity. Usually the hottest part of the day is avoided and matches are timed for evening kick-offs. In highly competitive international tournaments this is not always feasible and some teams from temperate climates are obliged to compete in conditions to which they are unaccustomed.

Altitude constitutes another environmental variable that can make supra-normal demands on soccer teams. This has applied to those teams who have competed at the two World Cup finals in Mexico in 1970 and 1986. It applies also to teams playing friendly or international qualifying matches at moderate altitude. Additionally, training camps for top teams are sometimes located at altitude resorts and this constitutes a particular novel challenge to sea-level dwellers.

The human body has mechanisms that allow it to acclimatize to some extent to environmental challenges. In the course of history it has evolved to match the environmental changes associated with the solar day. Consequently many physiological functions wax and wane in harmony with cyclical changes in the environment every 24 h. The sleep–wake cycle is dovetailed with alternation of darkness and light and the majority of the body's activities are controlled by biological clocks. These are disturbed when the body is forced to exercise at a time it is unused to, for example, after crossing multiple time zones to compete overseas. It is also disrupted if sleep is disturbed.

In this chapter, the major environmental variables that impinge on soccer play are considered. These include heat, cold, hypoxia, circadian rhythms and weather conditions. The biological background is provided prior to describing the consequences of environmental conditions for the soccer player.

12.1 Temperature

12.1.1 Thermoregulation

Human body temperature is regulated about a set point of 37°C. This refers to temperature within the body's core and is measured usually as rectal temperature, tympanic or oesophageal temperature. Oral temperature tends to be a little lower than these and is less reliable since the temperature within the mouth can be influenced by drinking cold or hot fluids and by the temperature of the air inspired.

For thermoregulatory purposes the body can be conceived as consisting of a core and a shell. There is a gradient of about 4°C from core to shell and so mean skin temperature is usually about 33°C. The temperature of the shell is more variable than the core and responsive to changes in environmental temperatures. Usually there is a temperature gradient from skin to the air and this facilitates loss of heat to the environment.

The human exchanges heat with the environment in various ways to achieve an equilibrium. The heat balance equation is expressed as:

$$M - S = E \pm C \pm R \pm K$$

where M = metabolic rate, S = heat storage, E = evaporation, C = convection, R = radiation and K = conduction.

Thermal equilibrium is attained by a balance between heat loss and heat gain mechanisms (Figure 12.1). Heat is produced by metabolic processes, basal metabolic rate being about $1 \, \text{kcal kg}^{-1} \text{h}^{-1}$. One kilocalorie (4.186 kJ) is the energy required to raise 1 kg water through 1°C. Energy expenditure during soccer might increase this by a factor of 15, with maybe only 20–25% of the energy expended reflected in external power output. The rest is dissipated as heat within the active tissues and as a result heat storage in the body increases. In order to avoid overheating, the body is equipped with mechanisms for losing heat. It also has built in responses to safeguard the thermal state of the body in circumstances where heat might be lost too rapidly to the environment, for example, in very cold conditions.

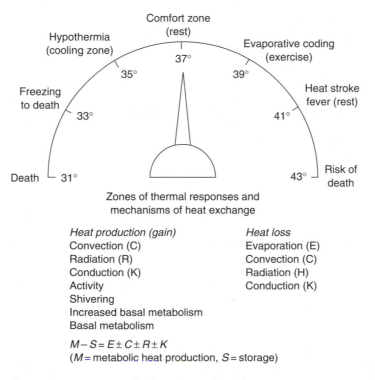

Figure 12.1 Heat loss and heat gain mechanisms.

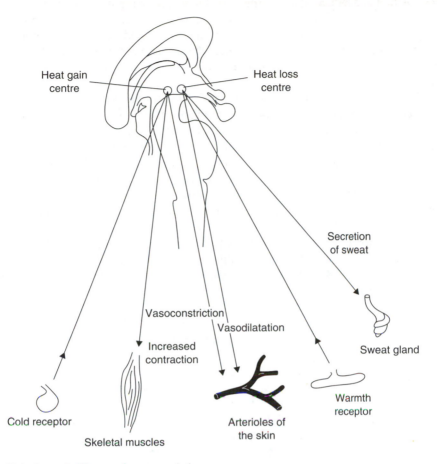

Figure 12.2 Control of human thermoregulation.

Body temperature is controlled by specialized nerve cells within the hypothalamus. The neurones in the anterior portion constitute the heat loss centre since they trigger initiation of heat loss responses.

These are effected by a redistribution of blood to the skin where it can be cooled and stimulation of the sweat glands to secrete a solution onto the skin surface where evaporative cooling can take place.

The hypothalamus is sensitive to the temperature of blood that bathes these cells controlling thermoregulatory responses. In addition to this direct information, the cells also receive signals from warmth and cold receptors located in the skin. In this way the heat loss and heat gain centres receive information about both the body's internal thermal state and environmental conditions (Figure 12.2).

12.1.2   *Exercise in the heat*

When exercise is performed, the temperature within the active muscles and core temperature both rise. When the exercise is carried out in hot conditions the skin temperature is elevated, reflecting the external challenge to the body. The hypothalamic response is represented by a diversion of cardiac output to the skin: the body surface can lose heat to the environment (by convection and radiation) due to the warm blood now being shunted through its subcutaneous layers. In strenuous exercise, such as intense competitive soccer, the cardiac output may be maximal or near it and the increased cutaneous blood flow may compromise blood supply to the active muscles. In such instances the soccer player will have to lower the exercise intensity, perhaps by taking longer recovery periods than normal or by running 'off-the-ball' less.

There are indications from motion analysis of players of the extent to which high environmental temperature affects their work-rates during matches. Ekblom (1986) reported that the distance covered in high-intensity running during match-play at an ambient temperature of 30°C was 500 m compared to 900 m when the temperature was 20°C. It is likely that this lowered work-rate reflects changes in the overall pace of the game. The exercise intensity and the level of play affect the magnitude of rise in the core temperature. Rectal temperatures averaging 39.5°C have been reported for Swedish First Division players in ambient temperatures of 20–25°C. The corresponding average for players of lower divisions was 39.1°C (Ekblom, 1986).

The distribution of blood to the skin is effected by means of dilatation of the peripheral blood vessels. There is a limit to the vasodilatation that occurs in thermoregulation. This is because increased vasodilatation reduces peripheral resistance and so causes a fall in blood pressure. The kidney hormone renin stimulates angiotensin which is a powerful vasoconstrictor and this response corrects a drop in blood pressure. The blood pressure decline is more of a risk in marathon running than in soccer play: whilst both entail an average loading of 75–80% $\dot{V}O_2$ max, the greater duration of marathon racing provides extra stress, probably with consequences for body water stores.

The sweat glands are stimulated when core temperature rises, loss of sweat by evaporation being the major avenue by which heat is lost to the environment during intense exercise. The glands respond to stimulation by noradrenaline and secrete a dilute solution containing electrolytes and trace elements. Heat is lost only when the fluid is vaporized on the surface of the body, no heat being exchanged if sweat drips off or is wiped away. When heat is combined with high humidity, the possibilities of losing heat by evaporation are reduced since the air is already highly saturated with water vapour. Consequently hot humid conditions are detrimental to performance and increase the risk of injury.

Soccer players may lose 3 l or more of fluid during 90 min of play in the heat. This is an average figure which varies with the climatic conditions and also between individuals. Some players may sweat little and will be at risk when competing in the heat due to hyperthermia. Those who sweat profusely may be dehydrated near the end of the game. A fluid loss of 13.1% body mass has been reported during a match at 33°C and 40% relative humidity. A similar fluid loss occurred when ambient temperature was 26.3°C but humidity was 78% (Mustafa and Mahmoud, 1979).

It is important that players are adequately hydrated prior to playing and training in the heat. Water is lost through sweat at a faster rate than it can be restored by means of drinking and subsequent absorption through the small intestine. Thirst is not a very sensitive indicator of the level of dehydration. Consequently players should be encouraged to drink regularly, about 200 ml every 15–20 min when training in the heat. The primary need is for water since sweat is hypotonic. Electrolyte and carbohydrate solutions can be more effective than water in enhancing intestinal absorption.

One important consequence of sweating is that body water stores are reduced. The body water present in the cells, in the interstices and in plasma seems to fall in roughly equal proportions. The reduction in plasma volume compromises the supply of blood available to the

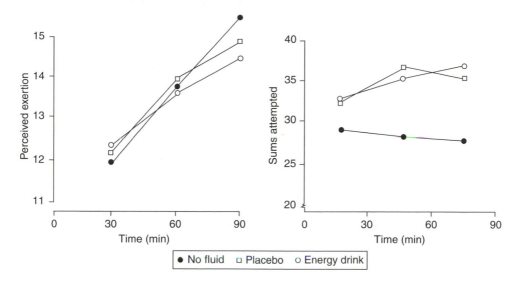

Figure 12.3 Rating of perceived exertion and the speed of adding under three experimental conditions: no fluid, a placebo and an energy drink.

active muscles and to the skin for cooling. Whilst the endocrine glands and kidneys attempt
to conserve body water and electrolytes, the needs of thermoregulation override these mechanisms and the athlete may become dangerously dehydrated through continued sweating.
The main hormones involved in attempting to protect against dehydration are vasopressin,
produced by the pituitary gland, and aldosterone secreted by the adrenal cortex which
stimulates the kidneys to conserve sodium.

Many components of soccer performance will be adversely affected once core temperature rises above an optimal level. This is probably around a body temperature of 38.3–38.5°C
(Astrand and Rodahl, 1986). Performance also deteriorates with progressive levels of dehydration. This drop in performance can be offset to some degree by fluid replacement. This
includes cognitive as well as physical and psychomotor aspects of skill. The data illustrated
in Figure 12.3 show how decision-making was best maintained when an energy drink was
provided to subjects compared with water only, which itself was superior to a trial when no
fluid was provided (Reilly and Lewis, 1985).

The thermal strain on the individual player is a function of the relative exercise intensity
(% $\dot{V}O_2$ max) rather than the absolute work-load. Therefore, the higher the maximal aerobic
power ($\dot{V}O_2$ max) and cardiac output, the lower the thermal strain on the player. A well-
trained individual has a highly developed cardiovascular system to cope with the dual roles
of thermoregulation and exercise. The highly trained individual will also acclimatize more
quickly than one who is unfit. Training also improves exercise tolerance in the heat but does
not eliminate the necessity of heat acclimatization.

12.1.3   *Heat acclimatization*

Acclimatization refers to reactions to the natural climate. The term *acclimation* is used to
refer to physiological changes which occur in response to experimentally induced changes in
one particular factor (Nielsen, 1994).

The main features of heat acclimatization are an earlier onset of sweating (sweat produced at a lower rise in body temperature) and a more dilute solution from the sweat glands. The heat-acclimatized individual sweats more than an unacclimatized counterpart at a given exercise intensity. There is also a better distribution of blood to the skin for more effective cooling after a period of acclimatization, although the acclimatized player depends more on evaporative sweat loss than on distribution of blood flow.

Heat acclimatization occurs relatively quickly and a good degree of adaptation takes place within 10–14 days of the initial exposure. Further adaptations will enhance the athlete's capability of performing well in heat stress conditions (Nielsen, 1994). Ideally, therefore, the athlete or team should be exposed to the climate of the host country for at least 2 weeks before the event. An alternative strategy is to have an acclimatization period of 2 weeks or so well before the event with subsequent shorter exposures nearer the contest as training is tapered before competition. If these are not practicable, attempts should be made at some degree of heat acclimatization before the athlete leaves for the host country. This may be achieved by pre-acclimatization.

1   Exposure to hot and humid environments, the player seeking out the hottest time of day to train at home.
2   If the conditions at home are too cool, players may seek access to an environmental chamber individually for periodic bouts of heat exposure. It is important that the players exercise rather than rest under such conditions. Repeated exposure to a sauna or Turkish bath is only partially effective. About 3 h per week exercising in an environmental chamber should provide a good degree of acclimatization (Reilly *et al.*, 1997).
3   The microclimate next to the skin may be kept hot by wearing heavy sweat suits or windbreakers. This practice will add to the heat load imposed under cool environmental conditions and induce a degree of adaptation to thermal strain.

On first exposure to a hot climate, players should be encouraged to drink copiously to maintain a pale straw-coloured rather than dark urine. They should drink much more fluid than they think they need since thirst is often a very poor indicator of real need. When they arrive in the hot country they should be discouraged strongly from sunbathing as this itself does not help acclimatization except by the development of a suntan which will eventually protect the skin from damage via solar radiation. This is a long-term process and is not beneficial in the short term, but negative effects of a sunburn can cause severe discomfort and a decline in performance. Players should therefore be protected with an adequate sunscreen if they are likely to be exposed to sunburn.

Initially, training should be undertaken in the cooler parts of the day so that an adequate work-load can be achieved and adequate fluid must be taken regularly. If sleeping is difficult, arrangements should be made to sleep in an air-conditioned environment but to achieve acclimatization the rest of the day should be partly spent exposed to the ambient temperature other than in air-conditioned rooms. Although sweating will increase as a result of acclimatization, there should be no need to take salt tablets, provided adequate amounts of salt are taken with normal food.

In the period of acclimatization, the players should regularly monitor body weight and try to compensate for weight loss with adequate fluid intake. Alcohol is inappropriate for rehydration purposes since it acts as a diuretic and increases urine output. Players are reminded to check that the volume of urine is as large as usual and that it is a pale straw colour rather than dark (de Looy *et al.*, 1988). Although there is no ideal measure of hydration status, the most suitable methods are osmolarity and conductivity (Pollock *et al.*, 1997).

12.1.4   Heat injury

Hyperthermia (overheating) and loss of body water (hypohydration) lead to abnormalities that are referred to as heat injury. Progressively they may be manifest as muscle cramps, heat exhaustion and heat stroke. They are observed more frequently in individual events such as distance running and cycling than in soccer but can occur in soccer matches or training sessions in the heat.

Heat cramps are associated with loss of body fluid, particularly in games players competing in intense heat (see Reilly, 2000). Although the body loses electrolytes in sweat, such losses cannot adequately account for the occurrence of cramps. These seem to coincide with low energy stores as well as reduced body water levels. Generally the muscles employed in the exercise are affected, but most vulnerable are the leg (upper or lower) and abdominal muscles. The cramp can usually be stopped by stretching the involved muscle, and sometimes massage is effective.

Heat exhaustion is characterized by a core temperature of about 40°C. Associated with this is a feeling of extreme tiredness, dizziness and breathlessness and tachycardia (increased heart rate). The symptoms may coincide with a reduced sweat loss but usually arise because the skin blood vessels are so dilated that blood flow to vital organs is reduced.

Heat stroke is a true medical emergency. It is characterized by core temperatures of 41°C or higher. Hypohydration – due to loss of body water in sweat and associated with a high core temperature – can be driven so far as to threaten life. Heat stroke is characterized by cessation of sweating, total confusion or loss of consciousness. In such cases treatment is urgently needed to reduce body temperature. There may also be circulatory instability and loss of vasomotor tone as the regulation of blood pressure begins to fail.

12.1.5   Competing in cold

Soccer in countries such as the United Kingdom is a winter sport and is often played in near-freezing conditions. Core temperature and muscle temperature may fall and exercise performance will be increasingly affected. Muscle power output is reduced by 5% for every 1°C fall in muscle temperature below normal levels (Bergh and Ekblom, 1979). A fall in core temperature to hypothermic levels is life-threatening and the body's heat gain mechanisms are designed to arrest the decline.

Among the responses to cold initiated by the posterior hypothalamus is a generalized vasoconstriction of the cutaneous circulation. This response is mediated by the sympathetic nervous system. Blood is displaced centrally away from the peripheral circulation and this increases the temperature gradient between core and shell. The reduction in skin temperature in turn decreases the gradient between the skin and the environment which protects against a massive loss of heat from the body. Superficial veins are also affected in that blood returning from the limbs is diverted from them to the vena comitantes that lie adjacent to the main arteries. In this way the arterial blood is cooled by the venous return almost immediately it enters the limb by means of counter-current heat exchange.

One of the consequences of the fall in limb temperature is that motor performance is adversely affected. In addition to the drop in muscular strength and power output as the temperature in the muscle falls, there is impairment of conduction velocity of nerve impulses to the muscles. The sensitivity of muscle spindles also declines and there is a loss of manual dexterity. For these reasons, it is important to preserve limb temperature in soccer players during competition. The goalkeeper in particular must maintain manual dexterity for handling the ball. Indeed much of the activity of the goalkeeper is spontaneous rather than directly imposed by demands of the game. The goalkeeper must remain alert during those periods when not directly involved in play in anticipation of the parts of the game when he or she is called upon.

Shivering is a response of the body's autonomic nervous system to the fall in core temperature. It constitutes involuntary activity of skeletal muscles in order to generate metabolic heat. Shivering tends to be intermittent and may persist during exercise if the intensity is insufficient to maintain core temperature. It may be evident during stoppages in play, especially when cold conditions are compounded by sleet.

Early symptoms of hypothermia include shivering, fatigue, loss of strength and coordination and an inability to sustain work-rate. Once fatigue develops, shivering may decrease and the condition worsens. Later symptoms include collapse, stupor and loss of consciousness. This risk applies more to recreational rather than professional soccer as some players might not be able to sustain a work-rate to keep them warm in extreme cold. In such events, the referee would be expected to abandon play before conditions became critical.

Cold is less of a problem than heat in that the body may be protected against exposure to ambient environmental conditions. The important climate is the microclimate next to the skin and this may be maintained by appropriate choice of clothing. Behaviourally, players might respond to cold conditions by maintaining a high work-rate. Alternatively, they may be spared exposure to the cold by conducting training sessions in indoor facilities where these are available.

Clothing of natural fibre (cotton or wool) is preferable to synthetic material in cold and cold-wet conditions. The clothing should allow sweat produced during exercise in these conditions to flow through the garment. The best material will allow sweat to flow out through the cells of the garment whilst preventing water droplets from penetrating the clothing from the outside. If the fabric becomes saturated with water or sweat, it loses its insulation and in cold-wet conditions the body temperature may quickly drop.

Players training in the cold should ensure that the trunk area of the body is well insulated. The use of warm undergarments beneath a full tracksuit may be needed. Dressing in layers is well advised: the outer layers can be discarded as body temperature rises and if ambient temperature gets warmer.

When layers of clothing are worn the outer layer should be capable of resisting both wind and rain. The inner layer should provide insulation and should also wick moisture from the skin to promote heat loss by evaporation. Polypropylene and cotton fishnet thermal underwear has good insulation and wicking properties and so is suitable to wear next to the skin.

Immediately prior to competing in the cold, players should endeavour to stay as warm as possible. A thorough warm-up regimen (performed indoors if possible) is recommended. It is thought that cold conditions increase the risk of muscle injury in sports involving intense anaerobic efforts; warm-up exercises may afford some protection in this respect. Competitors may need to wear more clothing than they normally do during matches.

Aerobic fitness does not necessarily offer protection against cold. Nevertheless, it will enable games players to keep more active when not directly involved in play and not increase the level of fatigue. Outfield players with a high level of aerobic fitness will also be able to maintain activity at a satisfactory level to achieve heat balance. On the other hand, the individual with poor endurance may be at risk of hypothermia if the pace of activity falls dramatically. Shivering during activity signals the onset of danger.

12.2   Altitude

12.2.1   *Physiological adjustments to altitude*

As altitude increases the barometric pressure falls. At sea level the normal pressure is 760 mmHg, at 1000 m it is 680 mmHg, at 3000 it is about 540 mmHg. High altitude conditions are referred to as hypobaric or low pressure and the main problem associated with this environment is hypoxia or a relative lack of oxygen.

Normally, the proportion of oxygen in the air is 20.93%, and the partial pressure of oxygen at sea level is 159 mmHg. The partial pressure of oxygen decreases with increasing altitude: this corresponds to the fall in ambient pressure since the proportion of oxygen in the air is constant. As a result there are fewer oxygen molecules in the air at altitude for a given volume of air. Less oxygen is inspired for a given inspired volume and this ultimately means a reduction in the amount of oxygen delivered to the active tissues.

As far as the uptake of oxygen into the body through the lungs is concerned, the important factor is the tension of oxygen in the alveoli (pO_2). Here the water vapour pressure is relatively constant at 47 mmHg as is the pCO_2 of 35–40 mmHg. The result of the fall in ambient pressure and consequently alveolar tension is that the gradient across the pulmonary capillaries for transferring oxygen into the blood becomes less favourable. Exercise that depends on oxygen transport mechanisms will be impaired at about 1200 m once desaturation occurs. This refers to the oxygen dissociation curve of haemoglobin (Hb) which is sigmoid-shaped and is affected by pressure. Normally the red blood cells are 97% saturated with O_2 but this figure falls when pO_2 levels drop at a point corresponding to this altitude (1200 m). The O_2–Hb curve is little affected for the first 1000–1500 m of altitude because of the flatness at its top. As the pressure drops further to reach the steep part of the curve, the supply of oxygen to the body's tissues is increasingly impaired. Nevertheless, at an altitude of 3000 m the arterial saturation is about 90%.

The immediate physiological compensation for hypoxia is an increase in ventilation. This is represented by an increased tidal volume (depth of breathing) and an increased breathing frequency. A consequence of this hyperventilation is that there is an increase in the CO_2 blown off from blood passing through the lungs. The elimination of CO_2 leaves the blood more alkaline than normal due to an excess of bicarbonate ions, CO_2 being a weak acid in solution in body fluid. Over several days the kidneys compensate by excreting excess bicarbonate, so returning the blood to the normal pH level. However, the body's alkaline reserve is decreased and so the blood has a poorer buffering capacity for tolerating additional acids (such as lactic acid diffusing from muscle to blood during exercise).

Once at altitude, there is an increased production of the substance 2,3-bisphosphoglycerate (2,3-BPG) by the red blood cells. This is beneficial in that it aids in unloading oxygen from the red blood cells at the tissues.

The oxygen-carrying capacity of the blood is enhanced by an increase in the number of red blood cells. This process begins within a few days at altitude and is stimulated by the kidney hormone erythropoietin. This hormone causes the bone marrow to increase red blood cell production: this requires that the body's iron stores are adequate and may indeed mean supplementation of iron intake prior to and during the stay at altitude. There is an apparent increase in haemoglobin in the first few days at altitude which reflects haemoconcentration due to a drop in plasma volume. Nevertheless there is a gradual true rise in haemoglobin which may take 10–12 weeks to be optimized. Even after a year or more at altitude the increases in total body haemoglobin and red cell count do not attain values observed in high altitude natives. As a result sea level natives will never be able to compete in aerobic events (including soccer) at altitude on equal terms with those born at altitude. They have to devise strategies to allow them to demonstrate their superior skills as well as prepare physiologically by acclimatizing.

12.2.2   Exercise at altitude

Soccer players will experience more difficulty in exercising at altitude compared with sea level in spite of the physiological adjustments to hypoxia that take place. Changes in maximum cardiac output and in the oxygen transport system lead to a fall in maximal oxygen uptake ($\dot{V}O_2$ max). At an altitude of 2300 m, corresponding roughly to Mexico City, the initial decline in $\dot{V}O_2$ max is about 15%. After 4 weeks at this altitude, there is an improvement in $\dot{V}O_2$ max but it still remains about 9% below its sea level value. For sea level dwellers the initial decline in $\dot{V}O_2$ max is 1–2% for every 100 m above 1500 m (Åstrand and Rodahl, 1986).

Soccer play is mostly at submaximal intensity, although periodically there are short maximal efforts. Maintaining a fixed submaximal exercise intensity is more stressful at altitude than at sea level. The highest level of endurance exercise that can be sustained is determined by the intensity above which lactate accumulates progressively in the blood. This 'lactate threshold' is lowered at altitude although the percentage of $\dot{V}O_2$ max at which it occurs is unaltered. In order to cope with the lack of oxygen, the active muscles rely more on anaerobic processes and so soccer players will need longer low-intensity recovery periods during match-play, following from their bouts of all-out high-intensity efforts.

Heart rate, ventilation and perceived exertion are all elevated beyond the normal sea level responses at any given submaximal exercise intensity. As a result the pace of tolerable exercise is reduced. Soccer players should be prepared to pace their efforts more selectively during matches at altitude. They will also need to accept a lower intensity during training sessions. This is especially important in the first few days at altitude.

Different individuals will be affected to varying degrees depending on factors such as level of aerobic fitness, prior acclimatization, previous experience of altitude. Physiological factors such as pulmonary diffusing capacity, total body haemoglobin, iron stores, nutritional state and so on may also determine why some individuals will suffer more at altitude than others.

Successful adaptation to altitude results in a decreased tachycardia in response to submaximal exercise compared with the heart rate on initial exposure. The heart rate response may approach sea level values after 3–4 weeks of exposure. Adaptations of skeletal muscles occur to aid their struggle against hypoxia. Improvements in maximum blood flow capacity and oxidative metabolism require a sojourn of many months at altitude. These long-term adaptations will not be of benefit to anaerobic processes. Sprinting ability may in fact be improved at altitude due to the reduced air resistance against which the body moves. Such conditions may be favourable for improving running speed. The buffering capacity of muscle is improved with a prolonged stay at altitude. This adaptation, along with changes in activities of enzymes associated with anaerobic glycolysis, complements the adaptations that occur in oxygen transport mechanisms.

12.2.3   Altitude sickness

The immediate and short-term adjustments which help the body adapt to altitude can have adverse side-effects. The most common problem is referred to as acute mountain sickness. This is characterized by headaches, nausea, vomiting, loss of appetite, sleep disturbances and irritability. These problems can be encountered at altitudes above 2000–2500 m.

The syndrome develops progressively, reaching a peak within about 48 h of initial exposure, and then disappears with adaptation. The problems are related to changes in the pH of the cerebrospinal fluid consequent to respiratory alkalosis and also increases in cerebral blood flow stimulated by hypoxia. Onset of acute mountain sickness may be sudden when ascent is rapid and exercise may increase the likelihood of its development. For this reason, intense training is discouraged for 3–4 days on going to altitude.

Individuals with low body stores of iron may experience difficulties at altitude once red blood cell production is stimulated by erythropoietin. This can be accentuated if appetite is affected by acute mountain sickness. At higher altitudes, neuropeptides released within the intestines cause a depression in appetite. Careful attention to diet is needed when going to altitude and during the immediate period of adaptation.

Attention is also directed towards adequate hydration. The air at altitude tends to be drier at sea level. More fluid is lost by means of evaporation from the moist mucous membrane of the respiratory tract. This loss is accentuated by the hyperventilation response to hypoxia. The nose and throat get dry and irritable and this can cause discomfort. It is important to drink more than normal to counteract the fluid loss. Indeed a rigorous regime of drinking fluids has been shown to offset the fall in plasma volume that is a characteristic response to altitude (Ingjer and Myhre, 1992).

It should be mentioned that the ambient temperature drops by about 1°C in every 300 m ascent. As a result some of the problems linked with cold environments are relevant considerations at altitude.

12.2.4 Soccer strategies

Acclimatization is imperative for a soccer team scheduled to compete at altitude. Major international tournaments have taken place at altitudes where aerobic processes are compromized. These have included two World Cups at Mexico City, the Olympic Games soccer tournament in Mexico in 1968 and the World Student games in 1979 at Mexico. Other countries play their home matches at altitude, including Colombia and Bolivia. In qualifying for the 1994 World Cup finals, Bolivia played its home matches at 2800 m, a factor that bestowed a considerable advantage to its players. Indeed, four of its top league clubs play at home at altitudes exceeding 3000 m and one of its clubs (El Alto) has a home ground at above 4000 m. Other South American clubs are disadvantaged when playing competitively at these levels.

Teams playing at moderate altitude may need to redistribute work-loads among players so that individuals can take longer recovery periods than normal. They may also need to time their offensive moves more effectively and concede possession to the opponents for longer than customary. Teams that rely on all-round work-rate from players, particularly in putting pressure on opposition players in possession of the ball, will need to modify their usual style of play. Occasionally, the direct style of play in quickly transferring the ball from defence to attack with long passes might prove effective.

The lowered air resistance at altitude alters ball-flight characteristics. Consequently, long kicks will travel further and shots at goal will travel faster. It seems important that all players should experience these conditions before actually competing in matches at altitude. This would be especially important for the goalkeeper and the strikers. There is no real method of simulating these conditions at sea level and so players have to accustom themselves to this aspect of skill at altitude.

Multi-venue soccer tournaments may entail qualifying matches or early rounds at different altitudes. This happened to some teams playing in the 1986 World Cup in Mexico, a number of matches being scheduled close to sea level. In such circumstances, it is difficult for the team management to make plans and generally preparations are made for the worst possible eventualities. Some flexibility is available in the choice of altitude for living accommodation and the team may descend to a lower altitude for specific strenuous training sessions. In this way the players can maintain a high standard of training stimulus and achieve a measure of acclimatization to altitude.

Many Olympic athletes use altitude training camps in the belief that the adaptations that occur will benefit subsequent performance at sea level. There are advantages and disadvantages to the practice. This ploy is unlikely to be of much help to soccer players whose competitive season tends to leave little room for such manoeuvres.

12.2.5   *Preparing for altitude*

Players scheduled to compete at altitude must consider the physiological consequences of such an engagement. Detailed preparatory recommendations have been outlined by de Looy *et al.* (1988) and much of the advice is appropriate to soccer teams.

It is not advisable to do strenuous training for at least 2–3 days until the stage of vulnerability to acute mountain sickness has passed. After that, prolonged training sessions should be reduced in intensity to the same perceptual load as at sea level; full work-outs are not advisable until 7–10 days after arrival. Recovery periods between intense short-term effort should be lengthened when intermittent exercise is being performed: this applies both to conditioning work and to games practice.

Rehydration following training at altitude must be complete as more fluids than normal may be lost during exercise. The diet should contain a greater than normal proportion of carbohydrate, especially in the first few days at altitude. This will compensate both for the increased reliance on glycogen as a fuel for exercise and for the fall in the tension of CO_2 in the blood consequent to hyperventilation.

About 14 days should be allowed before competition for acclimatization to altitudes of 1500–2000 m and 21 days before matches at 2000–2500 m. These periods may be shortened if the players have had previous exposures to altitude in their build-up to the tournament. Unacclimatized individuals need about 1 month to adapt to altitudes above 2500 m and may lose match fitness in the process. Fortunately, soccer play at this altitude is uncommon for sea level dwellers.

If it is impractical to stay at altitude for a long period before a competition, some degree of acclimatization may be achieved by frequent exposures to simulated altitude in an environmental chamber. Continuous exercise of 60–90 min, or 45–60 min of intermittent exercise performed four or five times a week at simulated altitude of 2300 m has shown good results in 3–4 weeks (Terrados *et al.*, 1988).

Portable simulators that induce hypoxia are available for wear as a back-pack. These lower the inspired-oxygen tension and accentuate exercise stress but also increase the resistance to breathing. There is no convincing evidence that they promote the kind of adaptations that are experienced at altitude or that result from prolonged exercise in a hypobaric chamber. Nevertheless, they may have psychological benefits for players in accustoming them to hypoxia. Portable simulators were used by the Danish soccer team, along with exercise tests in a hypobaric chamber (Bangsbo *et al.*, 1988) in preparation for the 1986 World Cup in Mexico. In more recent years, normobaric hypoxic chambers have become available to professional soccer clubs for training and rehabilitation purposes. It is not clear how effectively these are being employed.

12.3   Circadian variation

12.3.1   Circadian rhythm

Circadian rhythms refer to cyclical changes within the body that recur around the solar day. An example is the rhythm in core temperature (Figure 12.4) which shows a cycle every 24 h. A cosine wave can be fitted to the observations in rectal temperature and the time of peak occurrence identified. This is referred to as the acrophase and is usually found between 17:00 and 18:00 h. Many measures of human performance follow closely this curve in body temperature (Reilly *et al.*, 1993). These would include components of motor performance (such as muscular strength, reaction time, jumping performance) that are important in soccer play.

The other biological rhythm of major importance is the sleep–wake cycle. This is linked with the pattern of habitual activity, that is, sleeping during the hours of darkness and working or staying awake during daylight. Thus, there is a sharp contrast in arousal states between night time and day time. Arousal tends to peak just after mid-day at the time that circulating levels of adrenaline are at their highest. A team forced to compete at a time of day it would normally be inactive would not be well equipped, biologically or psychologically, to do so.

Circadian rhythms are described as endogenous or exogenous depending on the degree to which they are governed by environmental signals. These include natural and artificial light, temperature, type and timing of meals, social and physical activity. Endogenous rhythms imply internal body clocks, the suprachiasmatic nucleus cells of the hypothalamus being thought to be the site of control of circadian rhythms. Timekeeping functions have also been attributed to the pineal gland, its hormone melatonin and related substrates such as serotonin. Local timekeepers have also been isolated in both cardiac and skeletal muscle. It is likely that there is a family of clocks within the body which control a host of circadian

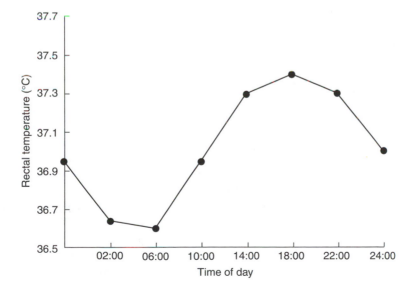

Figure 12.4 Circadian rhythm in rectal temperature.

rhythms and which are organized in a hierarchy. The most relevant of these for sports performance seem to be the body temperature curve and the sleep–wake cycle.

12.3.2   Sleep

Sleep is an enigma in the sense that it has never been conclusively explained why it is needed. One school of thought relates sleep to the restitution of the body's tissues. An alternative view is that the need for sleep is specific to nerve cells – the so-called brain restitution theory of sleep. Nevertheless, it is obvious that sleep is essential and this need is most apparent when sleep is deprived or disturbed.

Whilst the average sleep of a 20–30-year-old is about 7 h each night, there is a large variation between individuals both in the need for sleep and in the amount of sleep taken. Some athletes feel uneasy unless they sleep soundly for 8–9 h and place a priority on their sleeping arrangements. Brain states may be monitored during sleep by electroencephalography (EEG). Traces from EEGs demonstrate cycles of about 90 min, each cycle containing stages known as rapid eye movement (REM) and non-REM sleep. Non-REM sleep is further classified into stages 1–4. It is easy to awaken individuals from REM sleep but more difficult during non-REM sleep. Consequently, players who doze in the morning may slip into a further 90-min cycle of sleep which in all probability does them little good.

Professional soccer players tend to get adequate sleep when a complete week is considered. However, it is often inconsistent, players staying up late especially after an evening match. In such events it is difficult to get to sleep since catecholamine levels are elevated above normal and players still rethink details from the match completed earlier. Relaxation following a game may require a conscious behavioural strategy which differs from individual to individual.

Pre-match anxiety can disrupt sleep the night before playing. Complaints from players that they were unable to sleep are generally untrue as short periods asleep, albeit snatched unwittingly during the night, do provide a restorative function. In such cases a short nap during the day could be encouraged. A brief afternoon nap prior to an evening kick-off can promote a release from pre-match anxiety. There is a natural tendency to drowse in the mid-afternoon which is generally referred to as a 'post-lunch dip'.

Muscular performance may be unaffected by sleep loss, at least as shown in experiments of partial sleep deprivation where subjects are permitted only 2.5–3 h sleep a night (Reilly and Deykin, 1983). Complex tasks and decision-making, especially if demanded over a prolonged period such as 90 min, deteriorate with the duration of the task. Thus, concentration during a soccer match following nights of disrupted sleep requires a distinct motivational drive from the player concerned.

It has been possible to play soccer for days without sleep, although this was done indoors and at a low intensity. For over 91 h the level of play showed a cyclic change that corresponded to the circadian rhythm in core temperature of the players (Reilly and Walsh, 1981). The appearance of psychotic-like symptoms, particularly following the second night without sleep, makes meaningful play difficult under such circumstances. Clearly such a regime is not conducive towards serious soccer performance.

Effects of partial sleep loss accumulate and become increasingly evident with consecutive nights of progressive sleep disruption. Effects on training performance are more pronounced as the training session progresses in duration as it becomes difficult for the player to maintain motivation (Reilly and Piercy, 1994). A decline in the ability to concentrate is likely to promote errors in long training sessions with the consequence also of increased injury risk.

12.3.3   *Training and time of day*

The majority of motor performance measures demonstrate a peak in performance that occurs close to the acrophase of the circadian rhythm in body temperature. On this basis the ideal time for soccer play would be about 17:00–18:00 h, assuming the environmental temperature is comfortable. There is probably a window of some hours during the day when maximal performance can be achieved. This can be realized with appropriate warm-up and physical and mental preparation. Consequently, kick-offs at 15:00 and 19:30 hours do not necessarily entail sub-optimal performance, particularly as muscle and core temperatures rise during the course of match-play. Particular consideration to warm-up is needed in late kick-offs, say 20:00 h, in cold conditions.

There is often a mismatch between the time of training and the time at which matches are played. The majority of professional soccer teams train in the morning, starting at 10:00 or 11:00 h. Strenuous physical conditioning exercise is best conducted in the early evening, the time at which many amateur teams train. Joint stiffness is greatest in the morning and so special attention should be given to flexibility exercises in warming-up prior to morning training sessions. When players have to compete at a time of day to which they are unsuited, simply training at that time in the few days beforehand seems to be helpful.

Skills may be best acquired in mid-day sessions just as the curve in arousal approaches and reaches its high point. Consequently, there is a case for young professionals to have their skills work at light intensity in morning sessions. The more intense exercise can be retained for a later session following lunch and a rest for recovery.

12.3.4   *Jet lag*

Soccer players are sometimes called upon to travel vast distances to play in international team or club contests. They may also participate in closed-season tournaments or friendly matches overseas. Such engagements are made possible by the speed of contemporary air flight. Although international travel is commonplace nowadays it is not without attendant problems for the travelling footballer, which the team management and back-up staff should recognize in advance.

In the course of travelling abroad players encounter disruption of their regular routine. They may be particularly excited about the trip or may have had worries associated with planning for the departure. Vaccinations may be required, according to the country to be visited. The majority of top teams have arrangements overseen by their administrative and medical staff, as far as possible, in order to avoid otherwise inevitable embarrassments.

There are still routines to be faced – travelling to the airport of departure, checking in and going through security controls, possibly taking advantage of shopping facilities or coping with the frustration of delayed flights. Experienced management staff try to shield the players from such irritations. There is also a possibility of protracted disembarkation procedures on arrival and mix-ups with ground travel and accommodation. Having arrived safely at the destination, the player may be suffering travel fatigue, loss of sleep perhaps (depending on flight times), and a cluster of symptoms which have come to be known as jet lag.

Jet lag refers to the feelings of disorientation, light-headedness, impatience, lack of energy and general discomfort that follow travelling across time-zones. These feelings are not experienced with travelling directly northwards or southwards within the same time zone when the passenger simply becomes tired from the journey or stiff as a result of a long stay in a cramped posture. The feelings associated with jet lag may linger for several days after arrival and may be accompanied by loss of appetite, difficulty in sleeping, constipation and grogginess. In extreme cases the individual may even burst into tears when facing unanticipated difficulties. On the other hand, some people claim they never experience any problems and deny that the phenomenon of jet lag exists. Although there are undoubtedly individual differences in the severity of symptoms, many people may simply fail to recognize how they themselves are affected, especially in tasks requiring concentration and complex coordination.

Following a journey across multiple time zones the body's circadian rhythm at first retain the characteristics of their point of departure. However, the new environment forces new influences on these cycles, mainly the time of sunrise and onset of darkness. The body attempts to adjust to this new context but core temperature is relatively sluggish in doing so. As a rough guide, it takes about one day for each time zone crossed for body temperature to adapt completely. The individual may have difficulty in sleeping for a few days but activity and social contact during the day help in the adaptations of the arousal rhythm. Thus, arousal adjusts more quickly than does body temperature to the new time zone. Until the whole spectrum of biological rhythms adjusts to the new local time, thereby becoming re-synchronized, the performance of the soccer player may be below par.

Allowing for individual differences, the severity of jet lag is affected by various factors. In general, the greater the number of time zones travelled, the more difficult it is to cope. A 2-h phase shift may have marginal significance but a 3-h shift (e.g. British or Irish teams travelling to play European football matches in Russia or America or teams within the USA travelling coast to coast) will entail desynchronization to a substantial degree. In such cases the light times – time of departure and time of arrival – may determine how severe are the symptoms of jet lag that occur. It may also be wise to alter training times to take the direction of travel into account. Such a ploy was shown to be successful in American football teams travelling across time zones within the USA and scheduled to play at different times of day (Jehue *et al.*, 1993).

The severity of symptoms may be worse 2 or 3 days after arrival than on the day immediately following disembarkation. Symptoms then gradually abate, but may still be acute at particular times of day. There will be a window of time during the day when time of high arousal associated with the time zone departed from and the new local time overlap. This window may be predicted in advance and should be utilized for timing of training practices in the first few days at the destination.

The direction of travel also affects the severity of jet lag. It is easier to cope with flying in a westward direction compared to flying eastward. In flying westward, the cycle is lengthened and body rhythms can extend in line with their natural free-wheeling period of about 27 h and thus catch up. Our observations on travelling to Korea (9 h in advance of British Summer Time) and Malaysia (7 h in advance of British Summer Time) were that periods of 9 and 6 days, respectively, were inadequate for jet lag symptoms to disappear. In contrast, re-adaptation was more rapid on returning to Britain. However, when time zone shifts approach near maximal values – the minimum is a 12-h change – there may be little difference between eastward and westward travel and the body clock is likely to adjust as if the latter had occurred.

In the past, English League teams travelling to Japan to compete in the intercontinental cup have failed to allow time for jet lag symptoms decay. Not only did these representatives (Liverpool, Aston Villa, Nottingham Forrest) lose but their performances in the week following the return home were clearly compromised (Figure 12.5). Understanding the physiological processes involved in jet lag and the consequences of their disruption should encourage the use of practical measures to help the traveller to cope better.

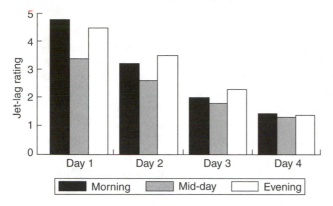

Figure 12.5 Mean jet lag ratings (scale 0–10) of soccer players after returning to England from Japan. Values had returned to zero on day 5. (Reproduced with permission from Reilly, 1993.)

British soccer teams travelling to Australia have used sleeping pills to induce sleep whilst on board. Although such drugs as benzodiazepines are effective in getting people to sleep, they do not guarantee a prolonged period asleep. They have been shown to be ineffective in a group of British Olympic athletes (Reilly *et al.*, 2001). Besides, they have not been satisfactorily tested for subsequent residual effects on motor performances such as soccer skills. They may also be counter-productive if administered at the incorrect time; for example, melatonin (which can act directly on the body-clock as well as being a hypnotic) had no benefit for travellers between the United Kingdom and Australia, a context which can elicit the most severe jet-lag symptoms (Edwards *et al.*, 2000). A prolonged nap at the time the individual feels drowsy (presumably at the time he/she would have been asleep in the time zone just departed from) simply anchors the rhythms at their former phases and so resists the adaptations to the new time zone.

On reaching the destination a key factor is to fit in immediately with the phase characteristics of the new environment. Players will already have worked out the local time for their disembarkation. There may be other environmental factors to consider such as heat, humidity or even altitude.

Having travelled westward players may be allowed to retire to bed early. Early onset of sleep will be less likely after an eastward flight. In this case, a light training session on that evening would be helpful in instilling local cues into the rhythms. There is some evidence that exercise does speed up the adaptation to a new time zone. Observations on professional players returning to England from the Far East showed that a light training session on the afternoon of arriving home was successful in alleviating jet-lag symptoms (Reilly, 1993). In contrast, training in the morning is not recommended after a long-haul flight to the East since it could act to delay the body clock rather than promote the phase adjustment required in this circumstance.

For the first few days in the new time zone, training sessions should not be all-out. Skills requiring fine coordination are likely to be impaired and this might lead to accidents or injuries if sessions with the ball are conducted too strenuously. Where a series of tournament engagements is scheduled, it is useful to have at least one friendly match during the initial period, that is, before the end of the first week in the overseas country.

In this period of adaptation a few caveats are noted. Alcohol taken late in the evening is likely to disrupt sleep and so is not advised. An alternation of feasting and fasting as recommended for commercial travellers in the USA is unlikely to gain acceptance among soccer players. Nevertheless, they could benefit from biasing the macronutrients in their evening meal largely towards carbohydrates. These would include vegetables with a choice of chipped, roast or baked potatoes, pasta dishes, rice and bread. These should include sufficient fibre to safeguard against constipation.

In the early days in the new country the players should be discouraged from taking prolonged naps. A nap at the time they would have been asleep had they stayed at home would make subsequent sleep more difficult and retard the adjustment of the major biological clocks to the new regimes.

By preparing for time zone transitions and the disturbances they impose on the body's rhythms, the severity of jet lag symptoms may be reduced. There has been little success in attempting to predict good and poor adaptors to long-haul flights. The fact that an individual escapes lightly on one occasion is no guarantee that the same individual will do so again on the next visit. Regular travellers do benefit from their experiences and develop personal strategies for coping with jet lag (Waterhouse *et al.*, 2002). The disturbances in mental performance and cognitive functions have consequences not only for players but also for management and medical staff travelling with the team, who by no means have immunity against jet-lag symptoms. Besides, the long periods inactive during the plane journey may lead to the pooling of blood in the legs and in susceptible individuals cause a deep-vein thrombosis. Consequently, regularly moving around the plane during the journey is advised.

12.4   Environmental monitoring

The environment in which the soccer player trains and competes has implications not just for performance but also for health and safety. The quality of the playing surface, for example, forces a choice of appropriate footwear so that performance can be executed without increased risk. Playing conditions sometimes exceed the bounds of safety and the match referee is entitled to declare the pitch unplayable.

A similar decision may be made in cases of air pollution (ozone, CO, SO_2 or Pb). Ozone concentration may exceed acceptable limits in some of the world's major cities (Athens, Mexico City, Seoul, Los Angeles) but generally is not a major problem for European players. The air is impurified in foggy conditions, but in such cases matches and training are usually curtailed for reasons of visibility.

Preparations for hot conditions entail choice of appropriate clothing. Light, loose clothing helps in creating convective air currents to cool the skin. Clothing of natural fibre such as cotton (or at least a cotton–polyester mix) is desirable under warm and radiant environmental conditions. In contrast, clothing with good insulation and preferably in layers helps maintain a warm microclimate next to the skin in cold conditions.

Calculating the risk of heat injury requires accurate assessment of environmental variables. The main factors to be considered are the dry bulb temperature, relative humidity, radiant temperature, air velocity and cloud cover. The most widely used index in sporting contexts is the wet bulb temperature, which takes both ambient temperature and humidity into account. The wind-chill index is employed in determining risk in cold conditions. Quite apart from the chilling effect of the wind, blustery conditions make ball flight more difficult to anticipate and skills become more erratic as a consequence.

The novel environmental challenge – hypoxia, temperature, travel, weather – calls for preparation on behalf of team management. Conditions may even change dramatically during the course of play. An awareness of the dynamic biological adjustments that the body makes means that adverse effects and discomfort associated with environmental variables can be countered to a large degree.

Summary

Soccer is played in a variety of challenging environments. Stresses may include heat, cold, altitude or disruption of the circadian body clock. Some account must be taken of the environment in which competition is scheduled. This may involve physiological preparation and changes in tactics may also be needed to enable players to cope.

References

Astrand, P.O. and Rodahl, K. (1986). *Textbook of Work Physiology*, McGraw-Hill, New York.

Bangsbo, J., Klausen, K., Bro-Rasmusen, T. and Larson, J. (1988). Physiological responses to acute moderate hypoxia in elite soccer players, in *Science and Football* (eds T. Reilly, A. Lees, K. Davids and W.J. Murphy), E & FN Spon, London, pp. 257–64.

Bergh, U. and Ekblom, B. (1979). Effect of muscle temperature on maximal muscle strength and power in human skeletal muscles. *Acta Physiologica Scandinavica*, **107**, 33–7.

Edwards, B.J., Atkinson, G., Waterhouse, J., Reilly, T., Godfrey, R. and Budgett, R. (2000). Use of melatonin in recovery from jet-lag following an eastward flight across 10 time-zones. *Ergonomics*, **43**, 1501–13.

Ekblom, B. (1986). Applied physiology of soccer. *Sports Medicine*, **3**, 50–60.

Ingjer, F. and Myhre, K. (1992). Physiological effects of altitude training on elite male cross-country skiers. *Journal of Sports Sciences*, **10**, 37–47.

Jehue, R., Street, D. and Huizengar, R. (1993). Effect of time zone and game time on team performance: National Football League. *Medicine and Science in Sports and Exercise*, **25**, 127–31.

de Looy, A., Minors, D., Waterhouse, J. *et al.* (1988). *The Coach's Guide to Competing Abroad*, National Coaching Foundation, Leeds.

Mustafa, K.Y. and Mahmoud, E.D.A. (1979). Evaporative water loss in African soccer players. *Journal of Sports Medicine and Physical Fitness*, **19**, 181–3.

Nielsen, B. (1994). Heat stress and acclimation. *Ergonomics*, **37**, 49–58.

Pollock, N.W., Godfrey, R.J. and Reilly, T. (1997). Evaluation of field measures of urine concentration. *Medicine and Science in Sports and Exercise*, **29**(Suppl 5), S261.

Reilly, T. (1993). Science and football: an introduction, in *Science and Football II* (eds T. Reilly, J. Clarys and A. Stibbe), E & FN Spon, London, pp. 3–11.

Reilly, T. (2000). Temperature and performance: heat, in *ABC of Sports Medicine* (eds M. Harries, G. McLatchie, C. Williams and J. King), BMJ Books, London, pp. 68–71.

Reilly, T. and Deykin, T. (1983). Effects of partial sleep loss on subjective states, psychomotor and physical, performance tests. *Journal of Human Movement Studies*, **9**, 157–70.

Reilly, T. and Lewis, W. (1985). Effects of carbohydrate feeding on mental functions during sustained physical work, in *Ergonomics International 85* (eds I.D. Brown, R. Goldsmith, K. Coombes and M.A. Sinclair), Taylor and Francis, London, pp. 700–2.

Reilly, T. and Piercy, M. (1994). The effect of partial sleep deprivation on weight-lifting performance. *Ergonomics*, **37**, 107–15.

Reilly, T. and Walsh, T. (1981). Physiological, psychological and performance measures during an endurance record for a 5-a-side soccer play. *British Journal of Sports Medicine*, **15**, 122–8.

Reilly, T., Atkinson, G. and Budgett, R. (2001). Effect of low-dose temazepam on physiological variables and performance tests following a westerly flight across five time zones. *International Journal of Sports Medicine*, **22**, 166–74.

184   *Thomas Reilly*

Reilly, T., Atkinson, G. and Coldwells, A. (1993). The relevance to exercise performance of the circadian rhythms in human temperature and arousal. *Biology of Sport*, **10**, 203–16.

Reilly, T., Maughan, R.J., Budgett, R. and Davies, B. (1997). The acclimatisation of international athletes, in *Contemporary Ergonomics 1997* (ed. S.A. Robertson), Taylor and Francis, London, pp. 136–40.

Terrados, N., Melichna, J., Sylven, C. *et al.* (1988). Effects of training at simulated altitude on performance and muscle metabolic capacity in competitive road cyclists. *European Journal of Applied Physiology*, **57**, 203–9.

Waterhouse, J., Edwards, B., Nevill, A., Carvalho, S., Atkinson, G., Buckley, P., Reilly, T., Godfrey, R. and Ramsay, R. (2002). Identifying some determinants of 'jet-lag' and its symptoms: a study of athletes and other travellers. *British Journal of Sports Medicine*, **36**, 54–60.

Part 3

Behavioural science and soccer

13 Coaching science and soccer

Andy Borrie and Zoe Knowles

Introduction

Everyone has at least a vague understanding of the term coaching and a personal concept of the kind of tasks that this activity entails. People also have personal perceptions of the kind of behaviour that they would expect an 'effective' coach to exhibit. It is arguable that within soccer the role of the coach or manager is valued to a greater extent than in perhaps any other sport. Certainly, coaches/managers are publically held accountable for poor performances or are highly praised, and paid, when trophies are won. Despite the eminent position of the coach within soccer, the complex process of coaching a team is still a phenomenon that is not clearly understood even by those who are practising coaches (Gould *et al.*, 1990). This lack of understanding is perhaps not surprising given that coaches can be involved in the establishment of basic skills in beginners, the provision of sound technical and tactical advice to intermediate performers or the planning and implementation of long-term training programmes with elite performers.

In recent years, the emphasis that many developed countries (e.g. USA, Canada, UK, Australia) have placed on improving the standards of sports performance within domestic and international competition has moved those countries to develop extensive coach education programmes. The profile of coaching has therefore risen sharply since the mid-1980s (Campbell, 1993) alongside which the volume of coach driven general literature regarding coaching practice has also risen. In much of this literature, coaching is characterized as being a multi-task occupation with the coach having to exercise a range of personal communication and management skills within a wide range of coaching roles (see Table 13.1).

Table 13.1 Coaching roles listed according to their source in the general coaching literature

Source	NCF (1986)	Pyke (1992)	Sabock (1985)
Coaching role	Instructor	Teacher	Teacher
	Teacher	Motivator	Guidance counsellor
	Motivator	Administrator	Diplomat
	Disciplinarian	Social worker	Organizer
	Manager	Scientist	Leader
	Administrator	Student	Dictator
	Social worker	Friend	Politician
	Scientist		Trainer
	Student		
	Friend		

The terminology used in the general coaching literature is often vague and the terms presented in Table 13.1 lack precise definition but the essential point about the potential range of roles is clear.

Many also believe that the role of the modern coach has now expanded far beyond the directing of practice sessions (Pyke, 1992; Launder, 1993; Woodman, 1993). It has been recognized that as standards of sports performance have grown so too has the need for more extensive and sophisticated support of the performer. The expanded elite coaching role involves taking responsibility for the performer outside as well as inside the practice/ competition environment and being aware of the performer's overall social and psychological development. Coaches are now frequently expected to take on almost any task that creates a better working environment for the performer.

The view of coaching as a multi-role activity clearly has a basis in reality but, and this should be a critical issue for scientists, the general coaching literature is currently based on anecdote not empirical research. The coach education systems that have arisen in the last decade are founded on what coaches think other coaches *ought* to know, not on sound research-based evidence.

Given the complexity of modern coaching, particularly in a major sport such as soccer, *only* a systematic and objective assessment of coaching will allow for the development of a clear understanding of the process. Scientists need to focus attention on systematically trying to identify the key components of coaching effectiveness so that coaches can prioritise their learning and maximize the time they have for developing their practice. Whilst it is not being claimed that science can ever provide a proven set of steps to universal coaching success, science should be able to illuminate critical facets of the process so that coaches and coach educators are better placed to develop the skills required.

This chapter will assess the coaching practice/process literature and draw links to soccer coaching with the intention of identifying key areas where science has clear messages for coaches and coach educators. The chapter will be split in to three sub-sections to provide a clear structure and help the reader to understand a diverse field of study. The sub-sections will be:

- simple process models and observational research;
- leadership concepts;
- knowledge and experience.

13.1   Simple models and observational research

Soccer coaching may range from simply correcting a child's attempt to kick a ball, through to developing cohesive attacking play with an international forward line. Regardless of the differences in these activities, some authors have suggested that there is a common process underpinning the work of all coaches (Fairs, 1987; Woodman, 1993). In such models, coaching has been represented as a series of unproblematic, sequential stages that a coach will pass through in the process of helping a player or team improve their performance (Franks *et al.*, 1986; Fairs, 1987). Figure 13.1 is derived from such models showing the stages usually presented as being the essential parts of coaching practice.

At one level, simplistic models have value in describing key facets of coaching. For example, such models clearly emphasize the cyclical aspects of coaching as well as presenting the, potentially, systematic nature of the process. These models also place great emphasis on the importance of data collection and interpretation as foundations of the coaching process

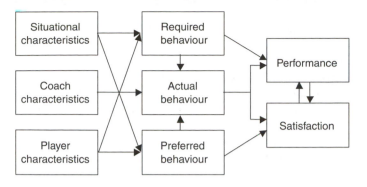

Figure 13.1 Simplistic model of the coaching process derived from the coaching literature.

with some authors, such as Fairs (1987) stating that collection of data is the most important stage within the process. The coach's observational skills are therefore presented as being an essential component of an effective coaching process and research on expert–novice visual search strategies would support this assertion. It has been shown that expert observers employ different visual search strategies to novices with the experts' observational skills allowing them to extract a greater amount of relevant information from their visual environment (Williams *et al.*, 1999). These findings support the concept that data collection is a key stage in determining the effectiveness of the coaching process and points towards the importance of developing observational skills in coach education programmes.

The model in Figure 13.1 also emphasizes the importance of teaching and instruction in the coaching process. This emphasis helps direct attention towards the large literature base covering coach behaviour in practice sessions. Coach instructional behaviour has been one of the most researched aspects of coaching practice over the last three decades gathering momentum in recent years (e.g. Soloman *et al.*, 1996; Bloom *et al.*, 1999; Spencer, 2001). This research has identified key points regarding effective behaviour when running a technical/tactical practice session. Douge and Hastic (1993) identified five principles of effective coaching from this literature base; they suggested that effective coaching requires:

- ordered management of the training environment;
- heavy emphasis on instruction behaviours;
- high frequency of correction and instruction comments;
- frequent use of questions and asking players for clarification;
- frequent provision of feedback and hustles.

Other strands of this research have emphasized the influence of the coach's behaviour within the practice session on players' evaluations of the coach. High levels of technical instruction, feedback, reinforcement and supportive behaviour have been associated with higher player perceptions of competence, enjoyment, self-esteem and self-confidence (Smith *et al.*, 1978, 1983; Smoll and Smith, 1993; Spencer, 2001). It is beyond the scope of this chapter to review this field in any depth but, when taken as a whole, there is much to tell us about what is required from the coach in terms of effective behaviour within practice sessions [for a more comprehensive review, see Abraham and Collins (1998) and Spencer (2001)].

Despite the value of observational studies of coaching, the coach's behaviour during practice is not the only component of effective coaching; many critical coaching behaviours occur away from the practice ground or the side of the pitch during a match. Consequently, observational research methods cannot fully assess all aspects of player–coach interaction or coach behaviour and principles of effective 'coaching' that are drawn from observational research may be more accurately described as principles of effective sports instruction not coaching. Similarly, models of coaching that reduce coaching intervention to the planning and delivery of practice sessions underplay large elements of the coach's expanded role.

Models such as the one contained in Figure 13.1 have been criticized as being episodic in nature when the reality of coaching is different (Lyle, 1999a). Coaching is a longitudinal activity with coaches operating within long-term plans that do as much to guide day-to-day practice as do the observations made from individual performances. Figure 13.1 does not allow for, or account for, any interaction between long-term and short-term goals or plans within the process. Neither does Figure 13.1 adequately account for the social dynamics of coaching such as the quality of coach–player relationships or the interplay between the coach and other support staff. All these elements are an important part of the overall effectiveness of the coaching process, particularly at an elite level, yet they are absent from the simple, cyclical models. An overemphasis on an episodic approach may hinder the development of coaching effectiveness (Cross and Ellice, 1997).

To summarize, simple models of the coaching process direct attention at some important facets that help coaches and coach educators to identify what is required for effective instructional behaviour. Such models do not account for the full complexity of coaching in that it is not episodic and the social dynamic of the coaching environment has an impact on practice.

13.2   Leadership research

An alternative to simple process models can be found in research where coaching has been conceptualised as a specific form of leadership. The most widely cited, and researched model, from this domain is the multi-dimensional model of sports leadership, proposed by Chelladurai and Carron (1978). Chelladurai's model was an attempt to apply concepts from general leadership research (e.g. business leadership), to sport and provides useful exemplar of the behavioural approach to leadership research. In the model presented in Figure 13.2, the main focus of attention is behaviour with leadership behaviours being directed towards the dual aims of achieving high performance whilst maintaining athlete satisfaction. These 'goals' are the product of three components of coach/leader behaviour: actual coaching behaviour, preferred behaviour and prescribed/required behaviour. These constructs are in turn influenced by three underpinning 'antecedents' of behaviour; the coach's characteristics, the players' characteristics and the situation.

The construct of required behaviour covers the behaviours that an organization will expect from a coach. For example, the Football Association will expect its coaches to maintain discipline in the England squad. Required behaviour is also influenced by the expectations of players. For instance, youth team players may expect their coach to spend time developing individual techniques rather than team tactics. The expectancies of players and organizations will have in turn been created by their respective underlying characteristics such as the age of the players or the culture within a club. Preferred behaviour represents

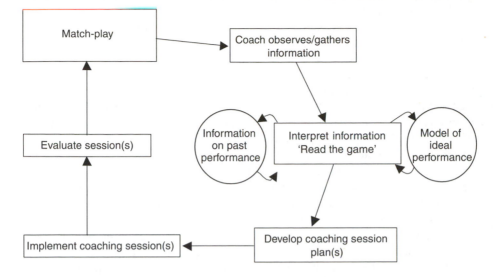

Figure 13.2 The multi-dimensional model of sports leadership from P. Chelladurai and A.V. Carron, *Leadership*. (© 1978 by the Canadian Association for Health, Physical Education and Recreation.)

what the players want from the coach, for example, players will have preferences for how much they want to be involved in decision-making about team strategy. The characteristics of the coaching environment will also have an influence on players' preferences for coaching behaviour; for instance, a recreational soccer club will probably not want its coach to run more than one practice session each week. Finally, the coach's actual behaviour will be a function of individual reading of expectation, preference, the coaching demands (i.e. technical/ tactical needs of the players) as well as underpinning characteristics such as personality, experience and so on.

Chelladurai hypothesized that the congruence between the three behaviour constructs determines performance and satisfaction. If the coach's behaviour pattern matches what is preferred and what is expected, then both performance and satisfaction should be maximized. If the coach responds to what is required but this does not match player expectation then performance might be optimal but player satisfaction will be low. Alternatively, if the coach matches player preferences but fails to meet required behaviour then satisfaction may be high but performance will suffer.

The multi-dimensional model has much to commend it and has justifiably been the subject of considerable research attention since its inception. The model, and its related research base, draws attention to the following key aspects of coaching:

- the importance of underpinning contextual issues in determining what effective coach behaviour will be in any given situation;
- coaching is an act where the coach aims to achieve some form of dynamic equilibrium, balancing out different, and potentially conflicting, inputs and demands from players and the situation;
- players' preferences for coach behaviour will have significant impact on their satisfaction;

- player characteristics, for example, player age and experience, will have an impact on preferences for coach behaviour (Chelladurai and Carron, 1983);
- situational factors such as national/organizational culture will influence player preferences for coach behaviour (Chelladurai *et al.*, 1988).

Whilst the model has been a useful addition to the coaching literature, its capacity to explain the process fully is limited. From a research perspective, almost all related work has focused on the lower portion of the model (i.e. the interaction of antecedent characteristics, player preferences, coach behaviour and satisfaction). The upper portion of the model, required/expected behaviour and performance, has not received research attention because of the complex difficulties one faces on trying to define what is required or expected in any given situation. Any attempt to investigate relationships amongst the constructs of the upper portion of the model will also be hindered by the fundamental problem of measuring coaching performance. Is a simple analysis of win/loss record an adequate measure of a coach's effectiveness? A coach can perform superbly but the team will still lose so how does one measure performance? These problems severely limit the usefulness of this model from both a research and practical perspective.

In addition, from the more practical perspective, the model also fails to consider or account for some key facets of coaching as recognized within the general coaching literature. First, the model presents a simplistic view of player satisfaction in coaching with no differentiation between long-term and short-term satisfaction. Practising coaches know that at times they may act/behave in a specific way because it is the 'correct' course of action even though it will be unpopular. For example, a coach may instigate a tough training schedule that causes dissatisfaction in a team but, once the schedule brings competitive success, satisfaction is enhanced. The model proposes a single and uncomplicated satisfaction construct but this does not match reality.

Second, and more importantly, the general coaching literature indicates quite clearly that coaching is a multi-role activity requiring a variety of roles across situations and over time. The model does not really allow for, or acknowledge, variation in roles with behavioural preferences being represented as relating to coaching in general and not to a specific role. However, practising coaches will recognize that player preference for specific forms of behaviour will be related to role. In a cup final in the tight time constraints between the end of normal time and the start of extra time, the players are unlikely to want the coach to engage in a very democratic, open team discussion of strategy. In contrast, when the coach is trying to organize travel arrangements for an away fixture, the players will probably want, and need, more open discussion of alternatives.

Overall leadership models, such as the one shown here, do highlight important factors that coaches and coach educators should consider in relation to coach effectiveness. However, leadership models that focus on behaviour, such as the multi-dimensional model, lack the complexity to account fully for coaching situations and practices. Given the complexity of the coaching environment, with its complex interactions between roles, behavioural preferences, individual characteristics and organizational expectancies, is it really feasible to think that a single behavioural model can account for all relevant factors? Would such a model ever be able to prescribe what behaviours are optimal for a specific coach in a specific coaching situation? This seems unlikely, therefore the question remains as to what is the most appropriate research paradigm that can yield models of coaching that are useful for both researcher and practitioner?

13.3   Knowledge-oriented research

In contrast to leadership models that focus on behaviour, more recent approaches to the study of coaching have focused on the issues of knowledge and decision-making. Since the early 1990s, a growing body of research has supported the view that expertise in coaching is based on the mental skills and knowledge that the coach has available rather than their behaviour in any given situation. Attention has therefore shifted from what coaches *do* to how coaches *think*.

The most widely used and cited model in this area is the Coaching Model (Côté *et al.*, 1995). This model was derived from a study of expert coaches and it proposes that coaching is driven by a coach's cognitive representation of what is required to develop a player or team. This cognitive representation, or schema, was termed a 'mental model of athletic potential' and, in the Coaching Model, the mental model determines the action that the coach will take at any given stage of the coaching process. The mental model also links together the main parts of the coaching process (training, competition and organization) and contextual factors (e.g. a player's age and ability level or the organizational culture of the club) in a single cognitive schema for potential coaching action. In essence the Coaching Model suggests that the way in which a coach mentally models the coaching process and the coaching environment is the foundation on which expertise is based.

The central tenet of the Coaching Model, that cognitive skills are at the heart of coaching expertise, is one that has considerable support in the literature. For example, studies of expert coaches and expert–novice differences in coaching have found that coaching expertise is linked to critical thinking and decision making skills (Strean *et al.*, 1997; Abraham and Collins, 1998). It has also been shown that expert coaches possess meta-cognitive forms of knowledge (i.e. they are able to draw together different knowledge sources to gain a better understanding of a situation and their position within it) (Salmela, 1995). It seems reasonable to assume that more integrated and comprehensive knowledge bases will contribute to more effective decisions regarding appropriate coaching action in any given situation. It has also been shown that the knowledge structures of expert coaches are vitally important in helping them to interpret the training environment correctly, anticipate the likely outcome of various coaching actions and make subsequent decisions regarding appropriate coaching interventions (Jones *et al.*, 1997; Saury and Durand, 1998; Lyle, 1999b).

Within the literature, there is growing agreement that the basis of coaching expertise is to be found in the coaches' use of knowledge to inform their decision making and subsequent action. This literature base suggests that coaching expertise is founded on the coaches' abilities to:

- perceive the most relevant information in a coaching environment;
- construct knowledge bases that link situational cues with coaching action and outcome;
- draw on a broad range of knowledge sources to help interpret the situation;
- form appropriate mental models of the situation to guide coaching action;
- predict the outcome of alternative courses of action with greater certainty.

One has to recognize that within this literature base issues of knowledge generation, knowledge construction, knowledge representation and decision-making are mixed together with little clarity or coherence. The coaching literature has not a common or agreed vocabulary for discussing issues relating to cognitive skills or knowledge and therefore it is difficult to draw more definite conclusions about the relative importance of these factors in determining coaching expertise. Nevertheless, if one accepts that coaching expertise is a matter of cognitive skill, a subsequent question has to be 'how does the expert coach develop such skills'? Fortunately for the coach educator, the literature does provide a relatively clear answer to this question.

The literature, almost universally, supports the idea that expertise in coaching is created through experience not through formal education courses. Douge (1987) described coaching as a 'learned trade' and this description has been supported in a number of studies. The work of Gould *et al.* (1990) with elite American coaches from a variety of sports, demonstrated that the coaches' primary means of knowledge development was through experience and interaction with other coaches. This finding has been reinforced by Salmela (1995) who, in a smaller but more in-depth study of elite coaches, found that the coaches had followed diverse and ad hoc learning pathways to reaching expert status with experience being the primary learning medium. It has also been shown that the personal knowledge bases created through experience play a central role in decision-making. Saury and Durand (1998) found that expert coaches used their own personal experience as performers as well as their experiences of past coaching situations to interpret what performers were experiencing at a given moment and what effect alternative actions would have on training.

The notion that experientially derived knowledge is critical to expert performance also finds support in the scientific study of other professions. Expertise in professions ranging from medicine to the armed services is grounded in the expert's capacity to use tacit knowledge that is derived from experience (Sternberg and Horvath, 1999). Tacit knowledge has been defined as 'procedural knowledge that guides behaviour but that is not readily available for introspection' (Sternberg and Horvath, 1999, p. 231). The emphasis in Sternberg and Horvath's work on tacit knowledge as being the basis of expertise has been echoed in the coaching literature. Several studies within coaching allude to the fact that expert coaches struggle to articulate the knowledge that they are using and that their knowledge structures are highly personal in nature (Salmela, 1995; Saury and Durand, 1998). It has been noted by various authors that expert coaching is not the overtly systematic process one might assume that it should be and that coaching action, especially in pressure situations, can seem to be intuitive and idiosyncratic (Cross, 1995; Bloom *et al.*, 1997; Saury and Durand, 1998; Lyle, 1999b).

There is a consensus within the literature that coaching expertise is grown out of experience. Experience allows the coach to develop complex, personalized knowledge structures, both conscious and tacit, that inform their future decision-making and action. Whilst all experts have experience, not all experienced coaches are experts. The issue for the coach educator is how does one learn from experience and how can this process be enhanced. The challenge for the educator of elite coaches becomes one of guiding experiential learning rather than creating educational programmes or packages.

Potential solutions to this challenge have been proposed by Saury and Durand (1998) and Lyle (1999b). In considering the development of expert coaches, these authors have advocated the use of structured learning opportunities such as:

- periods of coaching practice interspersed with periods of reflection;
- experience of the coaching environment within structured learning situations;
- simulation of real coaching experiences and exercises (e.g. a video of a real coaching situation with voice commentary from an expert coach);
- problem-based learning (i.e. asking developing coaches to solve specific problems that have arisen in expert coaching practice and guiding them to the solution).

Lyle also stated that 'many of these suggestions are best implemented by expert mentors' (p. 229). This statement is supported by the work of Bloom *et al.* (1998) who explored the importance of mentoring in development of coaches and performers. The authors found that the majority of expert coaches had been mentored within both their athletic and coaching careers and that they considered mentoring to have been an important ingredient in their experiential learning.

In addition to creating *in situ* learning opportunities and mentoring it has been shown that the development of reflective skills can enhance coach learning. Reflection and reflective processes have been recognized as being prerequisites to learning from experience across a range of disciplines including nursing (e.g. Ghaye and Lilleyman, 2000; Johns, 2000); education (e.g. Larrivee, 2000) and social work (e.g. Gardner, 2001). It is suggested therefore that coach education programmes should emphasize the development of reflective skills so that coaches are better able to learn from experience. The impact of reflective practice on coach development has been demonstrated by Knowles *et al.* (2001). When a structured reflective practice programme was combined with a period of mentored coaching experience, coaches were able to develop their reflective practice skills by generating a deeper understanding of their own practice and creating appropriate changes to that practice. Coach educators should look to develop learning skills such as reflection as well as deliver specific packages of knowledge.

The research that has been reviewed in this sub-section shows that expertise in coaching is largely based on the cognitive skills of the coach and that these skills are developed through experiential learning. The challenge for the coach educator is to enhance these processes by creating programmes that encompass real or simulated experience (through a practicum situation) whilst ensuring that the coach has the support and skills to learn from their experience.

Summary

Coaching is a diverse, dynamic and complex activity that encompasses many roles and requires many skills. Simple models cannot fully represent this process. However, these models do emphasize the importance of observation and instructional skills within the coaching process. In this respect, such models are of value to the coach and coach educator in that they draw attention to literature where science has defined clear principles for effective coaching practice.

In contrast to simple process models leadership models, such as the multi-dimensional model, focus on coaching as a specific form of sports leadership. The multi-dimensional model represents the complex interaction of player, coach and context of the coaching environment as well as differentiating between player satisfaction and performance as potential outcomes of the coaching process. By integrating these variables into one model, the multi-dimensional model captures more of the social dynamic of the coaching context than simple process models. However, it fails to account for the full variety of the coaching role described in the general coaching literature and does not account for the problematic interaction between player satisfaction and performance. Whilst one can draw some guiding principles for coach behaviour from the multi-dimensional model, and related research, it is not sufficiently robust to allow one to predict what is required for optimal coach behaviour in any given situation.

More recent research has shown that coaching expertise is not grounded in specific behavioural patterns but in the coaches' cognitive skills. Models such as Côté and co-workers'

Coaching Model emphasize the importance of knowledge construction in helping the coach to understand the coaching environment and make appropriate decisions regarding coaching action. The capacity of the coach to perceive relevant contextual cues, construct mental models of the situation and predict the outcome of alternative coaching action is what separates the expert from the novice. Furthermore, the literature also shows that such cognitive skills are grown from experience. The challenge of the coach educator is therefore one of constructing programmes that allow aspiring coaches to learn effectively from their experiences. The literature suggests that essential features of such programmes should be:

- periods of extended coaching practice;
- systematic mentoring during this practice;
- parallel development of the coaches' learning and reflective skills.

The scientific literature has much to say to both soccer coach and coach educator. The challenge in soccer, as in all sports, is to design and implement the type of coach education systems outlined above.

References

Abraham, A. and Collins, D. (1998). Examining and extending research in coach development. *Quest*, **50**, 59–79.

Bloom, G.A., Crumpton, R. and Anderson, J.E. (1999). A systematic study of the teaching behaviours of an expert basketball coach. *The Sport Psychologist*, **13**, 157–70.

Bloom, G.A., Durand-Bush, N. and Salmela, J.H. (1997). Pre and post competition routines of expert coaches of team sports. *The Sport Psychologist*, **11**, 127–41.

Bloom, G.A., Durand-Bush, N., Schinke, R.J. and Salmela, J.H. (1998). The importance of mentoring in the development of coaches and athletes. *International Journal of Sports Psychology*, **29**, 267–81.

Campbell, S. (1993). Coaching education around the world. *Sport Science Review*, **2**, 62–74.

Chelladurai, P. and Carron, A.V. (1978). *Leadership*. CAHPER, Sociology of Sport Monograph Series.

Chelladurai, P. and Carron, A.V. (1983). Athletic maturity and preferred leadership. *Journal of Sports Psychology*, **5**, 371–80.

Chelladurai, P., Imamura, H., Yamaguchi, Y., Oinuma, Y. and Miyauchi, T. (1988). Sport leadership in a cross-national setting: the case of Japanese and Canadian university athletes. *Journal of Sport and Exercise Psychology*, **10**, 374–89.

Côté, J., Salmela, J.H., Trudel, P., Baria, A. and Russell, S.J. (1995). The coaching model: a grounded assessment of gymnastics coaches' knowledge. *Journal of Sport and Exercise Psychology*, **17**, 1–17.

Cross, N. (1995). Coach effectiveness in hockey: a Scottish perspective. *Scottish Journal of Physical Education*, **23**(1), 27–39.

Cross, N. and Ellice, C. (1997). Coaching effectiveness and the coaching process: field hockey revisited. *Scottish Journal of Physical Education*, **25**(3), 19–33.

Douge, B. (1987). Coaching qualities of successful coaches. *Sports Coach*, **10**(4), 31–5.

Douge, B. and Hastie, P. (1993). Coach effectiveness. *Sport Science Review*, **2**(2), 62–74.

Fairs, J. (1987). The coaching process: the essence of coaching. *Sports Coach*, **11**(1), 17–19.

Franks, I.M., Sinclair, G.D., Thomson, W. and Goodman, D. (1986). Analysis of the coaching process. *Sports Science Periodical on Research and Technology in Sport*, January.

Gardner, F. (2001). Social work students and self awareness: how does it happen? *Reflective Practice*, **2**, 27–40.

Ghaye, T. and Lilleyman, S. (2000). *Reflection: Principles and Practice for Healthcare Professionals*. Quay Books, Salisbury, UK.

Gould, D., Giannani, J., Krane, V. and Hodge, K. (1990). Educational needs of elite US National, Pan-American and Olympic coaches. *Journal of Teaching in Physical Education*, **9**(4), 332–44.

Johns, C. (2000). *Becoming a Reflective Practitioner: A Reflective and Holistic Approach to Clinical Nursing, Practice Development and Clinical Supervision*. Blackwell Science, London.

Jones, D.F., Housner, L.D. and Kornspan, A.S. (1997). Interactive decision making and behaviour of experienced and inexperienced basketball coaches during practice. *Journal of Teaching Physical Education*, **16**, 454–68.

Knowles, Z., Gilbourne, D., Borrie, A. and Nevill, A. (2001). Developing the reflective sports coach: a study exploring the processes of reflective practice within a higher education coaching programme. *Reflective Practice*, **1**, 924–35.

Larrivee, B. (2000). Transforming teaching practice: becoming the critically reflective teacher. *Reflective Practice*, **1**, 293–308.

Launder, A. (1993). Coach education for the twenty-first century. *Sports Coach*, **16**, 2.

Lyle, J. (1999a). The coaching process: an overview, in *The Coaching Process: Principles and Practice* (eds N. Cross and J. Lyle), Butterworth Heinemann, Oxford, pp. 3–25.

Lyle, J. (1999b). Coaches' decision making, in *The Coaching Process: Principles and Practice* (eds N. Cross and J. Lyle), Butterworth Heinemann, Oxford, pp. 210–32.

NCF. (1986). *Coach in Action*. Springfeld Books, Leeds.

Pyke, F. (1992). The expanding role of the modern coach. *The Pinnacle*, **9**, 3.

Sabock, R.J. (1985). *The Coach*. Human Kinetics Books, Champaign, IL.

Salmela, J.H. (1995). Learning from the development of expert coaches. *Coaching and Sport Science Journal*, **2**(2), 3–13.

Saury, J. and Durand, M. (1998). Practical knowledge in expert coaches: on-site study of coaching in sailing. *Research Quarterly for Exercise and Sport*, **69**, 254–66.

Smith, R.E., Smoll, F.L. and Curtis, B. (1978). Coaching behaviours in Little League Baseball, in *Psychological Perspectives in Youth Sports* (eds F.L. Smoll and R.E. Smith), Hemisphere, Washington, DC, pp. 173–201.

Smith, R.E., Zane, N.S., Smoll, F.L. and Coppel, D.B. (1983). Behavioural assessment in youth sports: coaching behaviours and children's attitudes. *Medicine and Science in Sports and Exercise*, **15**, 208–14.

Smoll, F.L. and Smith, R.E. (1993). Educating youth sport coaches: an applied sport psychology perspective, in *Applied Sports Psychology: Personal Growth to Peak Performance* (ed. J.M. Williams), Mayfield Publishing, Mountain View, pp. 36–50.

Soloman, G.B., Striegal, D.A. and Eliot, J.F. (1996). The self-fulfilling prophecy in college basketball: implications for effective coaching. *Journal of Applied Sport Psychology*, **8**, 44–59.

Spencer, A.F. (2001). A case-study of an exemplary American College Physical Educator-tennis coach. *International Journal of Sport Pedagogy*, **3**, 1–27.

Sternberg, R.J. and Horvath, J.A. (1999). *Tacit Knowledge in Professional Practice: Researcher and Practitioner Perspectives*. Lawrence Erlbaum Associates, Mahwah, NJ.

Strean, W., Senecal, K.L., Howlett, S.G. and Burgess, J.M. (1997). Xs and Os and what the coach knows: improving team strategy through critical thinking. *The Sport Psychologist*, **1**, 243–56.

Williams, A.M., Davids, K. and Williams, J.G. (1999). *Visual Perception and Action in Sport*, E & FN Spon, London.

Woodman, L. (1993). Coaching: a science, an art, an emerging profession. *Sports Science Review*, **2**, 1–13.

14 Skill acquisition

*A. Mark Williams, Robert R. Horn
and Nicola J. Hodges*

Introduction

The acquisition of movement skills is fundamental to the development of skilful soccer players. Perhaps the primary role of the soccer coach is to help players develop these skills and improve as players. To achieve this goal, coaches need to have an understanding of how players learn and how practice should be structured and organized. Sports scientists can play an important role within this process by providing guidelines for coaches based on empirical research. As clubs and national associations continue to devote significant resources towards the development of elite soccer players, it is important that current practice is based on scientific evidence rather than on 'lay' opinion.

The aim in this chapter is to review current knowledge concerning how movement skills are acquired and refined, with particular reference to soccer. Initially, key terms such as skill, technique, learning and performance are defined and various measures of performance and learning evaluated. Next, current theoretical perspectives on motor learning are presented and the stages of skill learning described. The focus then turns to some of the key factors underpinning the design and implementation of effective instructional programmes. These include the role of instructions and demonstrations in conveying information about the skill to be learned, the importance of practice scheduling and the provision of feedback. The chapter is concluded by examining the potential benefits of guided practice versus discovery learning in the skill acquisition process.

14.1 Skill, technique, learning and performance

The term *skill* refers to a player's ability to select, organize, and execute an action, appropriate to a given situation in an effective, consistent and efficient manner. Implicit within this definition is the distinction between perceptual/cognitive and motor skills. The focus in this chapter is on motor skill learning and readers are directed elsewhere for literature pertaining to the acquisition of perceptual/cognitive skills (e.g. see Williams and Grant, 1999). Motor skills are distinct from perceptual/cognitive skills in that they require voluntary body and/or limb movement(s) to complete the task (Magill, 2001a).

Skill can be distinguished from *technique*, which refers to a basic motor action or movement pattern. Skill is the ability to use a technique appropriately and effectively at the right moment. Techniques form the building blocks upon which skill is developed. It is therefore essential that as a result of the skill acquisition process learners know how and when to apply techniques effectively.

The key features of skill are that it is:

- goal directed (players have a purpose and know what they are attempting to do);
- effective and reliable (the goal is achieved with maximum certainty and consistency);
- adaptable (to meet a variety of performance contexts);
- efficient (it uses no more effort than necessary); and
- learned (developed with practice and experience).

Learning is the process that underlies the acquisition and retention of skilled action. It is defined as '... a set of processes associated with practice or experience that leads to relatively permanent changes in the capability for movement' (Schmidt and Lee, 1999, p. 264). *Performance* simply refers to behaviour that can be observed. Learning is not directly observable, since it involves changes in central nervous system function, but it can be inferred by observing changes in performance over an extended period of time. Learning can only be deemed to have taken place if the improvement in performance is relatively permanent in nature. Temporary variations in performance generally occur for other reasons, such as changes in motivation, anxiety or fitness levels (e.g. fatigue) and are not truly representative of skill learning.

The distinction between performance and learning is important for scientists and practitioners since, as discussed at various points in this chapter, many of the typical interventions employed by coaches have differential effects on these two processes. Although players may have learnt a particular skill, they may not always perform to their potential because of the effect of various extraneous or performance variables such as fatigue or anxiety.

14.2   Measurement of skill learning

The coach needs to know whether learning has taken place and consequently, progress needs to be monitored and evaluated. Learning may be assessed through observation of the player in training sessions and matches or through more objective means such as soccer skill tests, quantitative performance statistics or qualitative video analysis. The typical approach is to plot improvement over time using a line graph or 'performance curve', as highlighted in Figure 14.1. An agreed measure of learning is required, such as performance on a passing skill test or the proportion of successful passes made during matches. It may also be useful for players and/or coaches to use performance diaries to record progress over time.

In order to measure learning accurately, retention or transfer tests must be used in association with performance curves. A retention test is when the player is re-tested after a period of time when the transient effects of performance have subsided. In contrast, a transfer test requires the learner either to perform the skill that has been practised in a new situation or to perform a new variation of the practised skill (i.e. passing over a different distance). Regardless of which method or combination of approaches is employed, it is difficult to determine accurately the amount that a player has learned.

14.3   Theoretical perspectives on motor learning

Various conceptual or theoretical models guide current research and coaching practice. Traditionally, most models of instruction have been grounded in cognitive psychology. The athlete is likened to a sophisticated computer that interprets (i.e. perceives) information for decision making prior to some form of response selection and execution. This 'information processing' model assumes that skilled performers accumulate through experience an extensive knowledge base that is stored in memory and which enables them to deal effectively

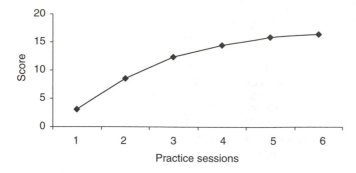

Figure 14.1 A line graph showing improvements in accuracy on a passing skill test. The shape of
the curve indicates a rapid improvement in performance at first followed by a general
'levelling off' during which improvements are relatively slow. This characteristic
of performance curves (i.e. steep at first and more gradual later) is one of the most
fundamental laws of practice.

with the complex demands of competition. As far as the execution of motor skills is
concerned, the model relies heavily on the existence of centrally stored representations of
movement (i.e. generalized motor programmes or 'schema' stored in memory, see Schmidt,
1975) as well as an ongoing role for both conscious and subconscious sensory feedback
mechanisms.

The assumption is that skilled players have well-established motor programmes with
invariant or non-changing features. These motor programmes enable movements to be carried out consistently and effectively with limited conscious control by sensory feedback
mechanisms. Novice performers have not acquired the necessary programmes or their representation of the movement is less defined and consequently, they are more likely to process
information consciously through various sensory loops (particularly vision). At a simplistic
level, as a player becomes more skilled there is a shift in control from attending to sensory
information about the movement to more automatic, less attention demanding performance.
At this more 'automatic' level, performance is generally more consistent due to the development and use of motor programmes which guide movement and allow sensory information
to be processed in a subconscious fashion.

In recent years, alternative theoretical models of motor control and learning have emerged
in which the performer is viewed less as a machine, but more as a complex, dynamical system. A greater emphasis is placed on the intricate link between the performer and his/her
environment (e.g. see Bernstein, 1967). Players are presumed to become more 'attuned' to
the perceptual demands of the task and environment, enabling effective and efficient performance (see Fowler and Turvey, 1978; Whiting, 1984). Scientists undertaking work from this
perspective have examined how coordinated movements, typically defined by the relationships between joints and/or units spontaneously emerge following a change in a specific
variable or control parameter. A good example would be changes in the phasing or coordination of the legs during the transition from walking to running as movement velocity (i.e.
the control parameter) increases. The emergent pattern of movement (i.e. running) is
assumed to represent an attractor point for stable movement behaviour at that speed. The key
issue is that changes in coordination are thought to self-organize without resorting to a
mechanism like a motor programme to control the movement (for a more detailed review,
see Davids *et al.*, 1994; Williams *et al.*, 1999). This self-organization is based, at least in

part, on the unique constraints imposed by the organism (performer), task and environment during goal-directed behaviour (Newell, 1996).

Although this chapter is written primarily from a cognitive psychology perspective, the important practical differences between the two conceptual models are briefly highlighted. In particular, the potential implications of the dynamical/ecological approach to coaching are discussed in light of recent calls for a more 'hands off' and less prescriptive approach to coaching (see Handford *et al.*, 1997). The emphasis on literature grounded in cognitive psychology is not intended to undervalue the importance of recent advances based on ecological approaches to skill acquisition, but merely a reflection of the far more extensive knowledge base that exists on motor learning from a cognitive perspective.

14.4   Stages of skill learning

It has been proposed by a number of researchers that individuals progress through different stages of learning a skill. These changes do not occur rapidly and the role of the coach is to help a player progress through the different stages. Fitts and Posner (1967) identified three principle stages of skill acquisition. The unique characteristics of each stage of learning are highlighted and implications for practice and instruction discussed throughout the remaining sections of this chapter.

14.4.1   Cognitive stage

This is the earliest stage of learning where the individual is preoccupied with trying to determine the basic characteristics and principles of the task. Players may think and verbalise about what to do and try to identify appropriate strategies to achieve the task (e.g. try different ways of kicking a ball). Performance is attention demanding and characterized by frequent, gross errors (e.g. miscuing or slicing the ball). Improvements at this stage are rapid, but performance fluctuates substantially as new strategies are tried and sometimes abandoned. Performers know that they are doing something wrong, if the ball fails to travel the required distance, for example, but they may not be aware of what they should be doing differently next time. Since the player relies heavily on the conscious evaluation and processing of sensory information, performance is often jerky, poorly timed and inconsistent.

14.4.2   Associative stage

The basic fundamentals or mechanics of the skill have been learned and consequently, the performer turns to the issue of how to perform the movement better. The learner begins to develop specific motor programmes for the actions and may need to divert less attention to processing the various sources of sensory input. The number and magnitude of errors are reduced and the performer concentrates on refining the skill. Actions are timed better and movement patterns are smoother and more efficient. Performers now begin to develop the ability to identify their own errors. This marks a definite change in the course of the learning process. At this stage, variability in performance begins to decrease and performance becomes more consistent. Only really skilled players actually move beyond this stage of learning (see Ericsson *et al.*, 1993).

14.4.3   Autonomous stage

At the advanced stage, learners can perform the skill without thinking about it, or at least thinking about it in a very different way to the beginner (i.e. it is automatic or autonomous).

During this more automatic stage, attention is devoted to strategic or stylistic aspects of performance. Players are able to detect their own errors and make appropriate adjustments to correct them. These adjustments may be immediate and on-line, requiring little conscious attention or occur as a result of in-depth evaluation of the situation after the skill has been performed. There is relatively little variability in performance and few errors. However, performance improvements will inevitably be very gradual and small. The developed skill or motor programme now only requires occasional modification and fine-tuning.

Alternative models of skill learning have been proposed by those working from a dynamical systems/constraints-based perspective (for a detailed review, see Handford *et al.*, 1997; Williams *et al.*, 1999). For example, Bernstein (1967) suggested that there are two sequential stages during the acquisition of a new movement pattern. Initially, the learner simplifies the movement by reducing or 'freezing' the motion around joints and by 'coupling' together different joint complexes (e.g. consider how the young child freezes the knee and ankle joints during the kicking action). As the learner progresses, the 'degrees of freedom' around the joints are released and eventually reorganized to ensure smooth, efficient and skilful movement (see Anderson and Sidaway, 1994; Williams *et al.*, 2002). More recently, Newell (1985, 1996) proposed that learners move through three distinct stages of learning termed: coordination, control and skill. During the coordination stage, the learner 'assembles' a new relative motion pattern that is subsequently scaled through a process termed control. This movement pattern becomes more fine-tuned following extensive practice in order to meet the accuracy or speed demands of the task and promote skilful behaviour.

14.5   Instruction and modelling of soccer skills

When introducing soccer skills, a crucial role for the coach is to provide the learner with the necessary information to perform the task. Such information may be provided verbally through the use of instructions or cues and/or visually via a demonstration. Although the old adage 'a picture paints a thousand words' suggests that demonstrations are likely to convey the most amount of information in a short period of time, this may not always be the case and in certain situations they may actually be detrimental to the learning process.

14.5.1   *Visual demonstrations and skill learning*

Demonstrations are presumed to be effective because they help players learn skills by acting as a model or visual template that they can copy. This process, termed observational learning, refers to a learner's tendency to watch the behaviour of others and to adapt his/her own behaviour as a result of the experience (Williams *et al.*, 1999). Whilst observing a model, the learner formulates a 'symbolic' representation or 'cognitive blueprint' of the activity in memory which serves to aid subsequent production of the skill. According to Bandura's (1986) Cognitive Mediation Theory, four inter-related component processes are implicated in the eventually learned behaviour. These are *attention* where the learner selectively attends to and extracts the distinctive features of the model. The coach may play an important role here in directing the learner to the appropriate information. *Retention* facilitates learning of the observed act by reconstructing and transforming it into a symbolic representation in memory. Cognitive activities such as rehearsal, labelling and organization are involved in the retention process. *Motor reproduction* uses the symbolic representation or schema to guide subsequent performance, supported by feedback and instruction. Finally, *motivation* mediates the decision to inhibit or reproduce an observed action.

Researchers have also attempted to determine the nature of the information picked-up by the learner when observing a model. For example, Scully and Newell (1985) argued that learners pick up the relative motions between the key body parts (i.e. the movement's topological characteristics) rather than specific information cues about the movement. According to this perspective, demonstrations should be most effective early in learning when the player is trying to acquire a new movement pattern (Magill and Schoenfelder-Zohdi, 1996). In this situation, the movement of various body parts in relation to each other defines the action pattern required for successful performance. Later in learning when the performer is trying to refine or 'scale' an existing movement pattern, demonstrations are presumed to be no more effective than verbal instructions. At this stage, continued practice on the task is by far the most important variable underlying skill acquisition (Scully, 1988).

Another recent suggestion is that very early in learning performers should be given the opportunity to practice the skill without any constraining information in the form of a demonstration or verbal instructions (Hodges and Lee, 1999; Hodges and Franks, 2001). Provided that the learners are aware of the desired outcome goal and receive feedback concerning goal attainment they may benefit from having the opportunity to try and adapt the movement to suit their own individual requirements. A demonstration may be provided after initial familiarity with the skill to help refine the movement.

14.5.2   Skill and status of the model

Selecting the most appropriate individual to perform the demonstration is an important consideration for the coach. Ideally, the model needs to be skilled and respected by the group (McCullagh and Weiss, 2001). A 'high-status' model causes learners to pay closer attention to the demonstration and provides an increased motivation to perform well, as compared to a 'low-status' model. The accuracy of the demonstration is also important since observational learning is based on direct imitation of the visually presented skill. If a specific technique is the goal of a particular practice session then it is important that the learners have a correct reference to base their subsequent attempts upon. If the coach is not technically proficient, then (s)he should consider asking a skilled member of the group to demonstrate.

Some authors (e.g. Adams, 1986; Pollock and Lee, 1992) have suggested that coaches should also employ models who are learning the skill. While highly skilled models provide a template for action, they do not provide information about error for the learner. Instead, observing a group member learning the task engages the learner in a problem solving process. Several studies indicate that learning models are at least as effective as skilled models (e.g. McCullagh and Meyer, 1997; Domingue and Maraj, 1998). The benefits of using learning models may be heightened if the observer not only watches the model, but also hears the feedback given by the coach. This latter technique may be particularly helpful if learners are grouped in pairs and one partner has the opportunity to observe whilst the other practices. If coaches provide different players with the opportunity to demonstrate the skill, this can also enhance self-confidence and motivate others in the group to try harder. Also, viewing different models performing the skill early in learning encourages the performer to try slightly different solutions to the movement problem and also enhances flexibility and adaptability.

14.5.3   Providing verbal instructions about the skill

For young learners a visual demonstration may be insufficient to acquire a new skill. Instead, a 'show and tell' approach is favoured in which demonstration is accompanied by verbal guidance. Several factors need to be taken into account when providing verbal instruction. In particular, instructions should be kept brief and simple because players have a limited capacity to absorb information, particularly when performing a new skill. The language employed should be non-technical and appropriate to the age level of the players. Verbal instruction should complement the demonstration by giving the players a general idea of how to perform the skill; it should give them the 'big picture' in as few words as possible. Verbal instruction and subsequent feedback should also be compatible (Hodges and Franks, 2001). If instructions are focused on a particular aspect of behaviour (e.g. keeping the head down during ball contact) then verbal prompts and ensuing feedback should direct attention towards this aspect of performance (Magill, 2001a).

Verbal instructions can sometimes have detrimental effects on learning (see Wulf and Weigelt, 1997; Hodges and Lee, 1999). Wulf and Weigelt (1997) found that 'expert strategy' instruction given to beginners when learning to move on a ski-simulator was actually detrimental to acquisition, as compared to a group who did not receive any technique instruction. The authors suggested that instructions and movement demonstrations direct attention inappropriately to the body, at the expense of a goal-related focus on the effects of the action (which might be encouraged by withholding instruction and promoting more discovery-learning conditions). This suggestion has subsequently been supported in a number of studies involving the tennis serve and putting in golf (e.g. Wulf *et al.*, 1999). In soccer, for example, watching a demonstration or receiving instruction concerning the angle of the hip, knee and foot when kicking the ball may increase the learner's attention to the body when performing the task. This internal focus of attention could be at the expense of goal attainment (i.e. an accurate pass).

14.6 Organizing practice sessions

Skilled performance in sport is dependent on years of deliberate and purposeful practice (Ericsson *et al.*, 1993), and soccer is no exception (Helsen *et al.*, 2000). Players must engage in practice that is designed to improve performance, and is effortful and high in concentration. Practice sessions need to be appropriately organized to ensure effective and efficient learning. Skills may be practised in many ways, for varying periods of time, under a range of conditions and according to different schedules. A key task for the coach is to manipulate these practice variables to meet the needs of players. Important issues include the optimal length and frequency of practice sessions and how practice sessions should be structured for effective learning.

14.6.1 Length and frequency of practice sessions

A suggestion is that a single practice session should last about an hour for young children and around an hour and a half for older players (Williams, 1998). The actual length of a practice session should depend on the nature of the skill(s) being practised and the work to rest period employed. Skills that are continuous and/or physically demanding require short sessions with frequent rest periods (i.e. distributed practice). However, most soccer skills are discrete actions that provide a natural rest between attempts and consequently, rest periods during practice need not be too frequent (i.e. massed practice), allowing coaches to make effective use of the time available. There is no empirical evidence to suggest that reducing rest time through massed practice degrades the learning of discrete skills (Lee and

Genevose, 1988). Although practising when fatigued may reduce performance, it does not appear to have a detrimental effect on learning. A key point for coaches is that practising a skill under the same or higher levels of stress than experienced during a match does not adversely affect skill learning. Players should practice under conditions that are similar to the actual game context so as to ensure effective transfer to the competitive environment.

As to the frequency of practice sessions, research indicates that daily, twice daily, or even more frequent sessions produce the greatest amount of learning (Schmidt and Wrisberg, 2000). Although there is certainly a degree of diminishing returns with extended practice periods, practising a skill beyond the amount needed to demonstrate satisfactory performance (i.e. overlearning) has a positive influence on the retention of motor skills (see Driskell *et al.*, 1992). However, school, family and work commitments may place limitations on the frequency of practice, whilst coaches must ensure that the risks of various overuse injuries as well as potential 'burnout' are monitored.

14.6.2 *Variability of practice*

An important consideration for coaches is whether consistent repetition is more effective for skill learning than practice involving some degree of variability. When learning to pass a ball, should players be required to pass the ball repeatedly back and forth over a consistent distance or should they vary, for example, the distance, angle and speed of the pass from one trial to the next? The former approach is referred to as 'constant or specific practice', whilst the latter is termed 'variable practice'. According to Schmidt's (1975) schema theory, variability within movement and context characteristics is important in producing an expansive, generative 'rule' to cope with a variety of similar but different instances.

Generally, the weight of research evidence supports the proposition that variable practice results in greater accuracy and consistency than specific repetition of the same skill (for a review, see Lee *et al.*, 1985; Van Rossum, 1990). This finding is more marked with children than adults (Wulf, 1991; Yan *et al.*, 1998). Coaches should therefore ensure that they vary practice by manipulating factors such as distance, speed, height or direction of the pass and that the practice session mimics the range of variations experienced during a match. However, coaches should not make the variation in practice so great that the task becomes a different skill to that originally practised. For example, with regard to the earlier example of instep passing, coaches should ensure that however they vary conditions, it is still the same skill or motor pattern that is practised. There is also a suggestion that variable practice is most effective when coupled with a random rather than blocked practice schedule as discussed in the next section (Lee *et al.*, 1985).

14.6.3 *Contextual interference effects: blocked versus random practice*

Closely related to variability of practice is the research on contextual interference. Contextual interference occurs when several skills are learnt within the same practice period (Battig, 1979). A low contextual interference practice schedule may involve practising one skill per session, or perhaps two separate skills (e.g. shooting and passing) in blocks of 20–30 min each (i.e. blocked practice). In contrast, much higher levels of contextual interference would arise if a variety of skills (e.g. shooting, passing, dribbling) were practised in a somewhat random manner throughout the session (i.e. random practice). In the most random practice schedule, a player never practices the same skill on consecutive trials. The degree of contextual interference can be considered as a continuum with a totally random order of skill practices at one end (high contextual interference) and blocks of practice on one skill at the other (low contextual interference; see Magill, 2001a). Small-sided games or match play typically require players to practice different skills from one trial to the next, whereas drill situations tend to involve repetitive practice of the same skill.

The research evidence indicates that a random or high contextual interference practice schedule is detrimental to short term performance, but, more importantly, is better for long term retention than blocked conditions (see Goode and Magill, 1986; Hall *et al.*, 1994). Although there is some evidence to suggest that blocked practice may be better very early in learning (Shea *et al.*, 1990), once a rough approximation of the movement is acquired coaches should try and avoid repetitious, blocked practice by presenting a variety of skills within the same session. The benefits of random practice also appear to be enhanced when skills differ more markedly (e.g. passing and heading may be more distinct than passing and shooting) (for an interesting discussion, see Brady, 1998).

Several hypotheses have been proposed to account for the positive effects of contextual interference on skill learning. The majority of these suggest that random practice schedules facilitate learning either by encouraging the performer to undertake more elaborate and distinctive processing or through the forgetting and subsequent recall of an action plan each time a skill is performed. Regardless of the ongoing controversy about the mechanisms underpinning the contextual interference phenomenon, the superiority of random over blocked practice is one of the most established findings in the motor learning literature.

14.6.4   *Easy-to-difficult progression*

An important characteristic of skill is that performance requires less attention or is more automatic as a result of continued practice on the task (Williams *et al.*, 1999). Coaches should therefore ensure that the level of difficulty of the practice is appropriate for the players under their guidance. This is particularly the case for beginners or low aptitude learners where the cognitive and attention demands of the task are likely to be very high. At this stage, coaches need to ensure that the task is not so difficult as to have a negative effect on learning and the players' commitment to practice. In contrast, at the other end of the continuum practice needs to be suitably demanding to promote further improvements in skill learning. Some techniques have been proposed in the literature to vary the attention demand on the learner, including segmentation and simplification (see Wightman and Lintern, 1985).

Segmentation or 'part practice' involves practising components of a skill in isolation before combining each part as learning progresses. The research evidence indicates that segmentation is most effective when tasks are high in complexity and low in organization (see Lee *et al.*, 2001). The organization of a task relates to the relationship or inter-dependency between its component parts, whereas task complexity refers to the number of parts or components to the skill. For example, controlling and passing a ball in soccer may be deemed to be a task that is low in organization and high in complexity. The separate components of ball control and passing would be highly complex for beginners and the separate components are not highly related since, for example, a child can have excellent ball control but poor passing skills. The suggestion is that the task may be broken down during practice into separate components involving ball control followed by the passing of a stationary ball. This approach has the advantage of reducing the attention demand of performing the whole task so that players focus attention on mastering each component separately. As learning progresses, these components can be combined so that the time-sharing skills necessary to coordinate the component activities can be acquired (see Wickens, 1989).

Simplification involves reducing the difficulty of either the whole skill or different parts of the skill. There are numerous examples of how coaches may simplify the demands of the task in order to facilitate learning. The majority of these involve the manipulation of time and space by varying the size of the playing area and/or the number of players involved. For instance, the attentional demand of trying to maintain possession of the ball when playing 3 versus 3 in a 10×10 m grid is much higher than practice involving 4 versus 1 players in the same area. In the latter example, the attentional demand may be increased by gradually introducing more opponents so that the learner can be placed under similar or even greater pressure than in a match situation. Innovative coaches use such techniques to ensure that practice sessions progress in an easy-to-difficult manner and learners are challenged to adapt to the increasing demands of practice and performance.

14.7 Providing augmented feedback

When performing a skill, learners receive sensory feedback about the correctness of the movement as well as about the success of their actions. With the possible exception of practice, this feedback is presumed to be the most important factor underpinning skill learning. Feedback can be provided intrinsically (i.e. via the performer's own sensory mechanisms) and/or extrinsically (i.e. from an external source). Information provided by a coach after the performance of a skill is typically referred to as terminal augmented feedback. When such information relates to the nature of the movement itself it is termed 'knowledge of performance'. Information about the outcome of the task, relative to some external goal, is termed 'knowledge of results'. This distinction is not always clear especially when the goal of the action is a particular movement pattern or technique.

Most empirical research has focused on the importance of knowledge of results during skill learning, primarily due to difficulties in attempting to provide knowledge of performance within a controlled experimental setting and in trying to measure changes in movement form and topology. In most instructional settings, learners are able to see whether they have been successful in achieving the intended goal and consequently, coaches are far more likely to provide learners with knowledge of performance about how the movement was performed rather than knowledge of results about the movement outcome (Magill, 2001b). Typical knowledge of performance statements for the soccer instep pass are presented in Table 14.1. Several factors relating to the provision of feedback have been examined in the literature such as its role and importance at each stage of learning as well as the optimal levels of precision and frequency for effective skill learning (for detailed reviews, see Swinnen, 1996; Magill, 2001a,b).

14.7.1 Importance of feedback during each stage of skill learning

Although learning can occur in the absence of feedback from a coach or significant others, the provision of augmented information leads to more efficient learning, ensures correct development of the motor skill, and better eventual performance. Players who receive constructive feedback also become more interested in the task, put more effort into practice, and persist longer after feedback is removed. As remarked on earlier, the provision of feedback is particularly important in the early stages of learning to guide error detection and correction. Beginners rely heavily on visual and verbal sources of information and consequently,

Table 14.1 A list of prescriptive KP statements for the soccer instep pass highlighting some of the typical errors in order of correction. (Adapted from Williams *et al.*, 2002)

Priority	KP statement
1	Place the support foot to the side of the ball
2	Strike the ball with the instep of the foot
3	Strike the ball in the centre
4	Keep your head down and look at the ball
5	Place the support foot approximately 30 cm away from the ball
6	Bring your knee over the ball
7	Follow through with the foot after contact
8	Take a longer last stride to the ball
9	Approach the ball with speed
10	Flex your knee as you bring your leg back
11	Use your arms for balance
12	That was correct

augmented feedback should be of most use at this stage. As learners progress, they develop the ability to detect and correct their own errors and are less dependent on prescriptive feedback. Skilled players often have a good indication of where they are going wrong and how they can attempt to rectify the problem on future attempts. It is important for the coach to develop good error detection and correction abilities in their athletes. Too much feedback can prevent learners from evaluating their own performance and cause them to become dependent on augmented feedback from the coach. The learner faced with the situation of bad performance (e.g. inability to kick straight) may lack the necessary skills to evaluate the performance and determine what needs to be changed to rectify errors. Coaches should therefore not be overly prescriptive. They should provide learners with the opportunity to process their own feedback and learn from their mistakes.

14.7.2   Precision of feedback

Feedback can be provided in a very detailed manner involving, for example, a range of verbal knowledge of performance statements, or it can be more coarse or qualitative in nature. Too much information is confusing and can actually hinder learning (Salmoni *et al.*, 1984). This is particularly the case with young children who are unable to simplify or 'round off' the information presented. Coaches should not attempt to rectify every error but should select one or two prescriptive statements and then present these in a simple and concise manner. Feedback should also be kept simple for adult learners early in learning, although more detailed feedback may be needed to facilitate skill acquisition as learning progresses (Magill and Wood, 1986). Once learners have acquired a rough 'ball park' idea of the movement, precise feedback is required to make the action more effective in achieving the desired outcome. The optimal level for the precision of feedback is dependent on a range of factors including the difficulty of the task, the age of the performers and their stage of learning (for a detailed review, see Magill, 2001a,b).

14.7.3   Frequency of feedback provision

While providing feedback on every trial appears to have a beneficial effect on performance, it is actually detrimental to skill learning (at least for discrete skills, see Swinnen, 1996). Providing feedback on every practice attempt can lead to information 'overload', result in over-dependence on augmented feedback, and prevent the learner from undertaking some degree of trial-and-error problem solving. The optimal frequency of feedback appears to be dependent on the player's stage of learning as well as the complexity or difficulty of the task. Early in learning or when practising difficult skills, players are likely to require more prescriptive feedback to improve performance (Wulf *et al.*, 1998). As skill develops, the frequency of feedback provision should be reduced or 'faded out' to encourage learners to detect and correct their own errors.

The frequency of feedback provision may be reduced using 'summary' or 'bandwidth' techniques. The former technique involves giving the learner a summary of performance after a certain number of practice attempts, whereas the latter involves establishing a performance-based bandwidth that provides a criterion for when feedback will or will not be given. The suggestion is that summary feedback should be given more frequently and/or the bandwidth for feedback provision should be wider early in learning or if the task is fairly complex (Guadagnoli *et al.*, 1996). One suggestion is that summary feedback be provided every five trials for complex skills and every fifteen or so attempts for simpler tasks (Schmidt and Wrisberg, 2000).

To prevent learners becoming overly dependent on feedback, the feedback provided should also be more descriptive than prescriptive as skill develops. Beginners require prescriptive information in order to correct the errors made during performance, whereas as learning progresses feedback should be more descriptive to encourage players to acquire the ability to detect their own errors (Magill, 2001a). A question and answer approach should also be encouraged so that learners' think about their own performance and develop a sensitivity to the 'feel' of movement (see Liu and Wrisberg, 1997). Coaches could ask players why they think the error occurred and how they might improve performance in subsequent practice.

14.8 Effective coaching: prescriptive coaching versus guided-discovery

A more 'hands-off' approach to instruction has recently been proposed in line with suggestions that skill learning is due to the natural emergence of movement solutions as a result of various constraints on action (see Newell, 1985). Constraints, which may be internal (e.g. physical characteristics) or external (e.g. rules, tactics, playing context) to the learner, are important influences, guiding and shaping emergent behaviours. According to the dynamical systems or the constraints-based perspective, movement patterns emerge as a consequence of the constraints placed on the learner, rather than the result of any pre-planned motor programme for the movement (Handford *et al.*, 1997; Williams *et al.*, 1999).

Coaching strategies that are overly prescriptive are likely to impose artificial constraints on learning, producing temporary and inefficient movement solutions. In the 'constraints-led' perspective, the role of the coach is to guide the learner to search, discover and exploit task solutions through a less prescriptive, more 'hands off' approach to instruction. The suggestion is that coaches should try to facilitate rather than discourage exploratory behaviour by manipulating the constraints of the learning environment. In this sense, coaching is about creating pressure for changes in movement form to emerge through self-exploration or 'discovery learning' (see Verijken and Whiting, 1990; Williams *et al.*, 1999).

Coaches could use conditioned games to elicit certain types of behaviours from players in soccer. One- and two-touch practices constrain the learner to search for available passing opportunities prior to receiving the ball, whilst games which allow goals to be scored solely from crosses encourage teams to employ width in offensive play as well as developing the skills of heading and volleying. Other constraints that may be employed to promote discovery learning include the implementation of tight time constraints, the restriction of space through use of playing areas or zones, and the selective use of opponents to manipulate 'pressure'. The process of manipulating constraints for the purposes of effective learning challenges coaches to be highly creative in designing games and practices that enable skills to emerge in learners (e.g. see Charlesworth, 1994; Thorpe, 1996).

Finally, it should be noted that whilst the process of manipulation constraints is typically associated with the dynamical systems approach, realistic practice conditions that encourage

Table 14.2 Some key 'dos' and 'don'ts' of instruction across the different stages of skill learning

Stage of learning	'Dos'	'Don'ts'
Cognitive	Provide a general idea of the movement by using verbal instructions and demonstrations Break skills down into constituent parts to simplify the task and then re-introduce them in a logical sequence or order (i.e. easy to difficult progressions such as initially learning to control the ball before concentrating on passing or introducing opponents gradually) Provide prescriptive feedback for error correction and motivation purposes Employ specific practice drills and low contextual interference conditions initially (i.e. only practice one skill per session)	Emphasize the outcome (i.e. stress the process of striking the ball well rather than where it goes) Concentrate too much on errors (i.e. reinforce those components of the task that are correct) Overload players with too much information
Associative	Encourage performers to evaluate their own performance Increase progressively the complexity of the task (e.g. introduce opponents, restrict time and space) Increase variability in practice as well as the amount of functional or contextual interference by practising more than one skill in a session (e.g. passing and shooting)	Give too much feedback
Autonomous	Use minimal intervention (i.e. encourage the player to evaluate his own feedback; use demonstrations less frequently) Deal only with highly specific components of the task Present the learner with complex, realistic and challenging practice (e.g. further restrict the time and space available to perform) Practice under realistic match conditions and encourage improvisation and adaptability through variety in practice Employ highly variable practice conditions and high contextual interference practice sessions (i.e. practice more than one skill in each session)	Assume that learning has stopped, continued intensive practice still remains essential to further develop skills (i.e. encourage 'overlearning')

subjective error detection and correction and promote variability and variety in movement solutions are recommended by the majority of researchers, irrespective of their theoretical stance.

Summary

In this chapter, some of the key principles underlying the acquisition of soccer skills were highlighted. In particular, efforts were made to outline potential implications for coaches in their endeavours to ensure that practice time is used appropriately. Due to the fact that the predominance of empirical research into motor learning has been conducted within the framework of information processing models, this chapter was written primarily from a cognitive perspective. We have attempted to illustrate also the potential implications of research conducted within the realms of dynamical systems and constraints-based approaches to learning. Within each section we have highlighted key principles and provided guidelines as to effective coaching practice. The main implications for coaches are summarized in Table 14.2.

The majority of motor learning research has been conducted within the laboratory such that controlled conditions of testing allow strong conclusions as to the effects of specific practice variables. It has been very difficult to assess the effectiveness of the various practice manipulations outlined in this chapter. Although some of the studies reviewed are more ecologically valid than others (i.e. have been examined in a practical setting), these types of studies typically fail to demonstrate concrete, repeatable learning effects as a result of difficulties in measurement and in isolating specific practice variables. In the practical setting the reality is that coaches typically employ a variety of practice techniques simultaneously and may vary these techniques to suit the individual. As a result of the difficulties in conducting scientific research when the practice variables are not manipulated in a consistent manner, or many variables are examined simultaneously, we do not have a clear understanding as to the cumulative or interactive effects of these practice variables. However, general practice principles have emerged throughout. These recommendations are intended to provide the coach with an idea of the type of instruction that is likely to be successful at each stage within the learning process. They are not intended to be applied too literally and clearly coaches need to employ 'craft knowledge' to adapt and modify these guidelines to suit the individual needs of the players under their care. Players learn skills at different rates and in varying ways and consequently, coaches need to be flexible and adaptable to ensure that soccer skills are learnt efficiently and effectively.

References

Adams, J.A. (1986). Use of the model's knowledge of results to increase the observer's performance. *Journal of Human Movement Studies*, **12**, 89–98.

Anderson, D.I. and Sidaway, B. (1994). Coordination changes associated with practice of a soccer kick. *Research Quarterly for Exercise and Sport*, **65**, 93–9.

Bandura, A. (1986). *Social Foundations of Thought and Action: A Social-Cognitive Theory*. Prentice-Hall, New York.

Battig, W.F. (1979). The flexibility of human memory, in *Levels of Processing in Human Memory* (eds L.S. Cermak and F.I.M. Craik), Erlbaum, Hillsdale, NJ, pp. 23–44.

Bernstein, N.A. (1967). *The Coordination and Regulation of Movements*. Pergamon Press, New York.

Brady, F. (1998). A theoretical and empirical review of the contextual interference effect and the learning of motor skills. *Quest*, **50**, 266–93.

Charlesworth, R. (1994). Designer games. *Sport Coach*, **17**, 30–3.

Davids, K., Handford, C. and Williams, A.M. (1994). The natural physical alternative to cognitive theories of motor behaviour: an invitation for interdisciplinary research in sports science? *Journal of Sports Sciences*, **12**, 495–528.

Domingue, J.A. and Maraj, B.K. (1998). Effects of model skill level in observational learning of the flying disc forehand. *Journal of Sport and Exercise Psychology*, **20**, S119.

Driskell, J.E., Willis, R.P. and Copper, C. (1992). Effect of overlearning on retention. *Journal of Applied Psychology*, **77**, 615–22.

Ericsson, K.A., Krampe, R.T. and Tesch-Römer, C. (1993). The role of deliberate practice in the acquisition of expert performance. *Psychological Review*, **100**, 363–406.

Fitts, P.M. and Posner, M.I. (1967). *Human Performance*. Brooks/Cole, Belmont, CA.

Fowler, C. and Turvey, M.T. (1978). Skill acquisition: an event approach with special reference to searching for the optimum of a function of several variables, in *Information Processing in Motor Control and Learning* (ed. G. Stelmach), Academic Press, New York, pp. 1–40.

Goode, S.L. and Magill, R.A. (1986). The contextual interference effect in learning three badminton serves. *Research Quarterly for Exercise and Sport*, **57**, 308–14.

Guadignoli, M.A., Dornier, L.A. and Tandy, R.D. (1996). Optimal length for summary knowledge of results: the influence of task-related experience and complexity. *Research Quarterly for Exercise and Sport*, **67**, 239–48.

Hall, K.G. and Magill, R.A. (1995). Variability of practice and contextual interference in motor skill learning. *Journal of Motor Behavior*, **27**, 299–309.

Hall, K.G., Domingues, D.A. and Cavazos, R. (1994). Contextual interference effects with skilled baseball players. *Perceptual and Motor Skills*, **78**, 835–41.

Handford, C., Davids, K., Bennett, S. and Button, C. (1997). Skill acquisition in sport: some applications of an evolving practice ecology. *Journal of Sports Sciences*, **15**, 621–40.

Helsen, W.F., Hodges, N.J., Van Winckel J. and Starkes, J.L. (2000). The roles of talent, physical precocity and practice in the development of football expertise. *Journal of Sport Sciences*, **18**, 75–90.

Herbert, E.P., Landin, D. and Solmon, M.A. (1996). Practice schedule effects on the performance and learning of low- and high-skilled students: an applied study. *Research Quarterly for Exercise and Sport*, **67**, 52–8.

Hodges, N.J. and Franks, I.M. (2001). Learning a coordination skill: interactive effects of instruction and feedback. *Research Quarterly for Exercise and Sport*, **72**, 132–42.

Hodges, N.J. and Lee, T.D. (1999). The role of augmented information prior to learning a bimanual visual-motor coordination task: do instructions of the movement pattern facilitate learning relative to discovery learning? *British Journal of Psychology*, **90**, 389–403.

Lee, T.D. and Genovese, E.D. (1988). Distribution of practice in motor skill acquisition: learning and performance effects reconsidered. *Research Quarterly for Exercise and Sport*, **59**, 277–87.

Lee, T.D., Chamberlin, C.J. and Hodges, N.J. (2001). Practice, in *Handbook of Sport Psychology*, 2nd edn (eds R.N. Singer, H.A. Hausenblas and C.M. Janelle), John Wiley and Sons, New York, pp. 115–43.

Lee, T.D., Magill, R.A. and Weeks, D.J. (1985). Influence of practice schedule on testing schema theory predictions in adults. *Journal of Motor Behavior*, **17**, 283–99.

Liu, J. and Wrisberg, C.A. (1997). The effect of knowledge of results delay and the subjective estimation of movement form on the acquisition and retention of a motor skill. *Research Quarterly for Exercise and Sport*, **68**, 145–51.

Magill, R.A. (2001a). *Motor Learning: Concepts and Applications*. McGraw-Hill, New York.

Magill, R.A. (2001b). Augmented feedback and skill acquisition, in *Handbook of Sport Psychology*, 2nd edn (eds R.N. Singer, H.A. Hausenblas and C.M. Janelle), John Wiley and Sons, New York, pp. 86–114.

Magill, R.A. and Schoenfelder-Zohdi, B. (1996). A visual model and knowledge of performance as sources of information for learning a rhythmic gymnastics skill. *International Journal of Sport Psychology*, **27**, 7–22.

Magill, R.A. and Wood, C.A. (1986). Knowledge of results precision as a learning variable in motor skill acquisition. *Research Quarterly for Exercise and Sport*, **57**, 170–3.

McCullagh, P. and Meyer, K.N. (1997). Learning versus correct models: influence of model type on the learning of a free-weight squat lift. *Research Quarterly for Exercise and Sport*, **68**, 56–61.

McCullagh, P. and Weiss, M.R. (2001). Modeling: considerations for motor skill performance and psychological responses, in *Handbook of Sport Psychology* 2nd edn (eds R.N. Singer, H.A. Hausenblas and C.M. Janelle), John Wiley and Sons, New York, pp. 205–38.

Newell, K.M. (1985). Coordination, control and skill, in *Differing Perspectives in Motor Learning, Memory and Control* (eds D. Goodman, R.B. Wilberg and I.M. Franks), Elsevier Science Publishing, Amsterdam, pp. 295–317.

Newell, K.M. (1996). Change in movement and skill: learning, retention and transfer, in *Dexterity and its Development* (eds M.L. Latash and M.T. Turvey), Lawrence Erlbaum Associates, Mahwah, NJ, pp. 393–430.

Pollock, B.J. and Lee, T.D. (1992). Effects of the model's skill level on observational motor learning. *Research Quarterly for Exercise and Sport*, **63**, 25–9.

Salmoni, A.W., Schmidt, R.A. and Walter, C.B. (1984). Knowledge of results and motor learning: a review and critical reappraisal. *Psychological Bulletin*, **95**, 355–86.

Schmidt, R.A. (1975). A schema theory of discrete motor skill learning. *Psychological Review*, **82**, 225–60.

Schmidt, R.A. and Lee, T.A. (1999). *Motor Control and Learning: A Behavioral Emphasis*. Human Kinetics, Champaign, IL.

Schmidt, R.A. and Wrisberg, C.A. (2000). *Motor Learning and Performance: From Principles to Practice*. Human Kinetics, Champaign, IL.

Scully, D.M. (1988). Visual perception of human movement: the use of demonstrations in teaching motor skills. *British Journal of Physical Education Research Supplement*, **4**, 12–14.

Scully, D.M. and Newell, K.M. (1985). Observational learning and the acquisition of motor skills: towards a visual perception perspective. *Journal of Human Movement Studies*, **11**, 169–86.

Swinnen, S.P. (1996). Information feedback for motor skill learning: a review, in *Advances in Motor Learning and Control* (ed. H.N. Zelaznik), Human Kinetics, Champaign, IL, pp. 37–66.

Thorpe, R. (1996). Telling people how to do things does not always help them learn. *Supercoach*, **8**, 7–8.

Van Rossum, J.H.A. (1990). Schmidt's schema theory: the empirical base of the variability of practice hypothesis. A critical analysis. *Human Movement Science*, **9**, 387–435.

Vereijken, B. and Whiting, H.T.A. (1990). In defence of discovery learning. *Canadian Journal of Sports Science*, **15**, 99–106.

Whiting, H.T.A. (1984). The concepts of adaptation and attunement in skill learning, in *Adaptive Control of Ill-defined Systems* (eds G. Selfridge, E.L. Risland and M.A. Arbib), Plenum, New York, pp. 187–205.

Wickens, C.D. (1989). Attention and skilled performance, in *Human Skills* (ed. D.H. Holding), John Wiley and Sons, Chichester, pp. 71–106.

Wightman, D.C. and Lintern, G. (1985). Part-task training for tracking and manual control. *Human Factors*, **27**, 267–83.

Williams, A.M. (1998). Practice, practice, practice. *Insight: the F.A. Coaches Association Journal*, **1**(4), 34–5.

Williams, A.M. and Grant, A. (1999). Training perceptual skill in sport. *International Journal of Sport Psychology*, **30**, 194–220.

Williams, A.M., Alty, P. and Lees, A. (2002). Effects of practice and knowledge of performance on the kinematics of ball kicking, in *Science and Football IV* (eds W. Spinks, T. Reilly and A. Murphy), Routledge, London, pp. 320–5.

Williams, A.M., Davids, K. and Williams, J.G. (1999). *Visual Perception and Action in Sport*, Routledge, London.

Wulf, G. (1991). The effect of type of practice on motor learning in children. *Applied Cognitive Psychology*, **5**, 123–34.

Wulf, G., Lauterbach, B. and Toole, T. (1999). Learning advantages of an external focus of attention in golf. *Research Quarterly for Exercise and Sport*, **70**, 120–6.

Wulf, G. and Weigelt, C. (1997). Instructions about physical principles in learning a complex motor skill: to tell or not to tell. *Research Quarterly for Exercise and Sport*, **68**, 362–7.

Yan, J.H., Thomas, J.R. and Thomas, K.T. (1998). Children's age moderates the effect of practice variability: a quantitative review. *Research Quarterly for Exercise and Sport*, **69**, 210–15.

15 Stress, performance and motivation theory

Martin Eubank and David Gilbourne

Introduction

The material in this chapter is designed around two contemporary blocks of theory. In section one, elements of the stress and performance literature are presented alongside contemporary perspectives offered by cognitive appraisal and coping theories. In section two, elements of achievement goal theory, cognitive appraisal and goal-setting are outlined from an integrated perspective. Finally, the chapter concludes with a number of practical suggestions that offer ideas for generating and maintaining an adaptive coping and motivational state.

When the many physical, psychological, environmental and other related demands are considered, it is not difficult to understand why competitive soccer places its participants under considerable stress. Furthermore, the prevalence and severity of such stressful encounters have been made greater by the ever-increasing profile of contemporary sport, and the importance, significance and value associated with success. Soccer is no exception, and today's players, coaches and managers are confronted with an array of demands in a number of different contexts. 'Facing up' to the demands of the game places a firm emphasis on an individual's capability to cope with them. Coping with stressful events is a natural and common occurrence, yet without this capability, the effect of stress on the emotional response and performance consequences for the player may be severe. Inability to cope with competitive demands and the psychological emotions that result are likely to inhibit (at best) and prevent (at worst) the success that the player is striving so hard to achieve. Stress is therefore a process that unfolds over time, a sequence of related events that lead to a particular end. As stress is a corollary of competitive soccer, the first part of this chapter explores these events, and examines the link between stress appraisals, emotional reactions and performance, and the concept of coping with stressful events in soccer.

The term 'motivation' is used in both everyday life and academic language. In the everyday context soccer players, coaches, managers and spectators may openly associate team or individual performance with different motivational states. So, depending on the level of performance, a player may be thought of as being high or low in motivation. The highly motivated player is frequently linked to a range of characteristics or attributes such as commitment, a willingness to work hard and a sense of dedication. In contrast, a player who is thought to lack motivation may appear to be lethargic and short on enthusiasm. In the context of soccer performance the term motivation may, therefore, be broadly associated with adaptive or maladaptive behaviour patterns. In a more academic context, sports psychologists may also be intrigued by the way a particular player is performing, but they are also interested in the *underlying mechanisms* that help to explain behaviour. To this end, various theoretical perspectives have been developed that help psychologists to understand motivation from this causal perspective. This work is considered in the latter half of the current chapter.

15.1   Psychological stress and performance

Stress is a process that is characterized by the interaction between individuals and their environment. Stress can therefore be defined as a substantial imbalance between demand and response capability, under conditions where failure to meet that demand has important consequences (Lazarus and Folkman, 1984; Lazarus, 1991). This process is illustrated in interactional models of the stress process that incorporate the link between competitive demands, their appraisal by the participant and the emotional responses and performance consequences that result (see Figure 15.1).

15.1.1   Understanding the stress process

Whilst the demands encountered in different environments may differ from one situation to the next, they represent an inevitable aspect of competitive soccer. If, for example, a list of the demands typically placed upon professional players was compiled, its length would be considerable! However, this stage of the process simply describes those demands that are present in the environment. The extent to which these competitive demands influence the emotional response is mediated by appraisal, as exposure to such demand is not considered stressful until the player makes a subjective interpretation of the situation. Cognitive appraisal is characterized by an evaluation of which factors are believed to hold personal significance (Lazarus and Folkman, 1984; Lazarus, 1991), and is arguably the most crucial stage of the entire stress process. The process of evaluation involves a primary appraisal of demand (i.e. what are the demands being placed upon me?) and a secondary appraisal of

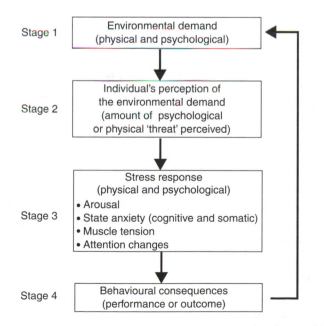

Figure 15.1 The stress process. (Adapted from Weinberg and Gould, 1999.)

Table 15.1 Major sources of stress (Adapted from Gould *et al.*, 1993a)

1 Physical demands on resources
   (e.g. the physical aspects of training and maintaining the required physical conditioning
   and health)
2 Psychological demands on resources
   (e.g. fear of failure, lack of confidence, pressure to perform well)
3 Environmental demands on resources
   (e.g. too much media exposure/hype, media criticism, excessive time demands)
4 Expectations and pressure to perform to a high standard
   (e.g. living up to expectations, fear of what others would think)
5 Significant other stressors
   (e.g. conflict with the coach, putting stress on family and friends)
6 Relationship issues
   (e.g. personal relationship problem, unfriendly competitor attitude, bad experience with sponsor)
7 Life direction concerns
   (e.g. figuring out the time to retire, post-sport career options)

resource (i.e. do I have the available resources to cope with them?), which implicates the 'imbalance' referred to within the definition of stress. The ability of the sporting environment to place individuals under considerable amounts of stress stems from the fact that primary appraised demands often outweigh secondary appraised resources.

Previous researchers (e.g. Scanlan *et al.*, 1991; Gould *et al.*, 1993a) have investigated the sources of stress identified by elite athletes. There are some consistent stress dimensions across a range of sports (Table 15.1).

Close examination of these dimensions also reveals that stress comes from a mixture of sport and life-specific contexts, which often interact to contribute to the stressfulness of the event.

15.2   Emotion and performance relationships

Competition emotions, and their relationship to performance, represent one of the major research avenues within sport psychology. Historically, anxiety has been considered to be one of, if not the major emotion in sport, and has been the most widely researched of the existing emotions. Anxiety is a multi-dimensional concept (Martens *et al.*, 1990), characterized as a negative psychological state that has a cognitive (mental) and a somatic (physiological) component which respond differently to the stressors within the environment. Based on this framework, different theoretical explanations for the anxiety–performance relationship have been proposed, with arguably the most viable of these being 'Catastrophe theory' (Hardy *et al.*, 1994; Hardy, 1996). We can all think of players who have been performing to the best of their ability, and then, for no apparent reason, exhibit a sudden and dramatic performance 'slump'. This slump can often be so severe that it takes considerable time for the individual to recover, if at all. This scenario is best described as a performance catastrophe, and it is from this that 'Catastrophe theory' takes its name.

15.2.1   *Performance catastrophes*

The catastrophe model is based on the premise that analysis of the anxiety concept should occur in three dimensions, and suggests that performance is related to the interaction of cognitive anxiety and physiological arousal. This interaction is likely to affect performance in

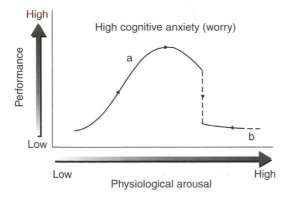

Figure 15.2 Catastrophe theory. (Adapted from Weinberg and Gould, 1999.)

some way, but the specific nature of the effect depends largely on the intensity of cognitive anxiety and physiological arousal at specific moments in time (see Figure 15.2).

A 'fine line' exists between the conditions required for optimal performance and those that will initiate a performance catastrophe. In the conditions of high cognitive anxiety that typically occur in soccer, increases in physiological arousal can have positive effects on performance. Once the optimal performance point is reached, a further increase in physiological arousal beyond the optimal 'threshold' is associated with a catastrophic drop in performance. The possibility of a catastrophe occurring is a frightening prospect, so understanding the conditions in which catastrophes happen may equip players to prevent its destructive effects.

15.2.2   Contemporary performance approaches

Significant shifts have recently occurred in relation to the conceptualization and measurement of competitive emotion and performance. In the case of performance, researchers have been criticized for continuing to adopt a 'narrowbanded' approach, examining the relationship of multiple anxiety variables to one global outcome measure of performance (e.g. game score, event time) rather than elements of the performance process. Outcome oriented anxiety–performance research is somewhat limited in its ability to facilitate understanding of the anxiety–performance relationship, as the mechanisms upon which anxiety exerts its effect have not being directly measured. It is therefore important to focus on the effect that anxiety has on some of the processes that constitute performance, rather than on the performance outcome itself. Recent research (e.g. Parfitt and Hardy, 1993; Parfitt *et al.*, 1995; Parfitt and Pates, 1999; Eubank *et al.*, 2000, 2002) has focused on the effect that anxiety has on task-relevant sub-components of performance. This includes those relating to perception (e.g. attentional narrowing and selectivity), working memory and processing bias. Anxiety has significant effects on these cognitive sub-components, illustrating the importance of empirical investigation into the mechanisms through which anxiety exerts its influence.

15.2.3   Contemporary emotion approaches

Anxiety has been viewed as the dominant emotion in sport, with research being based on the traditional assumption that anxiety is indicative of debilitative cognition that has negative effects on performance. Recent research (e.g. Jones *et al.*, 1994; Swain and Jones, 1996; Eubank and Collins, 2000) has shown such an assumption to be misleading, and contends that individual differences in anxiety states are not always reflected by differences in the intensity of response. In short, the positive and negative nature of cognition implicates the player's interpretation of anxiety, and not the intensity of the response to be a more meaningful measure in determining the effects of anxiety on performance. The importance of interpreting anxiety has a number of significant implications, not least the notion that anxiety can be 'treated' by restructuring rather than reducing it, and that such alterations in anxiety interpretation have a facilitative effect on self-confidence and performance (Hanton and Jones, 1999). In essence, 'the art is not to get rid of butterflies but to get them to fly in formation'. The control-process model of anxiety (Jones, 1995) provides a useful framework to address how such interpretations may be regulated. The model suggests that anxiety interpretation is dictated by expectancies of athletes that relate to their perceived ability to cope and attain goals, with the positive or negative nature of these expectancies being influenced by control. Perceptions of high control are accompanied by a positive expectancy to cope with the demands of the situation and attain the goals that have been set, which results in a facilitative interpretation of anxiety symptoms. Conversely, low perceived control is associated with negative expectations and debilitative symptom interpretation. Hence, the concepts of coping (Eubauk and Collins, 2000) and goal attainment (Jones and Hanton, 1996) are major mediators within the control process, and act as the mechanisms linking perceptions of control to facilitative interpretations of anxiety.

Despite these significant progressions, there is debate concerning the labelling of anxiety as a positive emotion. Do researchers accept that a 'positive anxiety' state for performance exists (Hardy, 1998) or should researchers reject this in favour of another affective state previously mis-labelled as anxiety? Lazarus (1991) and Burton and Naylor (1997) claimed that anxiety does not facilitate performance, as the current measurement of anxiety as facilitative or debilitative confounds anxiety (a negative emotion) with other positive emotions. The emotions of, for example, challenge, excitement and confidence, are, as Lazarus has suggested, positive in nature, yet have been mis-labelled as facilitative anxiety. This may contradict the use of a facilitative–debilitative classification of anxiety that attempts to combine positive and negative emotions that are actually discrete. It is not the facilitative–debilitative distinction that is the issue, but rather that such a distinction is not relevant for all emotions. Examples of 'facilitators' being excited and/or motivated but also anxious appear to exist (Jones and Hanton, 2001). The competitive psychological state is a highly complex simultaneous experience of positive and negative feelings.

As there are many other pleasant and unpleasant emotions that play an important role in performance, a number of researchers (Kerr, 1997; Hanin, 2000; Jackson, 2000) have questioned why sport psychology is so preoccupied with anxiety, as there is little justification for restricting investigations in this way. The current consensus would appear to be that whilst anxiety is an important ingredient of psychological state, it may not be the central variable that researchers assume it to be. The examination of the anxiety response alone cannot fully explain the affective experience of players, emphasizing a need to address the performer's own idiosyncratic labelling of the experienced symptomatology. This issue reflects a distinct shift away from anxiety and towards emotion (Cerin *et al.*, 2000; Hanin, 2000; Lazarus, 2000), where contemporary models such as 'Individualized Zones of Optimal Functioning' (Hanin and Syrja, 1995) and 'Flow State' (Jackson, 1996) may have much to offer. This change of direction represents a need to understand players' specific emotional experiences and their effect on performance. The challenge for researchers is to explore the complex emotional states that reflect the idiosyncratic emotional experience and vocabulary of the player, examine the sets of basic emotions experienced throughout competition and focus on individual differences and factors determining them.

15.3   Coping with stress and anxiety

According to Anshel *et al.* (2001), stress can be appraised as harmful, threatening or challenging, which influences the perceived intensity and importance of the stressor and therefore the choice of coping style. Coping is 'a conscious process of constantly changing cognitive and behavioural efforts to manage specific external and/or internal demands or conflicts appraised as taxing or exceeding one's resources' (Lazarus and Folkman, 1984). Such a process involves a particular coping style, which is reflected by a dispositional tendency to use certain types of coping strategies across time (Anshel *et al.*, 2001). These styles have a primary focus (e.g. problem and emotion focused, approach and avoidance focused), and can take various forms (e.g. cognitive and behavioural, engagement and disengagement, adaptive and maladaptive). The coping strategy that is then employed is related to the use of one or more cognitive or behavioural attempts at reducing perceived stress intensity. There is debate as to the 'compatibility' of problem and emotion (Lazarus and Folkman, 1984) versus approach and avoidance (Anshel, 1996) labels to characterize coping style. In a review of coping with stressful events in sports, Anshel *et al.* (2001) attempted to simplify this focus and form by arguing that there are four basic coping styles: (1) Approach Behavioural (actions to try to resolve a stressful situation); (2) Approach Cognitive (thoughts to improve emotional status and reduce stress intensity); (3) Avoidance Behavioural (actions to physically remove the athletes from the stress source that can be either adaptive (functional) or maladaptive (dysfunctional)); and (4) Avoidance Cognitive (thoughts to mentally remove the athletes from the stress source). Anshel *et al.* (2001) also argued that the use of such styles may be influenced by a number of factors that include time, controllability and confidence. When the player is confident and there is time to address a stress source that is controllable, approach strategies are appropriate and are likely to achieve best results. Conversely, when the player's confidence is low and there is insufficient time to address a stress source that is uncontrollable, avoidance strategies may be more appropriate and productive.

An understanding of a dynamic approach to coping suggests that individuals have a repertoire of options available to them from which they can build what they believe to be the most effective strategy depending upon the nature of the situation. Gould *et al.* (1993a,b) found coping strategies of elite athletes to be represented by 'thought control', 'task focused', 'behavioural' and 'emotional control' strategies that were used to deal with particular stressors. These findings reflect a dynamic and complex coping process that uses both problem- and emotion-focused strategies, often in combination, rather than being limited to a single strategy approach (see Table 15.2).

The acquisition and integration of different coping styles and strategies may also reflect the facilitative and debilitative interpretation of anxiety symptoms (Eubank *et al.*, 2000). From a strategic perspective, the positive expectancy of being able to cope is evidenced by the greater number of familiar and comfortable coping resources possessed by the facilitator. In contrast, debilitators run into coping-related problems because their coping resources are relatively limited. Hence, when in highly demanding environments, they are often at a loss as to how they should cope with the situation, and the coping strategies that they do possess often become unworkable.

Table 15.2 Most frequently cited coping strategies associated with specific stress sources (Adapted from Gould *et al.*, 1993c)

Stress source	Coping strategy general dimension
Physical demands	Rational thinking and self-talk Pre-competition mental preparation and anxiety Management Changing to healthy eating attitudes/ behaviour Training hard and smart
Psychological demands	Pre-competition mental preparation and anxiety Management Positive focus and orientation Training hard and smart
Environmental demands	Time management and prioritization Isolation and deflection
Expectations and pressure to perform to a high standard	Positive focus and orientation Training hard and smart Rational thinking and self-talk Social support Pre-competition mental preparation and anxiety Management
Relationship issues	Positive focus and orientation Social support Striving for positive working relationship Isolation and deflection Rational thinking and self-talk
Life-direction concerns	Time management and prioritization Rational thinking and self-talk
Uncategorized stress sources	Reactive behaviours Social support Isolation and deflection

15.3.1 A strategic approach to coping

Clearly, a broad spectrum of problem and emotion-focused coping strategies for use in different situations to control different sources of stress is required. Such an individual-specific 'goodness of fit' between primary appraised demands and secondary appraised coping resources facilitates the selection of an appropriate coping strategy for every situation encountered. This issue illustrates the crucial importance of identifying the specific sources of the stress experienced. In terms of approach coping, Anshel *et al.* (2001) suggested that players should initially engage in coping attempts that prevent stress appraisals by subjectively interpreting objectively stressful events as benign or irrelevant. This strategy requires considerable levels of mental toughness. If stress is appraised, then a challenge interpretation should be adopted, whereby players deal with, learn, or benefit from the stressor and take the opportunity to attain growth and demonstrate mastery for mental gain and profit. Such an appraisal is compatible with an approach-coping style. Finally, it is accepted that appraisal of stress as having the potential for harm, loss and/or threat will happen and can be appropriate. The use of personal resources (mental skills) to practice meeting and overcoming threatening events can enable future psychological and behavioural adjustments to take place, so that if and when the stressor re-occurs, equivalent potential for harm, loss and/or threat is removed. Eventually, the stressor could be re-appraised as a challenge and ultimately be prevented.

15.4 Motivation and soccer

As with the stress and performance literature there are many conceptual contributions to the broad topic of motivation. Indeed, motivation is best viewed as an umbrella term under which many different theoretical positions exist. One of these, 'achievement goal theory', has been selected to form the primary conceptual base for discussion alongside the concepts of cognitive appraisal and self-confidence. Although these three motivational constructs are often presented separately in the literature, in the present chapter they will be viewed from an integrated perspective.

Achievement goal theory is a social–cognitive approach to achievement motivation (Newton and Duda, 1999). Studies in this area emerged initially from school-based research and sport psychologists have since utilized achievement goal theories to help them understand and explain behaviour within sports settings. Biddle (1999) argued that the motivation-based research has been driven, in the main, by achievement goal theory. This block of literature has focused primarily on the way individuals define success and can be divided into dispositional or situational-based inquiry. Achievement 'dispositions' are discussed through the terms task and ego orientation. Task-oriented individuals are thought to define success through the application of 'self-referenced' criteria such as personal improvement; as an example, soccer players might reflect on how their capacity to meet the physical demands of the game has improved over recent months. In contrast, an ego orientation is associated with a more interpersonal focus. So, players may judge how good or bad they are by referring to the performance of others (Newton and Duda, 1999). In this case a desire to win or outperform relevant others is the benchmark of success. For example, a player might contrast his/her physical improvements in contrast to peers in the same team or league (Duda, 1987). More generally, task-orientation and success in sport have been correlated with the endorsement of effort and persistence, whilst an ego orientation is associated more with success being driven by 'chance' factors and 'social approval' (Lochbaum and Roberts, 1993).

In contrast, literature which is more focused on the achievement qualities inherent in the situation emphasizes the importance of an individual's perception of what is termed the 'motivational climate' (Ames, 1992). The concept of a motivational climate, in a definitional sense, mirrors the dispositional material as situations are also thought to contain task or ego-type cues. Apart from the fact that the term 'mastery' is used to depict what dispositionally we understand as 'task', the core themes (and indeed findings) are similar. The separation between literature that has a dispositional or situational focus can sometimes obscure the view that situational cues and dispositional potential interact to mediate the 'goal-involvement' states experienced by players. Goal-involvement has been described by Treasure *et al.* (2001) as an 'interpretive lens' through which performance information is understood. It follows that this experience is qualitatively different when task or ego states are operational.

Over the years, dispositional research has linked task orientation with positive motivational states, such as the demonstration of persistence (Duda, 1988, 1989) and effort (Duda, 1988). Conversely, an ego orientation (particularly in combination with perceptions of low confidence or low actual ability) is associated with lower levels of persistence and a lack of effort (Duda, 1992). Biddle (1999) argued that dispositional research consistently demonstrates that high task orientation (either on its own or) in combination with high ego orientation is a motivationally positive profile. This view has been underscored by Georgiadis *et al.* (2001). They noted that a high task, high ego profile was associated with both intra-personal and normative thinking. They argued that this mind set was the most motivationally adaptive for sport participation. The notion of adaptation can be linked to the potential for any one's cognitive state to be flexible. As noted earlier, this fluctuation is thought to depend on the perception of momentary situational cues and the dispositional tendencies of the individual (Treasure *et al.*, 2001). It has already been suggested that the adaptive nature of ego-involvement is largely dependent upon an individual's perception of his or her own ability. Put simply, interpersonal comparison is not thought to be problematical when capacity to perform at the required level is perceived to be high (Hardy, 1997). The logic here is that players, who believe they possess a high ability to meet the demands of the task in maybe a physical, tactical and technical sense, are more likely to experience success (given they are indeed in possession of the required skills) than players who possesses a perception of a low ability. Consequently, interpersonal feedback from players with high ability should be more agreeable and so levels of perceived self-confidence are more easily sustained.

Although achievement goal theory has established that individuals are likely to demonstrate *both* task and ego tendencies (Georgiadis *et al.*, 2001), research over the years has tended to focus on populations who are either highly task- or highly ego-oriented. For example, researchers have explored links between goal-dispositions and sportsmanship attitudes (Duda *et al.*, 1991) and the relationship between dispositional factors and perceptions of success criteria in significant others (Escarti *et al.*, 1999). These issues draw attention to the notion that goal-orientations, as well as influencing specific behaviours in specific situations, are also expected to relate to a coherent set of world views (Treasure *et al.*, 2001). In this context, Kavussanu and Roberts (2001) considered 'moral functioning' (such as unsportsmanlike behaviour and judgments about the legitimacy of injurious acts) within college basketball. They noted that, compared to the female participants, male participants reported higher level of ego orientation, lower task orientation, lower levels of 'moral functioning', demonstrated greater approval of 'unsportsmanlike' behaviour, and were more likely to see injurious acts as legitimate. The general tone of these findings is repeated across the achievement goal literature with highly ego oriented individuals reporting more hierarchical and cynical world views than those held by their high task counterparts. Little is indicated in the literature concerning 'how' these (often) incompatible world views are simultaneously accommodated by someone with a high task, high ego profile.

15.4.1   Critiques of achievement goal theory: linking goal-involvement to goal-setting

Harwood *et al.* (2000, 2001) proposed a third 'goal-dimension' which merits consideration. They argue for task and ego involvement states to be separated into a tripartite formula which includes task involved *process*, task involved *product* and ego involvement. Task involved process is an intra-individual state that relates to trying hard and mastering a task through the execution of effort. Task involved product is also intra-individual in focus and is associated with trying to surpass previous performance levels. Finally, ego orientation is a normative cognition and a full explanation of this involvement state has been outlined earlier.

The tripartite aspect of the critique has been selected due to the applied nature of the proposition and the potential links between the three goal dimensions and elements of the goal-setting literature. There are conceptual distinctions and points of synthesis existing between achievement goal involvement states and goal-setting processes. Involvement states are understood to represent a particular *cognitive focus* arrived at through the appraisal-led interplay between dispositional potential and the antecedent properties that exist within the environment. As noted earlier, achievement goal theorists refer to the environmental element in the above equation as the motivational climate. In contrast, goal-setting is generally seen to be an applied 'technique' (Gilbourne, 1996), which *also* helps to *focus cognitions* in a particular direction (Weinberg, 1994; Gilbourne and Taylor, 1998). Goals, in this context, may be linked to an individual's aspirations to achieve a particular outcome in either the short, medium or long term. From a research perspective, the goal-setting literature offers numerous examples of people setting *their own* goals in a spontaneous way. In such cases the nature of the goals that individuals set may be reasonably linked to aspects of their achievement goal-profile and also to the antecedents present in the environment (Weinberg, 1994). The saliency of a particular goal state (be it depicted as a broad goal involvement state or a focus on a specific goal-setting strategy) may also change. In that sense, cognition is viewed as dynamic rather than fixed (Lazarus, 2000). For example, players' cognitive focus can be expected to alter as they evaluate the impact that their behaviour is having on the situation.

When the language and concepts that surround goal involvement states and goal-setting are brought together they suggest a certain interdependence. For example, if we use the tripartite terminology suggested by Harwood *et al.* (2001), an individual who is presently task-involved (process) may be expected to deploy goal-setting strategies that are broadly classifiable as being 'process goal' in nature (i.e. intra-individual, based on sustaining attitudinal or technical quality). Similarly, someone who is task-involved (product) may tap into performance-type goals (objectives that relate to prior intra-individual standards). Ego involvement could be intuitively, but less convincingly, linked to outcome goals (such as winning). Suggesting associations between goal involvement states and goal-setting processes can create controversy. Whilst the conceptual context of these criticisms are wide ranging, independent observers may conclude that Duda (1997) and Treasure *et al.* (2001) see little connection between goal-involvement states and goal-setting cognitions. In the discussion to follow, a more inclusive perspective on this topic is sketched out.

Cognitive appraisal processes suggest that thinking states, feeling states and acting states are cyclical, dynamic and interdependent (Lazarus, 2000). Our inclusive agenda are based on the notion that the appraisal mechanism accommodates achievement goal concepts and goal-setting processes. For example, it seems reasonable to assume that particular goal-setting strategies can influence the tone of an achievement context and so encourage players to engage specific goal-achievement states. Indeed, this point is made unequivocally in the writing of Ames (1992). She noted that a mastery climate would need to house antecedents that encouraged task involvement. To this end, Ames (1992) outlined practical guidelines which aim to bring a task climate into being and, in doing so, overtly refers to task 'friendly' goal-setting processes such as establishing personal targets, recognizing effort and rewarding personal development. In contrast, it is not unreasonable to propose that goals which emphasize winning (outcome goals) would contribute towards interpersonal cognitions, engender an ego oriented motivational climate and so prompt ego involvement.

Of course, these strategies are not the only signature of a particular goal-involvement state (Duda, 1997). As an example, the grouping (organization of a motivational climate) and cooperation within and across groups (the behaviour demonstrated within the situation) reflects the intention of the organizers and the willingness of the participants to adhere to a particular goal involvement state. Duda (1997) observed that the creation of a motivational climate through the 'manipulation of goals' is only one strategy amongst many. Nevertheless, the direction and tone of any achievement setting can be influenced by the highlighting of different goal attainment processes and Ames (1992) suggested a pivotal role for goal-setting in that regard.

If the cognitive linkage between goal involvement and goal-setting is viewed from the opposite perspective, then it also seems plausible that goal-setting strategies may be used by players as a function of their particular goal-involvement state. In this view, goal-setting frameworks may not be the antecedent mechanism that leads to a particular goal involvement state, they could rather be strategies that *result from* a player's goal involvement. In this instance, a player who is 'task involved *product*' may formulate performance type goals. In contrast, at any one moment a player may be ego involved and so interpersonally engaged. Given this latter state of cognition the player may formulate outcome type goals and adopt behaviours that seek to satisfy that perspective.

Although conceptually helpful, it is artificial to segregate possible cause and effect relationships between goal-involvement and goal-setting strategies. When a cognitive appraisal model of cognition, affect and behaviour is 'layered' onto the goal-involvement and goal-setting literature, both goal-involvement and goal-setting led conditions could be seen to be activated *differentially* as cycles of cognition spiral on. These spirals of cognition could also be viewed as short term, such as within the time frame of a specific competition, or long term, say over a competitive season. So, as events (antecedents) in the situation unfold, a player may (in a volitional or non-volitional manner) switch goal involvement.

As goal-involvement states are likely to change, it seems logical to suppose that a particular goal involvement may be not always helpful or unhelpful to performance. The key in this dynamic is the self-confidence of the player (an issue discussed earlier) and the nature of the feedback. Hardy (1997) noted that ego involvement, when perceived ability is high and feedback is positive, can be productive. Harwood *et al.* (2000) also referred to this phenomenon and the following example lends from their tripartite-based thinking. Let us consider the cognitions deployed by a player who is failing to manage the demands of a competitive season. An ego-involved state would provide negative information as the failing player would perceive himself or herself as being outperformed by others. It may also be unproductive to deploy task oriented product cognitions as the player may also sense that present performance falls below those enjoyed in the past (again feedback is negative). Finally, even task process cognitions may contribute to a sense of failure. For example, if effort has been maintained and adaptive approaches to mastery have been consistently deployed and *still* task product and ego-based cognitions show no progress is being made, then (as time progresses) a player is likely to become aware that the efforts are not bearing fruit. This final phase is likely to be more longitudinal in nature. Overall and over time these cognitions are likely to reduce self-confidence. The task process element of the scenario is unlikely to be accepted so readily by Treasure *et al.* (2001) who claimed that task-type cognitions act to buffer negative feedback. Appraisal theory would lead many to agree that, in the short term, this may be so. The cyclic nature of appraisal, however, would also lead some to question about how long a full commitment to effort can be maintained in the face of little progress.

It is also possible to sketch out a more positive scenario. In a review of a disappointing season a male player may feel that compared to his team-mates he lacks strength (an ego-involved cognition). This may encourage him to seek guidance from fitness trainers and nutritionalists who encourage him to follow a strict dietary and training regimen over the closed season. This process may be structured around task product and task process strategies (i.e. a personal goal programme). As the player works hard at his new training (task process), he begins to see progress as body fat levels decrease and weight-training loads increase (task product). These results further reinforce task process cognitions which in turn help to ensure that positive task product information is sustained. When all players return from the close season the situational antecedents include much more opportunity for the player to utilize interpersonal information. When the player subsequently engages in ego-involved cognitions, the feedback is much more positive as his increased strength allows comparisons with his peers to be more favourable. The gross impact of the above set of actions and cognitions is likely to be increased self-confidence.

The links between achievement goal theory and the goal-setting literature draw from both established and contemporary literature. The conceptually integrated approach outlined here is viewed as one way through which greater understanding of motivational processes may be achieved.

Finally, the achievement-goal literature does suggest a number of practical ideas for practitioners who wish to establish task motivational climates (Ames, 1992). Some of these have been sketched out earlier and further ideas are outlined in the practical section to follow.

15.5   Practical suggestions

The material presented in the first section of this chapter documents the importance of stress factors to soccer performance, the inter-relationships between appraisal, emotion and coping being emphasized. Within this conceptual equation, coping would appear to be the critical component as all emotional states have the potential to be adaptive when the ability of the player to cope is high. Adopting strategies that help to develop coping skills in players is a core consideration. Stress management techniques must help players to create an optimal state for every situation they encounter. The decision as to which technique to implement in specific situations is therefore crucial. The practical suggestions that follow encourage coping to be developed through a number of stress management skills:

- Players should engage in coping attempts that prevent stress appraisals by subjectively interpreting objectively stressful events as benign or irrelevant. For example, the opposition could be interpreted as threatening and cause high levels of anxiety, but players should be encouraged to ignore or accept the opposition as something that is outside of their control.
- Players should engage in coping attempts that turn potential threats into an actual challenge. For example, the importance of the game could be interpreted as threatening and cause negative emotions, but players should be encouraged to deal with such stressors by re-interpreting them, thus taking the opportunity to attain mental growth and demonstrate mastery.
- Players should be encouraged to develop robust mental skills to meet and overcome threatening events in order to eliminate potential for psychological harm and performance loss. For example, the need to live up to the expectations of others could be interpreted as threatening and cause anxiety. The player needs mental skills that can be matched to the form of anxiety being experienced.
- If the player needs to control cognitive anxiety, a cognitive treatment such as meditation may be selected. This technique focuses on breathing and uses a 'mantra' (i.e. key-word) to maintain/regain composure and concentration.

- If the experience of anxiety is somatic, a somatic treatment such as progressive neuromuscular relaxation (PNR) may be adopted. This technique also has a breathing component, and involves tensing and relaxing specific muscle groups to develop the player's awareness of the difference between tension and relaxation in the muscles.
- These same techniques could be used in different ways to induce appropriate emotional states, which serve to 'psyche-up' the player and instill a readiness to perform.
- When a player's anxiety response has both a cognitive and somatic component, a multi-modal package of techniques that allows the player to deal with the two types of anxiety simultaneously needs to be utilized.

The material presented in the second half of the chapter stresses the importance of motivational factors to soccer performance, where the inter-relationships between self-confidence, goal-involvement and on-going cognitive appraisal have been emphasized. Within this conceptual equation, self-confidence would appear to be the critical component as all goal-involvement states have the potential to be adaptive when self-confidence is high. As a consequence, adopting strategies that help to develop self-confidence in players is important. The practical suggestions that follow encourage self-confidence to be developed through goal-setting processes. These ideas also encourage goal-setting to be viewed as an antecedent mechanism which encourages particular goal-involvement states to be active:

- Encourage players to become active agents in their own development so that they establish goals which are personally relevant, challenging but attainable.
- Within any goal-setting structure stress the importance of effort and ensure that personal progress is given appropriate recognition.
- Allow the overall goal-setting structure to embrace a mix of intra-personal and inter-personal or outcome type goals.
- Always link process goals to the attainment of outcome goals. For example, a team may want to win (outcome) so the actions that can make this happen (process) need to be emphasized. Similarly, a player who aspires to play in the first team (outcome) will need to focus on the elements of his/her game which presently prevent selection (process).
- Encourage players to understand the costs and benefits of engaging (volitionally or non-volitionally) in different goal-involvement states. For example, if an ego state consistently leads to negative emotions, then it might be beneficial for this cognition to be replaced by a more intra-personal focus.

Summary

This chapter has considered elements of the stress and performance and achievement goal literature and has sought to review these varied conceptual positions from an integrated perspective. In this regard, psychological concepts have been presented as interrelated rather than independent phenomena, which all draw from recent research and conceptual critique. For example, the emerging debate on emotional diversity and the dynamic impact of appraisal on the deployment of coping mechanisms are central to psychological stress and performance in soccer. The cyclical interactions between stress appraisals, emotional reactions and soccer performance and the concept of coping with stressful events in sport operate continuously to make the stress process a transient one, where the way in which individuals appraise, respond and behave in response to environmental stressors is constantly changing. Appraisal is presented as the catalyst for this change, as without changes in appraisal, our responses and behaviours to the situations we encounter would always be the same. Whilst the basic stages of the stress process are well established and understood the concepts that relate to them are still evolving. In a similar manner, there are new lines of thinking within the motivation literature. It is necessary to investigate the diversity of these contemporary views (in conceptual and applied terms). These conceptual discussions have implications for those who participate in soccer. With this in mind, a list of practical ideas, outlined to help soccer coaches and players deal with the demands of the game have been presented.

References

Ames, C. (1992). Achievement goals, motivational climate and motivational processes, in *Motivation in Sport and Exercise* (ed. G.C. Roberts), Human Kinetics, Champaign, IL, pp. 161–76.

Anshel, M.H. (1996). Examining coping style in sport. *Journal of Social Psychology*, **136**, 311–23.

Anshel, M.H., Kim, K.W., Kim, B.H., Chang, K.J. and Eom, H.J. (2001). A model for coping with successful events in sport: theory, application and future directions. *International Journal of Sport Psychology*, **32**, 43–75.

Biddle, S.J.H. (1999). Motivation and perceptions of control: tracing its development and plotting its future in exercise and sport psychology. *Journal of Sport and Exercise Psychology*, **21**, 1–23.

Burton, D. and Naylor, S. (1997). Is anxiety really facilitative? Reaction to the myth that cognitive anxiety always impairs sport performance. *Journal of Applied Sport Psychology*, **9**, 295–302.

Cerin, E., Szabo, A., Hunt, N. and Williams, C. (2000). Temporal patterning of competitive emotions: a critical review. *Journal of Sports Sciences*, **18**, 605–26.

Duda, J.L. (1987). Towards a developmental theory of children's motivation. *Journal of Sport Psychology*, **9**, 130–45.

Duda, J.L. (1988). The relationship between goal perspectives and persistence and intensity among recreational sport participants. *Leisure Sciences*, **10**, 95–106.

Duda, J.L. (1989). Relationship between task and ego orientation and the perceived purpose of sport among high school athletes. *Journal of Sport and Exercise Science*, **11**, 318–35.

Duda, J.L. (1992). Motivation in sport settings: a goal perspective analysis, in *Motivation in Sport Settings* (ed. G. Roberts), Human Kinetics, Champaign, IL, pp. 57–92.

Duda, J.L. (1997). Perpetuating myths: a response to Hardy's Coleman Griffiths Address. *Journal of Applied Sport Psychology*, **9**, 303–9.

Duda, J.L., Olson, L.K. and Templin, T.J. (1991). The relationship of task and ego orientation to sportsmanship attitudes and the perceived legitimacy of injurious acts. *Research Quarterly for Exercise and Sport*, **62**, 79–87.

Escarti, A., Roberts, G.C., Cervello, E.M. and Guzman, J.F. (1999). Adolescent goal orientation and the perceptions of criteria of success used by significant others. *International Journal of Sport Psychology*, **30**, 309–24.

Eubank, M.R. and Collins, D.J. (2000). Coping with pre-and in-event fluctuations in competitive state anxiety: a longitudinal approach. *Journal of Sports Sciences*, **18**, 121–31.

Eubank, M.R., Collins, D.J. and Smith, N.C. (2000). The influence of anxiety direction on processing bias. *Journal of Sport and Exercise Psychology*, **22**, 292–306.

Eubank, M.R., Collins, D.J. and Smith, N.C. (2002). Anxiety and ambiguity: it's all open to interpretation. *Journal of Sport and Exercise Psychology*, **24**, 239–53.

Gilbourne, D. (1996). Goal setting during injury rehabilitation, in *Science and Soccer* (ed. T. Reilly), E & FN Spon, London, pp. 185–200.

Gilbourne, D. and Taylor, A.H. (1998). From theory to practice. The integration of perspective theory and life development approaches within an injury specific setting. *Journal of Applied Sport Psychology*, **10**, 124–39.

Georgiadis, M., Biddle, S.J.H. and Auweele, Y.V. (2001). Cognitive, emotional and behavioural connotations of task and ego goal orientation profile: an ideographic approach using hierarchical class analysis. *International Journal of Sport Psychology*, **32**, 1–20.

Gould, D., Jackson, S.A. and Finch, L.M. (1993a). Sources of stress in national champion figure skaters. *Journal of Sport and Exercise Psychology*, **15**, 134–59.

Gould, D., Eklund, R.C. and Jackson, S.A. (1993b). Coping strategies used by U.S. Olympic wrestlers. *Research Quarterly for Exercise and Sport*, **64**, 83–93.

Gould, D., Finch, L.M. and Jackson, S.A. (1993c). Coping strategies used by national champion figure skaters. *Research Quarterly for Exercise and Sport*, **64**, 453–68.

Hanin, Y.L. (2000). Successful and poor performance and emotions, in *Emotions in Sport* (ed. Y.L. Hanin), Human Kinetics, Champaign, IL, pp. 157–87.

Hanin, Y. and Syrja, P. (1995). Performance affect in junior ice hockey players: an application of the individual zones of optimal functioning model. *The Sport Psychologist*, **9**, 169–87.

Hanton, S. and Jones, G. (1999). The effects of a multimodal intervention program on performers: II. Training the butterflies fly in formation. *The Sport Psychologist*, **13**, 22–41.

Hardy, L. (1996). Testing the predictions of the cusp catastrophe model of anxiety and performance. *The Sport Psychologist*, **10**, 140–56.

Hardy, L. (1997). Three myths about applied consultancy work. *Journal of Applied Sport Psychology*, **9**, 277–94.

Hardy, L. (1998). Response to the reactants on three myths about applied consultancy work. *Journal of Applied Sport Psychology*, **10**, 212–19.

Hardy, L., Parfitt C.G. and Pates, J. (1994). Performance catastrophes in sport: a test of the hysteresis hypothesis. *Journal of Sports Sciences*, **12**, 327–34.

Harwood, C., Hardy, L. and Swain, A. (2000). Achievement goals in sport: a critique of conceptual and measurement issues. *Journal of Sport and Exercise Psychology*, **22**, 235–55.

Harwood, C., Hardy, L. and Swain, A. (2001). Persistence and effort in moving achievement goal research forward: a response to Treasure and colleagues. *Journal of Sport and Exercise Psychology*, **23**, 330–46.

Jackson, S.A. (1996). Toward a conceptual understanding of the flow experience in elite athletes. *Research Quarterly for Exercise and Sport*, **67**, 76–90.

Jackson, S.A. (2000). Joy, fun and flow state in sport, in *Emotions in Sport* (ed. Y. Hanin), Human Kinetics, Champaign, IL, pp. 135–55.

Jones, J.G. (1995). More than just a game: research developments and issues in competitive anxiety in sport. *British Journal of Psychology*, **86**, 449–78.

Jones, J.G. and Hanton, S. (1996). Interpretation of competitive anxiety symptoms and goal attainment expectancies. *Journal of Sport and Exercise Psychology*, **18**, 144–57.

Jones, J.G. and Hanton, S. (2001). Pre-competitive feelings states and directional anxiety interpretations. *Journal of Sports Sciences*, **19**, 385–95.

Jones, J.G., Hanton, S. and Swain, A.B.J. (1994). Intensity and interpretation of anxiety symptoms in elite and non-elite sports performers. *Personality and Individual Differences*, **17**, 657–63.

Kavussanu, M. and Roberts, G.C. (2001). Moral functioning in sport. An achievement goal perspective. *Journal of Sport and Exercise Psychology*, **23**, 37–54.

Kerr, J.H. (1997). Up for the match? Experiencing arousal and emotion in sport, in *Motivation and Emotion in Sport* (ed. J.H. Kerr), Psychology Press, Sussex, pp. 89–113.

Lazarus, R. (2000). How emotions influence performance in competitive sports. *The Sport Psychologist*, **14**, 219–29.

Lazarus, R.S. (1991). *Emotion and Adaptation*. Oxford University Press, New York.

Lazarus, R.S. and Folkman, S. (1984). *Stress, Appraisal and Coping*. Springer Publishing Company, New York.

Lochbaum, M.R. and Roberts, G.C. (1993). Goal orientations and perceptions of the sport experience. *Journal of Sport and Exercise Psychology*, **15**, 160–71.

Martens, R., Vealey, R.S. and Burton, D. (1990). *Competitive Anxiety in Sport*. Human Kinetics, Champaign, IL.

Newton, M. and Duda, J.L. (1999). The interaction of motivational climate, dispositional goal orientations and perceived ability in predicting indices of motivation. *International Journal of Sport Psychology*, **30**, 63–81.

Parfitt, G. and Hardy, L. (1993). The effects of competitive anxiety on memory span and rebound shooting tasks in basketball players. *Journal of Sports Sciences*, **11**, 517–24.

Parfitt, G., Hardy, L. and Pates, J. (1995). Somatic anxiety and physiological arousal: their effects upon a high anaerobic, low memory demand task. *International Journal of Sport Psychology*, **26**, 196–213.

Parfitt, G. and Pates, J. (1999). The effects of cognitive anxiety and self-confidence on components of performance during competition. *Journal of Sports Sciences*, **17**, 351–6.

Scanlan, T.K., Stein, G.L. and Ravizza, K. (1991). An in-depth study of former elite figure skaters: III. Sources of stress. *Journal of Sport and Exercise Psychology*, **13**, 103–20.

Swain, A.B.J. and Jones, J.G. (1996). Explaining performance variance: the relative contribution of intensity and direction dimensions of competitive state anxiety. *Anxiety, Stress and Coping*, **9**, 1–18.

Treasure, D.C., Duda, J.L., Hall, H.K., Roberts, G.C., Ames, C. and Maehr, M.L. (2001). Clarifying misconceptions and misrepresentations in achievement goal research in sport: a response to Harwood, Hardy and Swain. *Journal of Sport and Exercise Psychology*, **23**, 317–29.

Weinberg, R.S. (1994). Goal-setting and performance in sport and exercise settings: a synthesis and critique. *Medicine and Science in Sports and Exercise*, **26**, 469–77.

Weinberg, R.S. and Gould, D. (1999). *Foundations of Sport and Exercise Psychology*. Human Kinetics, Champaign, IL.

16 Soccer violence

Benny Peiser and John Minten

Introduction

Football in its varying forms is, without doubt, the most popular sport in the world. It is also the sport most discussed, studied, evaluated and scientifically analysed. Whilst rugby or American football are almost exclusively played in Western and Commonwealth countries, soccer is by far the biggest global sport. As a result, the world-wide football business is considered to be the most lucrative sport industry, estimated to be worth in the region of £150 billion per year (Syzmanski and Kuypers, 1999).

Currently, the *Federation Internationale de Football Associations* (FIFA) represents 204 national soccer associations which in total constitute around 220 million active members (both male and female). According to FIFA, there are some 150 million active players. Altogether, 1.3 million referees officiate weekly at soccer matches for some 600 000 soccer clubs which represent some 4 million soccer teams world-wide. Together with an estimated 200–300 million additional soccer players who are not organized in clubs or associations affiliated to FIFA, in particular pupils and students involved in school sport, there are an estimated 400–500 million active soccer players world-wide. This impressive number indicates that some 10% of the world's population are, in one way or another, participating in soccer.

Soccer is the only sport which attracts almost the entire world population. For example, the 1998 World Cup final was watched by well over 3 billion people around the globe. It has become a multi-billion business, thereby transforming a nineteenth-century elite sport into the single biggest leisure industry in the world. The universal appeal, popularity and economic weight of soccer has also led to the application of scientific research into an ever growing number of questions and problems.

16.1 Soccer violence: a brief history

The origins of modern soccer are rooted in pagan blood rituals. Ancient ball games such as the Persian *buzkashi*, the Mesoamerican *peloya*, or the Roman *harpastum* were extremely violent and cruel contests which involved killings, blood sacrifices and serious injuries. When, during the fourth century AD, Christianity became the official state religion of the Roman Empire, all pagan rituals, in particular Greek and Roman athletic and gladiatorial contests, were radically suppressed. The majority of these traditional games were associated with heathen and sacrificial cults which Christianity tried to eradicate (Peiser, 1997). The violent nature also contradicted the very basis of Judeo-Christian ethics. Jews and Christians were uncompromising in their attitude toward pagan spectacles. They attacked the very nature of these cruel combat sports, finding them incompatible with the idea of the holiness of life. The belief that God made man according to His image and likeness contradicted the participation in physical activities which involved the infliction of pain and injuries upon other living beings. Despite the continuous efforts of the church, kings and magistrates throughout history to ban these games, neither the various types of mob football nor ancient blood sports could be entirely erased. The pagan pastimes proved to be so popular that even the most powerful authorities finally gave up their battle against unlawful sports.

In particular, the repeated attempts by English lawmakers during the Middle Ages to outlaw soccer once and for all failed. Between the fourteenth and seventeenth centuries, football was banned on more than 30 occasions. In most of these cases, football was outlawed because of its inherent violence. Medieval football was a wild and brutal game played according to informal rules which allowed a high level of tolerated physical violence (Dunning and Sheard, 1979). During the nineteenth century, British pedagogues finally abandoned their struggle against the brutal sport. Since they could not deny its popularity, they now joined in. As violence and rebellion in England's public schools rose steadily, school masters discovered that football could be functionalized according to their own interests as a safety valve and means of controlling extremely violent pupils. Instead of banning the sport outright, new regulations were introduced in order to give it a more organized and less brutal character. These developments finally constituted the basis of modern soccer (Dunning and Sheard, 1979).

The codification and reformation of the rules of the game during the nineteenth century led to a reduction of the most blatant brutality of traditional football. Yet, assaults and grievous harm were still legitimate as long as physical violence occurred within the rules of the game. Only during the last decades of the twentieth century have many attempts to cleanse soccer further from its violent traditions led to a slow but gradual decline in the level of aggression and hostility.

16.2   What is soccer violence?

Violence can be defined as any form of behaviour which inflicts pain, harm or injury on another living being or oneself, thereby violently disturbing the homeostasis of the victim. Participants in all codes of football normally accept a certain degree of painful and injurious attacks on their bodies as legitimate. Bodily contact is acknowledged in the laws of the game of soccer. All significant body contacts that occur in tackles and collisions are deemed legal according to the rules of the game. Physical violence in soccer is frequently the result of reckless or intentional behaviour. Because of a positive relationship with victory and success, 'tough play' is demanded by fans, coaches and managers. Each player learns from an early age how to use various parts of the body in order to tackle, block and foul. Thus, both minor and extreme forms of violence are still inherent in today's soccer and resulting injuries are regarded as normal by-products of the game.

There is little doubt that many societies throughout the world have become more violent and brutalized during the last 100 years or so. Two world wars, numerous genocides and countless regional and civil wars have documented only too dramatically that civilizations and highly developed cultures can regress into barbaric states. In recent decades, almost all Western democracies have seen an increase of crime and anti-social behaviour. Against this background, the number of violent incidents in sports, in particular in soccer, seem to have equally mushroomed.

Violence in soccer is of utmost interest in the social and sports sciences. Its social and political repercussions are not just of academic interest. Football-related violence has been the focus of sociological, moral and judicial debate for many years. Contact sports, particularly American football, rugby and soccer, have all tolerated dangerous behaviour by players contrary to the spirit and letter of the rules (Mueller and Blyth, 1988). Despite the fact that research in the field of soccer violence has quickened its pace in the course of the past decade, little has been done to combat the violence itself.

16.3   The present state of soccer violence

Soccer is undoubtedly one of the most aggressive of today's sports both on and off the pitch. Hooliganism or spectator violence has long been an international predicament, causing riots and destruction throughout the world. Consequently, there has been a growing interest in the social aspects of soccer-related violence. It is not only spectator violence, however, which is an increasing world-wide nuisance; according to many observers, violent and unruly behaviour are also increasing on the field of play. Violence in soccer receives publicity through spectators and the media. Because various forms of aggressive behaviour in soccer are not only legitimate and tolerated but also encouraged and applauded, they reinforce the perception of a fierce and sometimes brutal game.

Since the 1970s, there has been a steady increase in scientific research on soccer violence (Bakker *et al.*, 1993). The growing concerns about intolerable levels of violent behaviour demonstrate the relevance of these kind of investigations and the need for significant changes. The past two decades have also brought an apparent upsurge in the level of brutality, violence and death in many other areas of contact sports.

Statistics show that there has been a dramatic increase in cautions and dismissals from play in the English leagues. The average number of illegitimate physical tackles in one professional soccer match is about 30 fouls per game. If we multiply the average foul rate by 16 000 (i.e. the number of professional matches governed by the Football Association in 2002), we can estimate a total of almost 480 000 incidents per year of actual bodily harm.

16.4   Perspectives on soccer violence

16.4.1   The medical perspective

The risks of injury to individuals in general and of soccer injuries in particular have been neglected for a long time. As a result of violence, indifference or foul play in soccer, millions of people each year require medical care after accidents or acts of physical violence. At a time when economic crises are jeopardizing efforts to improve the health of nations around the globe, violence-related injuries cost the world community almost US$ 500 billion in medical care, sick pay and lost productivity every year (World Health Organization, 1993).

The level of soccer violence can best be demonstrated by analysing injury rates. Soccer at both the professional and amateur level is a moderately risky sport resulting in a high rate of injuries. Whilst in soccer about 75% of all players sustain one or more injuries per year, the injury rate in rugby and American football is nearly 100% (Atyeo, 1979). It is estimated that the risk of sustaining a serious injury in American football is five times higher than in soccer.

Soccer injuries have become an increasing interest in sports medicine. It is estimated that 50–60% of all sports injuries and up to 10% of all hospital-treated injuries in Europe are due to soccer (Franke, 1980). In soccer, there are fewer injuries inflicted during training sessions than there are during real matches. The vast majority of soccer injuries result from collisions, direct kicks, tackles, hits by a kicked ball or from falling.

On average, 30–40% of traumatic soccer injuries are caused by foul play. In many cases these injuries are self-inflicted because the player committing the foul may sustain the more serious injury. Many soccer injuries are preventable to a large extent if significant rule changes were to be made. Medical researchers are therefore recommending more preventive measures. The development of a preventive strategy has to take into account the compulsory use of protective equipment (Ekstrand *et al.*, 1983) and significant changes in the rules of the game. These changes would help to reduce the high levels of violence and injurious conduct. More importantly, they would reduce the level of legitimate bodily contacts which are responsible for the bulk of serious injuries.

16.4.2 The psychological perspective

Does participating and watching soccer lead to a reduction of aggression and violence or does it in reality cause increased hostility? The most popular misconception regarding soccer violence holds that aggression on and off the field acts as a social and individual safety valve. According to the so-called catharsis theory of aggression, expressions of antagonistic behaviour in soccer reduce the overall level of social aggression among players and spectators. It is suggested that aggression not acted out in soccer (or other sports) will otherwise burst out in a much worse form of anti-social or even criminal behaviour (Bennett, 1991; Russell, 1993; Peiser and Madsen, 1995).

Accordingly, participation in soccer is widely regarded as a socially adequate means of discharging pent-up hostility and frustration, thereby improving the peaceful and lawful nature of society. Empirical research, however, shows clearly that participation in soccer has often quite the opposite effect (Coakley, 2000). Far from reducing the level of aggression, playing or watching soccer regularly reinforces aggressive impulses and occasionally leads to violent behaviour on and off the field. Due to its inherent confrontational structure and antagonistic dynamics as a mock battle, soccer creates its own, inherent aggression. Sports psychologists have found that watching the varying forms of football significantly increased the levels of agitation and aggression (Russell *et al.*, 1988/89; Young and Smith, 1988/89). During the past 20 years, none of the major psychological studies on football and sport violence found any support for the catharsis theory (Bakker *et al.*, 1993). Furthermore, since aggression is an essential prerequisite of victory in soccer, its catharsis is neither intended nor desired.

Obviously, soccer players and spectators are very concerned with victory. Given that the odds of losing are high, this mind-set regularly results in frustration, and frustration often enhances aggression and violence. Psychological studies of soccer have shown that observations of foul play or bodily harm on the field commonly produce high levels of emotional arousal and in consequence lead to a significant increase in hostility among players and spectators. As a result, many social scientists consider spectator violence a direct response to the violence observed on the soccer field (Russell, 1993).

According to social learning theory, aggressive behaviour is mainly learned through observing and imitating aggressive conduct of others. Observing soccer violence often results in imitation of such behaviour (Young and Smith, 1988/89). Soccer undoubtedly provides legitimate and socially acceptable possibilities for learning aggressive and violent behaviour. Since many players act as popular role models, their violent conduct is reproduced by other players and observers. Learning to play soccer, therefore, in most cases involves also imitating aggressive and foul play. This is part of the socialization process and is, in many cases, supported by coaches, parents and peers alike.

Violence in soccer is also related to the great insecurity associated with participating in a highly competitive team sport. Players are often willing to use violent behaviour to 'prove' themselves. Violence can easily become a means for players to prove their worth and establish membership on their teams. Soccer violence can also become a way to reaffirm manhood and injuries often become symbols of courage if the injured player endures the pain and remains in the game. According to Coakley (2000), all forms of football are socially constructed in ways that not only ritualized belligerence, but tie the expression of aggression to certain forms of masculinity. Being able to be tough and play violently has become part of gaining respect 'as a man' within many male groups.

16.4.3   *Soccer violence and the media*

From the perspective of trying to reduce the amount of violent behaviour, the relationship between the communication media and player violence may be viewed as a controversial one. On the one hand, communications in the media can be viewed as condemning acts of player violence. Indeed, the media have often made the disciplinary committees of the governing bodies in soccer aware of incidents of serious violent play missed by the referee in what is becoming increasingly known as 'trial by video'. On the other hand, it may be fair to say that violence on the pitch is often hyped up by television and newspapers commentators in an effort to attract more viewers and readers. After all, the communication media have always been aware that violence 'sells'.

Television experts do this by means of slowing down, 'zooming' in on and frequently replaying the more serious of acts of violence. Moreover, forthcoming matches are often billed as violent battles between 'tribes' or players in order to draw the attention of soccer fans some of whom look forward to these clashes with relish. The newspapers have been known to create melodramatic headlines and have even branded soccer matches between rival countries as a replication of previous wars. In recent years, the mass media have increasingly condemned both player and spectator violence and instead have focused on the entertaining aspects of the game not least because the relationship between the media and soccer is a two-way process with both needing each other to survive (Gratton and Taylor, 2000).

16.4.4   *The legal perspective*

What are the legal consequences violent soccer players have to face? In the first place, the structure of soccer as a contact sport makes harmful bodily contacts inevitable (Finn, 1994). In contrast to non-contact sports, the rules of soccer legitimize and endorse a whole range of actual bodily harm, so that the border to criminal violence is often unclear. Furthermore, even legal experts have struggled to determine what exactly constitutes 'violence' in soccer (Grayson, 1999). Conn (1999) pointed out that no matter how severe the offence, soccer violence is traditionally seen as separate from 'real' violence. Thus, the courts have usually been reluctant to deal with cases pertaining to violent acts in soccer. In recent years, however, law courts have to deal with an increasing number of cases given that soccer players may well be liable for injuries that result from intentional or reckless tackles.

Most dangerous assaults and battery, such as direct hits and kicks against an opponent, are strictly prohibited in soccer. Actual bodily harm and serious injury, however, constitute legitimate bodily contacts as long as they occur within the rules of the game. According to criminal law, actual bodily harm is defined as any hurt or injury intended to interfere with the health or comfort of the victim. The pain or injury does not have to be serious or permanent but must be more than trifling. Pain or discomfort resulting from bruises or swellings can be sufficient evidence of an assault occasioning actual bodily harm. Grayson (1999), a sports law specialist, has criticized the general tolerance towards sports violence. In his view, violent breaches of sporting laws and rule of play, condoned or inadequately disciplined by over-tolerant administrations, coaches and referees, have created a misconception that in sport the criminal law stops at the borders of the soccer pitch. In recent years, however, the Football Association and the legal profession have progressively moved towards a consensus where players who cause serious injuries through intentional or reckless challenges should be brought before the courts (Grayson, 1999).

16.5   Rule changes and disciplinary action

In 1994, recognizing that violent play was not only inadequately punished but also responsible for soccer injuries which were ending the careers of many professional players prematurely, FIFA persuaded the IFAB to make a number of amendments to the laws of the game (Elleray, 1998). As a result of these changes, referees were instructed to take more stringent action against 'reckless challenges' by sanctioning them with yellow and red cards (Rollin, 1999). More specifically, any 'challenge from behind with little or no attempt to play the ball' was to be sanctioned with an immediate red card as this offence was now deemed as serious foul play which endangered the safety of an opponent (Grayson, 1999). As a result of FIFA's reforms, more players have been booked or sent off during matches than before (Elleray, 1998). It is still uncertain, however, if this has led to a significant reduction in soccer player violence across different countries and in particular at the lower levels of play. In addition, it is also unclear if the reason for more cautions and dismissals is due to the more rigorous enforcement of the rules by the referees or a result of increased numbers of fouls.

The introduction of the new rules regarding foul play was initially met with great scepticism. The high number of bookings and cautions during matches led to criticism from all quarters that the referees were being too strict in the enforcement of the rules and that the entertainment and enjoyment of the match was being jeopardized (Elleray, 1998). Quite the reverse attitude has been taken by the governing bodies of soccer (such as UEFA and FIFA). These organizations have been concerned that referees are being too lenient and therefore routinely monitor referees to ensure that their directives are being followed adequately (Elleray, 1998). If the directives are not followed sufficiently, the referee may be demoted to less important fixtures (Goldman, 2000).

Whilst violence against players has been highlighted, the issue of violence against match officials also warrants mention. In September 1998, Paolo Di Canio (then at Sheffield Wednesday) pushed referee Paul Alcock to the floor after being sent off. The Football Association (FA) acted swiftly in punishing him with a fine of £10 000 along with a 11-match ban. The problems with threatening behaviour and dissent displayed at the top levels of the game are reflected in the rising imitation against referees at the grass root levels of the game. The problem is not confined to verbal abuse. Whilst Premiership managers and players routinely question decisions and call for professional referees with technological assistance, referees at the grass roots are regularly physically assaulted. As a result of the disenchantment with refereeing, the FA is confronted with a serious shortage of people willing to take up refereeing at all levels (Conn, 2000).

Prior to the commencement of the 2000–01 season, the FA introduced a new scheme to reduce player violence by introducing a disciplinary panel. The panel is composed of independent FA councillors who punish players for certain offences and those missed by the referee. The panel also has the power to overrule the FA's disciplinary committee in handing out punishments for violent behaviour. The panel issues punishments for cautions and dismissals incurred by players in the form of fines and suspensions. Fines, particularly at the top level in the Premier League, may be viewed as trivial for players already earning high salaries and are therefore not likely to curb violent misconduct on the pitch. Suspensions, on the other hand, may be regarded by some players as a welcome rest between matches. Consequently, there is some concern that the present disciplinary procedures present in English soccer are not too effective in dealing with violent play at the top level.

16.6   Hooliganism

16.6.1   Soccer hooliganism

Social disorder associated with soccer has been present since its codification at the end of the nineteenth century. However, from the 1960s, hooliganism in the UK and other European countries became perceived as a major social problem. After the Hillsborough disaster in 1989 – at which 96 Liverpool fans were killed in a crush on the terraces – and the subsequent Taylor Report (1990), a new era arrived with reduced levels of hooliganism and a more positive discourse was established around soccer. While there is disagreement about the extent to which the problem of hooliganism has been eradicated, it is clear that the experience of fans today is very different from that of 15 years ago. Nevertheless, hooliganism remains of interest to academics and politicians alike, especially following any high profile incident.

16.6.2   Nature and development of hooliganism

Soccer hooliganism consists of not only the swearing and unruly behaviour which often occurs in other sporting situations. It also involves more serious pitch invasions, the fighting in and around soccer grounds and clashes with the police in city centres or in transit to games.

In the pre-war period, disorder often took the form of attacks on officials and players and vandalism. Dunning *et al.* (1988) claimed that all modern manifestations of hooliganism were to be found occurring in this period, including fights between rival fans, and trouble at away matches. The perception that hooliganism began in the 1960s may be due, in part, to the fact that television began to show both the games and the hooligan acts for the first time. This increased the exposure of hooliganism and shaped public attitudes towards it, though in the 1960s there actually was an increase in the prevalence and a change in the nature of hooliganism.

By the 1970s and 1980s soccer hooliganism had come to be perceived as a major social problem. During this period the phenomenon changed, not least because of the attempts to solve the problem. In the 1960s, it was largely confined within the soccer grounds themselves, with concerns centring on pitch invasions, fights between fans and the 'taking of ends'. As the strategies employed by the police became more sophisticated and grounds became segregated, hooliganism was much more likely to occur in the surrounding neighbourhoods, and in transit to games making it more difficult to contain and causing greater inconvenience for the surrounding populace.

Against the background of major domestic problems there was a growing concern regarding the behaviour of British soccer fans abroad. This initially was focused on English club sides with trouble reported in a variety of European cities in 1974, 1977, 1980 and 1981 (Williams *et al.*, 1984). This problem reached its nadir in May 1985 at the Heysel Stadium disaster in Brussels. In this incident, there were many deaths among fans in the stadium awaiting the start of the European Cup Final between Liverpool and Juventus. There was little protest from a British Government elected on a 'law and order' ticket when, as a result of the Heysel incident, English clubs were banned from European competitions.

There was also concern in the 1980s about the rise of so-called super-hooligan groups. These were much more sophisticated in their organization, such as the 'Inner City Firm' of West Ham United and the 'Headhunters' of Chelsea. This led to the formation of specialist police sections and undercover operations, which had limited success in combating these groups. It appeared at this time that there was a rise in the number of middle class hooligans, wearing smart designer clothes. The vast majority of arrests, though, were still of working class youths, the clothes more of a statement of 'hardness' than an indication of class position (Murphy *et al.*, 1990).

After the ban on English clubs competing in Europe, attention was focused on the behaviour of English fans supporting their national team abroad. Throughout the 1990s, when domestic hooliganism was undoubtedly less prevalent, events surrounding international tournaments and matches (e.g. the Republic of Ireland versus England match in Dublin in 1994) continued to make headlines. This concern about English fans abroad has persisted into the new Millennium and prompted the Home Office to form a Working Group to explore soccer disorder. There is no doubt that other countries in Europe have had equivalent, or at times worse, problems of soccer hooliganism. Despite this, English fans were perceived, at least, to be the worst in Europe and hooliganism became known as the 'English Disease'. Williams (2001) suggested that it is important to understand the underlying causes of this international disorder, since nearly all of the most recent major incidences at international matches have involved English fans, either as instigators or as targets.

16.6.3   Social theories of hooliganism

There has been much written in the academic literature, and many explanations of soccer hooliganism have been put forward over the past three decades with early work including that of Taylor (1971a,b, 1982), Clarke (1978) and Marsh *et al.* (1978). Armstrong (1998) in an ethnographic study of Sheffield United fans described hooligans in terms of their seeking 'social dramas', giving them a sense of excitement, honour and belonging. Social–psychological perspectives explain hooliganism in terms of the search for experiences not present in everyday life; sharing emotions of joy and sadness, excitement and risk. These theories address why acts of hooliganism occur. Yet any adequate theory has to explain why this kind of hooliganism predominantly has been associated with soccer, and why it involves young, working class males. Theories need to locate the hooliganism phenomenon in a context of time and place.

Influential in the study of soccer disorder has been the work of researchers at the Sir Norman Chester Centre for Football Research at Leicester University (Williams *et al.*, 1984; Dunning *et al.*, 1988; Murphy *et al.*, 1990). The theory of Dunning *et al.* (1988) is based on an historical analysis of the development of soccer and associated acts of disorder. Hooliganism has gone through cycles, with many examples before First World War, less between the wars, and a escalation from the 1960s to the 1980s. What accounts for this variation is the involvement in soccer of the lower working classes. Soccer hooliganism is the ritualistic and real violence that comes from 'rough' working class masculinity, a 'violent masculine style'. The lower working class began to perceive soccer games as somewhere that violence could be expressed, which was reinforced by the media, starting an 'amplification spiral'. The more the media highlighted the problem the more it became defined to the hooligans as a place to express this behaviour.

The post-Hillsborough period of fashionable, business-orientated soccer has certainly seen a reduction in the incidence, and the reporting of, hooliganism. It is clear that since the early 1990s it is in no one's interests to focus on negative images of the sport. Redhead (1997) proposed that we have entered an era of post-fandom, where the traditional view of 'fans' no longer holds and soccer 'consumption' has changed significantly. Has soccer hooliganism died out? No, but it has certainly changed in its form, marginalized and less visible.

16.6.4   *Government reports and attempts to eradicate hooliganism*

Attempts to eradicate soccer hooliganism have included segregation, heavy policing, closed circuit television (CCTV) and monitoring, all-ticket matches, stricter penalties, restrictions on alcohol and, more recently, international banning orders. Academics, although not able to agree on causes of hooliganism, are critical of many attempts that have failed to eradicate the problem.

Some of these measures have had a major impact on soccer in Britain. Table 16.1 outlines the major Government reports into soccer hooliganism and the main points of interest and recommendations.

The most recent of the official investigations into soccer hooliganism was the report of the Working Group on Football Disorder, chaired by Lord Bassam (2001). This was

Table 16.1 Summary of major government reports into soccer and soccer violence

Report	Year	Main points
Chester	1966	Wide-ranging report on state of the game including soccer hooliganism
Harrington	1968	Recommended many, now common, methods of deterrence, for example, segregation, all-ticket matches, stricter penalties
Lang	1969	Stronger and more sophisticated policing. Improvements in grounds and tighter operation by clubs
McElhone	1977	Similar recommendations to previous reports with emphasis on restrictions on alcohol
Smith	1978	First real report from a social science perspective. Noted failure of past measures
Teasdale	1984	Better implementation of existing strategies and careful timing of fixtures, introduction of CCTV, membership schemes, better international liaison
Popplewell	1986	Considered Bradford fire and hooliganism. Proposed searches before entry, additional powers for police and specific offences
Taylor	1990	Report into Hillsborough disaster. Many proposals related to safety. All-seater stadia, removal of perimeter fencing, suspension of membership schemes
Bassam	2001	Working Group on Football Disorder. Numerous recommendations on tackling racism, improved communications, stewarding and preparation for tournaments. Proposed formation of a new England Members Club

prompted by the trouble surrounding EURO 2000 in Belgium and the Netherlands, though the focus of the report was wider and considered issues on how 'to make football a more positive experience for all'. Recommendations were numerous and included measures to improve communications, management of disorder overseas, stewarding and preparation for tournaments. It also proposed the formation of a new England Members Club and strategies for tackling racism as well as measures to broaden the appeal of the domestic game and help grassroots soccer.

The Football (Disorder) Act 2000 established international banning orders for some English fans. This has been controversial in that it has lowered the threshold of evidence required for a ban and has been criticized by the civil liberty group Liberty. There has also been concern as to the effectiveness of such bans in targeting hooligans.

Serious hooliganism in the domestic Premier League has, arguably, been eradicated and attendance has risen again following the low point of the 1985–86 season. Periodically, incidents give the authorities cause for concern, as do incidents in the lower divisions where the management of crowds is less well developed and new technologies are used less by 'organized' hooligans.

16.6.5 *Stopping soccer hooliganism?*

Since the late 1980s, the UK has taken a lead in the development of control measures to deal with hooliganism. While police responses have been regularly reactive rather than proactive, new strategies of crowd control have been profoundly influenced by technological advances. More importantly, though, the police have introduced less confrontational crowd control tactics which have proven more effective and are more appreciated by soccer fans (Marsh *et al.*, 1996).

When looking at the improvements in the context of reducing crowd violence, many researchers have highlighted the effectiveness of a clean and tidy stadium as an instrument of affirmative crowd behaviour (DeBenedette, 1988). After the publication of the Taylor Report, all-seater stadia were installed at all First Division grounds. The idea of changing the spectator into the customer with improved facilities all round seemed the way forward. Along with CCTV surveillance equipment, the general conditions of the grounds have improved significantly with new and refurbished facilities helping to create a more pleasant and welcoming atmosphere. This has proved a drastic change from the treatment of the fans in the past when perimeter fences were considered by many as a mark of contempt for the fans by those in authority.

The recommendation of the Taylor Report to improve the condition of professional soccer grounds was welcomed by both authorities and fans in the sense that they provided higher levels of revenue, lower levels of hooliganism and much better comfort. It is clear that the British soccer stadium has come a long way since the tragedies of the 1980s. It is fair to state that they rank among the world's best with regard to comfort, class and facilities.

Just as important were the changes introduced in the way soccer crowds were policed by law enforcement agencies. The discovery that crowd behaviour is heavily influenced by the relations between the crowd and police has encouraged developments of new technologies and strategies that manage this relationship more proficiently (Durrheim and Foster, 1999). In the past, poor relations between the police and soccer fans were well documented and regularly attributed as a partial cause for the severity of the hooliganism problem. Sadler (1992), for instance, looked at the benefits and drawbacks of the use of a specialist police force to deal with the problem of hooliganism. He concluded that although a specialist force would avoid the need to train all police officers in crowd control methods and would hence be more cost effective, there would be a risk that such an elite force may over-react to an otherwise non-threatening situation. As a result, fans may be aggravated into further violence.

Various researchers have highlighted that some of the causes for crowd conflict among soccer fans can be attributed to the police acting too hastily, as a result of lack of experience or due to a mutual lack of communication between the police and the soccer fans (Stott and Reicher, 1998). Rushed action or inconsistent operations undertaken by the police can easily become counter-productive and may act in a way that fuels more violence (Kerr, 1994).

As a direct result of social research into crowd control of soccer fans, the policing of matches in the UK has improved significantly in recent years. Improvements have also been made so far as trained stewards are taking on an increasing role in crowd management. Other improvements, such as the installation of CCTV along with the successful establishment of a National Football Intelligence Unit show that crowd control in British soccer has come a long way over the past decade. These changes have certainly helped to improve the protection and satisfaction of the average soccer fan.

Summary

This chapter focused on some fundamental problems of soccer violence. In particular, the psychological, medical and legal perspectives of player violence were discussed. This interdisciplinary review also examined the socio-political issues surrounding the problem of soccer hooliganism. In addition, it probed recent advances in crowd management and assessed effort to reduce outbreaks of crowd violence in soccer.

References

Armstrong, G. (1998). *Football Hooligans: Knowing The Score*. Berg, Oxford.

Atyeo, D. (1979). *Blood and Guts. Violence in Sports*. Paddington Press, New York.

Bakker, F.C., Whiting, H.T.A. and van der Brug, H. (1993). *Sport Psychology. Concepts and Applications*. John Wiley, New York.

Bassam, Lord. (2001). *Report of the Working Group on Football Disorder*. HMSO, London.

Bennett, I.C. (1991). The irrationality of the catharsis theory of aggression as justification for educators' support of interscholastic football. *Perceptual and Motor Skills*, **72**, 415–18.

Clarke, J. (1978). Football and the working class: tradition and change, in *Football Hooliganism: The Wider Context*. (ed. R. Ingham), Inter-Action Imprint, London, pp. 37–60.

Coakley, J. (2000). *Sport in Society: Issues and Controversies*, 6th edn. McGraw-Hill, Boston, MA.

Conn, J.H. (1999). Professional player on player violence: another plaintiff failure in redress by the McKichan Court. *Journal of Legal Aspects of Sport*, **9**(2), 63–74.

DeBenedette, V. (1988). Spectator violence at sports events: what keeps enthusiastic fans in bounds? *Physician and Sportsmedicine*, **16**(3), 202–11.

Dunning, E. and Sheard, K. (1979). *Barbarians, Gentlemen and Players. A Sociological Study of the Development of Rugby Football*. Martin Robertson, New York.

Dunning, E., Murphy, P. and Williams, J. (1988). *The Roots of Football Hooliganism: An Historical and Sociological Study*. Leicester University Press, Leicester.

Durrheim, K. and Foster, D. (1999). Technologies and social control: crowd management in liberal democracy. *Economy and Society*, **28**(1), 56–74.

Ekstrand, I., Gillquist, I. and Liljedahl, S.-O. (1983). Prevention of soccer injuries. *American Journal of Sports Medicine*, **11**, 116–20.

Elleray, D. (1998). *Referee!* Bloomsbury, London.

Finn, G.P.T. (1994). Football violence. A societal psychological perspective, in *Football, Violence and Social Identity* (eds R. Giulianotti, N. Bonney and M. Hepworth), Routledge, London, pp. 91–127.

Franke, K. (1980). *Traumatologie des Sports*. Georg Thieme Verlag, Stuttgart.

Goldman, R. (2000). Referees and the laws of the game, in *Rothmans Football Yearhandbook 2000–2001* (ed. J. Rollin), Headline, London.

Gratton, C. and Taylor, P. (2000). *Economics of Sport and Recreation*. E & FN Spon, London.

Grayson, E. (1999). *Sport and the Law*, 3rd edn. Butterworths, London.

Kerr, J.H. (1994). *Understanding Soccer Hooliganism*. Open University Press, Buckingham.

Marsh, P., Rosser, E. and Harre, R. (1978). *The Rules of Disorder*. Routledge & Kegan Paul, London.

Marsh, P., Fox, K., Carnibella, G., McCann, J. and Marsh, J. (1996). *Football Violence in Europe*. The Amsterdam Group. http://www.sirc.org/publik/football_violence.html.

Mueller, F.O. and Blyth, C.S. (1988). Forty years of head and cervical spine fatalities in American football: 1945–1984, in *Science and Football* (eds T. Reilly, A. Lees, K. Davids and W.J. Murphy), E & FN Spon, London, pp. 224–9.

Murphy, P., Williams, J. and Dunning, E. (1990). *Football on Trial*. Routledge, London.

Peiser, B. (1997). "Thou shalt not kill:" the Judeo-Christian basis of the civilizing process. *The Sports Historian* **17**, 93–108.

Peiser, B. and Madsen, T. (1995). Physical play, foul play and violence in football: comparative analysis of violent play among professional soccer players in Italy and England, in *Science and Football III* (eds T. Reilly, J. Bangsbo and M. Hughes), E & FN Spon, London, pp. 188–200.

Redhead, S. (1997). *Post-Fandom and the Millennial Blues*. London, Routledge.

Rollin, J. (1999). Discipline problems in football, in *Sport and the Law* (ed. E. Grayson), Butterworths, London, pp. 559–62.

Russell, G.W. (1993). *The Social Psychology of Sport*. Springer, New York.

Russell, G.W., Di Lullo, S.L. and Di Lullo, D.D. (1988/89). Effects of observing competitive and violent versions of a sport. *Current Psychology*, Winter, 312–21.

Sadler, S. (1992). Crowd control: are there alternatives to violence? Community disorder and policing. *Conflict Management in Action*, **12**, 107–28.

Stott, C. and Reicher, S. (1998). How conflict escalates: the inter-group dynamics of collective football crowd 'violence'. *Sociology*, **32**, 353–77.

Szymanski, S. and Kuypers, T. (1999). *Winners & Losers: The Business Strategy of Football*. Viking, London.

Taylor, I. (1971a). Soccer consciousness and soccer hooliganism, in *Images of Deviance* (ed. S. Cohen), Penguin, Hamondsworth, pp. 134–64.

Taylor, I. (1971b). Football mad: a speculative sociology of football hooliganism, in *Sociology of Sport: A Collection of Readings* (ed. E. Dunning), Frank Cass, London, pp. 352–77.

Taylor, I. (1982). The soccer violence question. Football hooliganism revisited, in *Sport Culture and Ideology* (ed. J. Hargreaves), Routledge & Kegan Paul, London, pp. 152–96.

Taylor, The Rt. Hon. Lord Justice. (1990). *The Hillsborough Stadium Disaster: Final Report*. HMSO, London.

Williams, J. (2001). Who you calling a hooligan?, in *Hooligan Wars. Causes and Effects of Football Violence* (ed. M. Perriman), Mainstream Publishing, Edinburgh, pp. 37–53.

Williams, J., Murphy, P. and Dunning, E. (1984). *Hooligans Abroad*. Routledge & Kegan Paul, London.

World Health Organization (1993). *Handle Life with Care. Prevent Violence and Negligence*. WHO, Geneva.

Young, K. and Smith, M.D. (1988/89). Mass media treatment of violence in sports and its effects. *Current Psychology*, Winter, 298–311.

Part 4

Match analysis

17 Notational analysis

Mike Hughes

Introduction

A considerable amount of effort has been devoted to establishing objective forms of analysis and demonstrating their importance in the coaching process. There are difficulties facing any single individual attempting to analyse and remember objectively the events occurring in complex team games, such as soccer. One of the main solutions to these inherent problems has been to use notational analysis systems. Coaches, scouts and managers have designed and developed systems for gathering information which over the past three decades have been improved by both coaches and sports science researchers, to the point where the design of the systems has become an end in itself. The aim of this chapter is not only to review the data that have been produced, but also assess the major innovations and developments in the systems and to examine some of the practical uses of notation in soccer.

17.1 Historical perspective

The earliest recorded form of music notation was conceived in the eleventh century (Hutchinson, 1970; Thornton, 1971), although it did not become established as a uniform system until the eighteenth century. Historical texts give substantial evidence pointing to the emergence of a crude form of dance notation much later, in about the fifteenth century. Thornton (1971) stated that the early attempts at movement notation may well have 'kept step' with the development of dance in society, and as a consequence the early systems were essentially designed to record particular movement patterns as opposed to movement in general.

Dance notation actually constituted the 'starting base' for the development of a general movement notation system. Arguably the greatest development in dance notation was the emergence of the system referred to as 'Labanotation' or 'Kinetography-Laban', so-called after its creator, Rudolph Laban in 1948. Laban highlighted three fundamental problems encountered in the formulation of any movement notation system:

1 recording complicated movement accurately;
2 recording this movement in economical and legible form;
3 keeping abreast with continual innovations in movement.

These three fundamental problems left dance in a state of flux, incapable of steady growth for centuries (Hutchinson, 1970). In almost the same way, these problems still beset analysts today.

The next 'step' in the development of movement notation came with the conception of another form of dance notation, Choreology, by Benesh and Benesh (1956). In this form of notation, five staves formed the base or matrix for the human figure, that is,

All notation was completed on a series of these five line grids with a complex vocabulary of lines and symbols.

The major underlying disadvantage of the Benesh and Laban methods of notation in terms of sport is that they are both primarily utilized for the recording of patterns of movement rather than its quantification. Later researchers attempted to develop a system of movement notation based entirely on the mathematical description of movement in terms of the degrees of a circle in a positive or negative direction. As with the systems of Labanotation and Choreology, this system did not allow the description of movement in terms familiar to sport or everyday life.

Movement notation developed primarily in the field of expressive movement, gradually diversified into game analysis, specifically sport. Ensuing research proved severely limited both in variety and detail, and was at a fairly global and unsophisticated level (Sanderson and Way, 1977).

Hand notation systems are in general very accurate but they do have some disadvantages. The more sophisticated systems involve considerable learning time. In addition, the amount of data that these systems produce can involve many hours of work in processing them into forms of output that are meaningful to the coach, athlete or sports scientist. Even in a game like squash, the amount of data produced by the method of Sanderson and Way required 40 h of work to process one match.

Computerized notation systems have enabled these two problems, in particular the data-processing load, to be tackled in a positive way. Used in real-time analysis or post-event in conjunction with video recordings, they enable immediate, easy access to data. They also enable the sports scientist to present the data in graphical form more easily understood by the coach and athlete. The increasing sophistication, and reducing cost, of video systems have greatly enhanced the whole area of post-event feedback, from playback with subjective analysis by a coach to detailed objective analysis by means of notation systems.

The four major purposes of notation are:

1   analysis of movement;
2   tactical evaluation;
3   technical evaluation;
4   statistical compilation.

Team sports have the potential to benefit immensely from the development of computerized notation. Purdy (1977) suggested that the sophisticated data manipulation procedures available would aid the coach in efforts to improve performance. Many of the traditional systems outlined above are concerned with the statistical analysis of events which previously had to be recorded by hand. The advent of on-line computer facilities overcame this problem, since the game could then be digitally represented first, by means of data collection directly onto the computer, and then later documented via the response to queries pertaining to the game.

The information derived from this type of computerized system can be used for several purposes:

- immediate feedback;
- development of a database;
- indication of areas requiring improvement;
- evaluation;
- as a mechanism for selective searching through a video recording of the game.

All of the above functions are of paramount importance in improving the coaching process, the initial raison d'etre of notational analysis. The development of a database is a crucial element, since it is sometimes possible to formulate predictive models as an aid to the analysis of different sports, subsequently enhancing future training and performance. Both types of notation, manual or computerized systems, have their advantages and disadvantages. Hand notation systems are cheap and accurate, if fully defined operationally and used correctly. The disadvantages of these types of systems are that the time required for data processing can be very long. When the systems become more sophisticated, in order to gather data in more complex analyses of match-play, which is often the case with a team game such as soccer, then considerable learning and training time is necessary in order to ensure accuracy and reliability of the operator.

One advantage of using computers is that the data analysis, once the software has been designed, written and tested, takes only as long as the computer and/or printer takes to process the output. Second, the learning time required to use notation systems can be reduced considerably, especially with the variety of advances in special keyboards, digitization pads, graphical user interfaces and voice interactive systems that have become available for computer hardware. Finally, the rapid developments of computer graphics, word-processing, database and multi-media packages to present the analyses in clear and appealing ways have enabled athletes and coaches to understand complex data. This potential has only just begun to be tapped. The disadvantages of computerized systems are that they are expensive and can be less accurate than hand notation systems, unless very carefully designed and validated.

Skilful management of data input and output can make the use of the systems, and the understanding and assimilation of the data, easier for the coach or player. The advances made in tackling these problems in computerized notation analysis provide a useful structure with which to explore the developments in this field.

17.2   The development of sport-specific notation systems (hand notation)

Notation systems were commercially available for analysis of American football as early as 1966, and the Washington Redskins was one of the first sports teams to use such a system. American football is the only sport that has, as part of its rules, a ban on the use of computerized notation systems in the stadium. How this rule is being enforced is not clear. All clubs do claim to use similar hand notation systems, the results of which are transferred to computer after the match. Clubs exchange data, just as they exchange videos, on opponents. Because of the competitive nature of this, and other 'big money' sports, actual detailed information on the results of these analyses is seldom available.

The first publication of a comprehensive sport notation system in Britain was that by Downey (1973), who developed a detailed system which allowed the comprehensive notation of lawn tennis matches. Detail in this particular system was so intricate that not only did it permit notation of such variables as shots used, positions and so on, but it also catered for type of spin used in a particular shot. Downey's notation method has served as a useful base for the development of systems for use in other racket sports, specifically badminton and squash.

17.3   Performance indicators in soccer

17.3.1   *Performance indicators*

If two players, A and B, have 4 and 6 shots on target respectively, it is not appropriate to report that player B is having the better game. What are the respective totals of shot attempts? Player A could have had 4 shot attempts, while player B had 12 shot attempts, thus resulting in shooting indices of 4/4 and 6/12 shots on target per attempt, respectively. Even this could be further analysed – how many shooting opportunities did each player have? Player A could have had a total of 12 opportunities but decided to pass 8 times instead of shooting, player B shot on all 12 of the possible opportunities that were presented. Does this now indicate that player B was having the better game? Further analysis of the game could show that the passing options adopted by player A were deemed better tactically for the team. This would lead to analysis at further levels and so on. As noted above, simple analysis of data induces a simple interpretation, which is not always fitting in sport. The analyses are always dependent upon how we define the 'performance indicators' by which we are judging performance of a team or individual.

Through an analysis of game structures and the performance indicators used in recent research in performance analysis, Hughes and Bartlett (2002) defined basic rules in the application of performance indicators to sport. In every case, success or failure in a performance is relative, either to your opposition, or to previous performances of the team or individual. To enable a full and objective interpretation of the data from the analysis of a performance, it is necessary to compare the data collected to aggregated data of a peer group of teams, or individuals, which compete at an appropriate standard. In addition any analysis of the distribution of actions across the playing surface must be normalized with respect to the total distribution of actions across the area.

Performance indicators, expressed as non-dimensional ratios, can have the advantage of being independent of any units that are used; furthermore, they are implicitly independent of any one variable. Mathematics, fluid dynamics and physics in general have shown the benefits of using these types of parameters to define particular environments. They also enable an insight into differences between performers that can be obscure in the raw data. The particular applications of non-dimensional analysis are common in fluid dynamics, which offers empirical clues to the solution of multi-variate problems that cannot be solved mathematically. Sport is even more complex, because of the result of interacting human behaviours. Applying simplistic analyses of raw sports data can be highly misleading.

For the different types of games considered, Hughes and Bartlett (2002) contended that the classification of the different action variables being used as performance indicators follow rules that transcend the different sports. The selection and use of these performance indicators depend upon the research questions being posed, but it is clear that there are certain guidelines that will ensure a more clear and accurate interpretation of these data. These are summarized below.

17.3.2   Match classification

Always compare observations with opponents' data. Where possible, the comparison should be with aggregated data from peer performances of both your own team and that of your opponents.

17.3.3   Technical/tactical

The technical and tactical variables should be treated in a similar way. Always normalize the action variables with the total frequency of that action variable or, in some instances, the total frequency of all actions, and present these data with the raw frequency, and/or processed, data.

Most notational analysts have not followed these simple guidelines to date. The utility of performance analysis could be considerably enhanced if its practitioners agreed and implemented such conventions in the future.

17.4   Hand notation systems in soccer

17.4.1   Motion analysis and match analysis

The definitive motion analysis of soccer, using hand notation, was by Reilly and Thomas (1976), who recorded and analysed the intensity and extent of discrete activities during match-play. They combined hand notation with the use of an audio tape recorder to analyse in detail the movements of English First Division soccer players. They were able to specify work-rates of the players in different positions, distances covered in a game and the percentage time in each of the different ambulatory classifications (Figure 17.1). They also found that typically a player is in possession of the ball for less than 2% of the game. Reilly (1990) has continually added to this database enabling him to define clearly the specific physiological demands in soccer, as well as other football codes. The work by Reilly and Thomas has become a standard against which other similar research projects can compare their results and procedures. A detailed analysis of the movement patterns of the outfield positions of Australian professional soccer players was completed in a similar study to that above (Withers *et al.*, 1982). The data produced agreed to a great extent with those of Reilly and Thomas (1976); both studies emphasized that players cover 98% of the total distance in a match without the ball, and were in agreement in most of the inferences made from the work-rate profiles.

An alternative approach towards match analysis was exemplified by Reep and Benjamin (1968), who collected data from 3213 matches between 1953 and 1968. They were concerned with actions such as passing and shooting rather than work-rates of individual players. They reported that 80% of goals resulted from a sequence of three passes or less. Fifty per cent of all goals came from possession gained in the final attacking quarter of the pitch.

Bate (1988) found that 94% of goals scored at all levels of international soccer were scored from movements involving four or less passes, and that 50–60% of all movements leading to shots on goal originated in the attacking third of the field. Bate explored aspects of chance in soccer and its relation to tactics and strategy in the light of the results presented by Reep and Benjamin (1968). It was claimed that goals are not scored unless the attacking team gets the ball and one, or more, attacker(s) into the attacking third of the field. The greater the number of possessions a team has, the greater chance it has of entering the attacking third of the field, therefore creating more opportunities to score. The higher the

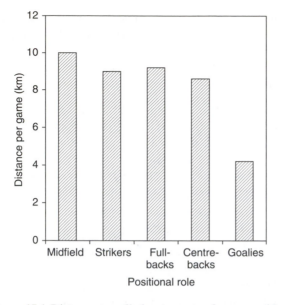

Figure 17.1 Distance travelled per game of soccer with respect to the position of the player. (Data from Reilly and Thomas, 1976.)

number of passes per possession, the lower the total number of match possessions, the total number of entries into the attacking third, and the total chances of shooting at goal. Thus, Bate rejected the concept of possession football and favoured a more direct strategy. He concluded that to increase the number of scoring opportunities a team should play the ball forward as often as possible; reduce square and back passes to a minimum; increase the number of long passes forward and forward runs with the ball and play the ball into space as often as possible.

These recommendations are in line with what is known as the 'direct method' or 'long-ball game'. The approach has proved successful with some teams in the lower divisions of the English League. It is questionable whether it provides a recipe for success at higher levels of play.

Harris and Reilly (1988) considered attacking success in relation to team strategy and the configuration of players around the point of action, by concentrating mainly upon the position of attackers in relation to the defence and overall the success of each attacking sequence. This was a considerable departure from many of the systems previously mentioned which have tended to disseminate each sequence into discrete actions. Harris and Reilly provided an index describing the ratio of attackers to defenders in particular instances, while simultaneously assessing the space between a defender and an attacker in possession of the ball. These were analysed in relation to attacking success, whereby a successful attack resulted in a goal, an intermediate attack resulted in a non-scoring shot on goal, and an unsuccessful attack resulted in an attack ending without a shot. Successful attacks tended to involve a positive creation of space, where an attacker passes a defender – an unsuccessful attack involved a failure to use space effectively due to good organization of defensive lines.

These examples represent the different purposes of notational analysis. More recent research using hand notation tends to use a data gathering system and then process the data in a computerized database. Pettit and Hughes (2001) used a hand notation system to analyse all the matches from the 1998 World Cup, through the aid of a database into which the data were entered. The system was designed in an order, like a flow chart so as each action occurred the operator imputted the data from field to field, for example, first the time was inputted then the event that led to the cross, crossed from and to, and so on. If a shot was taken the data were added, otherwise the process was started again to input the data for the next cross. Abbreviations were used to help speed up the process of inputting the data. The system, designed to analyse crossing and shooting, was based on that used by Partridge and Franks (1989a,b). All 64 matches from the 1998 World Cup were notated post-event over a period of 90 min plus injury time, although extra time and penalty shootouts were omitted from the analyses. The time the cross occurred, events leading up to, team, area crossed from, area crossed to, type of cross, in front or behind the defence, result of cross, if applicable: whether or not a pass was made, number of passes in sequence, shot type, height of shot, direction in relation to goalkeeper, speed and intent of shot, contact, outcome and possession, and so on were analysed, which enabled the frequency of the actions to be recorded. On these criteria, results for the 1986 and 1998 World Cup finals were compared.

17.4.2 Penalty kicks

Penalties are now a subject of myth, romance, excitement, dread, fear and pressure – depending upon whether you are watching or taking them. They have either helped the careers of soccer players or destroyed them. Yet little research has been completed on penalty kicks. Using a hand notation system Hughes and Wells (2002) notated and analysed 129 penalties with an intention to examine:

- the time in preparing the shot;
- the number of paces taken to approach the ball;
- the speed of approach;
- the pace of the shot; and
- its placement and the outcome.

In addition, the actions of the goalkeeper were notated – position, body shape, movements as the player approached, his first movements and the subsequent direction, the outcome. A summary of their findings are presented below.

- One in five penalties were saved (20%; 3/15), one in 15 missed (7%; 1/15) and three in four resulted in a goal (73%; 11/15).
- Players using a fast run up had 25% of their efforts saved, because the player then tried either 50 or 75% power.
- Best success ratios are from an even run up of 4, 5 and 6 paces.
- There is no laterality in the success ratios – left footers and right footers have the same success rates.
- No shots above waist height were saved.
- In every case, the goalkeeper moved off the line before the ball was struck.
- Although there is only a small dataset, the goalkeepers who stood still while the striker approached the ball, had the best save and miss ratios.

It is evident that hand notation systems provide a detailed record of behaviour during soccer match-play. It is possible also to derive theories of play from such analysis. A complete theory must take the opposition into consideration, and this requires highly sophisticated modelling which is beyond the scope of this chapter.

17.5   Computerized notation systems

17.5.1   Limitations

Using computers does introduce extra problems in notational analysis, of which the system users and programmers must be aware. Increases in possibilities of error are enhanced by either operator errors, or hardware and software errors. Any system is subject to the limitations of human perception where the observer misunderstands an event, or incorrectly fixes a position. The computer–operator interface can result in the operator thinking the correct data are being entered when this is not the case. This is particularly so in real-time analysis. Hardware and software errors are introduced by the machinery itself, or the programs of instructions controlling the operation of the computer. Careful programming can eradicate this latter problem. To minimize both of these types of problems, careful validation of computerized notation systems must be carried out. Results from both the computerized system and a manual system should be compared and the accuracy of the computerized system quantitatively assessed.

Although computers have only recently impinged on the concept of notational analysis, this form of technology is likely to enhance manipulation and presentation of data. Its benefits are due to the ability of computers to process large amounts of data easily and the improved efficiency of the graphics software.

17.5.2   Data entry

A fundamental difficulty in using a computer is entering information. The traditional method employs the QWERTY keyboard, but studies using this method to enter data about soccer have shown that this can be a lengthy and monotonous task.

The time spent by three professional soccer players in different match-play activities was calculated by Mayhew and Wenger (1985). The analysis was completed post-event from videotapes, using a specially designed computer program. The results indicated that soccer is predominantly an aerobic activity, with only 12% of game time spent in activities that would primarily stress the anaerobic energy pathways. The mean time of 4.4 s for such high-intensity work indicated that the lactacid acid energy supply system was the anaerobic system of primary importance. The interval nature of soccer was partly described, and suggestions for the design of soccer-specific training programmes were offered. Considerable keyboard skills, and time, were required to enter the data into the computer. The work did not really extend in any way the previous efforts of Reilly and Thomas (1976).

An alternative to the specially designed keyboard is the use of digitization pads. In England, researchers worked with the 'Concept Keyboard', in Canada they used the 'Playpad', but both systems were similar. These are touch-sensitive pads that can be programmed to accept input to the computer, via the pad, over which specially designed 'overlays' can be placed. This enabled a pitch representation, as well as action and player keys to be specific and labelled (Figure 17.2). The digitization pads considerably reduced the time to learn the system, and made the data input quicker and more accurate. The system enabled an

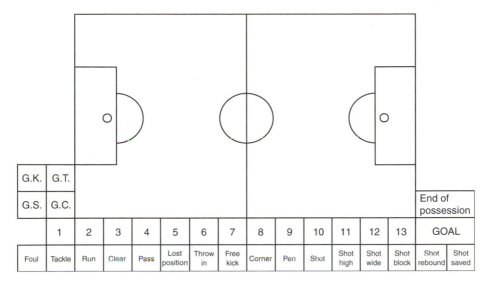

G.K.	G.T.														
G.S.	G.C.													End of possession	
	1	2	3	4	5	6	7	8	9	10	11	12	13	GOAL	
Foul	Tackle	Run	Clear	Pass	Lost position	Throw in	Free kick	Corner	Pen	Shot	Shot high	Shot wide	Shot block	Shot rebound	Shot saved

Figure 17.2 The overlay for the Concept Keyboard for data input for notational analysis of soccer. (Reproduced with permission from Hughes *et al.*, 1988.)

analysis to be performed of patterns of play, both at team and player levels, and with respect to match outcome.

Hughes *et al.* (1988) used the concept keyboard and hardware system developed by Church and Hughes (1987), but with modified software, to analyse the 1986 World Cup finals. Patterns of play of successful teams, those teams that reached the semi-finals, were compared with those of unsuccessful teams (i.e. teams that were eliminated at the end of the first rounds). The main observations are as follows:

1   Successful teams played significantly more touches of the ball per possession than unsuccessful teams.
2   The unsuccessful teams ran with the ball and dribbled the ball in their own defensive area in different patterns to the successful teams. The latter played up the middle in their own half, the former used the wings more.
3   This pattern was also reflected in the passing of the ball. The successful teams approached the final sixth of the pitch by playing predominantly in the central areas while the unsuccessful teams played significantly more to the wings.
4   Unsuccessful teams lost possession of the ball significantly more in the final one sixth of the playing area both in attack and defence.

Partridge and Franks (1989a,b) used a similar system with a Playpad to enter their data (the overlay for which is shown in Figure 17.3) and produced a detailed analysis of the crossing opportunities from the 1986 World Cup. They carefully defined how they interpreted a cross, and gathered data on the following aspects of crosses:

1   build up of attacks;
2   area of build up;
3   area from which the cross was taken;

4   type of cross;
5   player positions and movements;
6   specific result of the cross;
7   general result, if the opportunity to cross was not taken.

Fifty games were analysed from videotape, using specifically designed software on an IBM XT microcomputer that enabled each piece of information relating to crossing opportunities to be recorded and stored. The program recorded the time at which all actions took place during the match, for extracting visual examples post-analysis, in addition to the usual descriptive detail about the matches, that is, venue, teams. A second program was written to transform and download these data into dBASE III+. Afterwards, this database was queried to reveal selected results. Partridge and Franks (1989a) related their results to the design of practices to aid players understand their roles in the successful performance of crossing in soccer; this work was then used as an example of how notation analysis can aid the coaching process.

Similar systems have now been updated to enable a graphical user interface to input the data, thus eliminating the need for a concept keyboard or digitization pad. These systems use on-screen graphics that are controlled and interfaced by means of the computer mouse. They were used to investigate the effect of the changes in the rules of soccer in Premier Division matches in England in the 1992–93 season. It was found that the time-wasting of goalkeepers

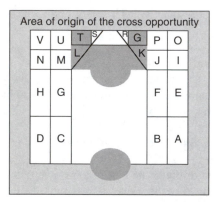

Figure 17.3 The overlays for the Playpad for data input for notational analysis of crosses by players in soccer. (Reproduced with permission from Partridge and Franks, 1989a.)

was significantly reduced, that the midfield was more congested as a result of fewer back passes, and that although there were more errors in defence, there were not any more goals or goal-scoring possibilities.

The use of digitization pads has considerably eased some of the problems of data entry, but one of the more recent innovations in input is the introduction of voice entry of data into the computer. It has been demonstrated that this type of system can be used by the computer 'non-expert' for notation systems, and that, as prices come down and hardware efficiency and flexibility improve, this method of data entry will be the future of interfacing sports data with computers.

Data input has moved from the use of difficult QWERTY keyboards and complex learning skill problems, through special keyboards specifically task-designed, concept keyboards, digitization pads, voice interactive systems and graphical user interfaces in the attempt to simplify the task and make it more accurate. As the technology improves, the easier it will become to access the computer and all the vast potential it offers to the player, coach and sports scientist.

17.5.3   *Data output*

In practical applications to sport, it is imperative that the output from notation systems is immediate and, perhaps more important, clear, concise and to the point. The first systems produced tables of data, often incorporated with statistical significance tests that were difficult for the non-scientist to understand. Some researchers attempted to tackle the problem before the advent of computerized graphics, but it was debatable whether the type of presentation was easier to understand than the tables of data. Frequency distributions across graphical representations of the playing area offer a compact form, which coaches of different nationalities have found easy to assimilate.

Yamanaka *et al.* (1993) took this a step further in presenting their data (see Figure 17.4). They demonstrated the ethnic differences in international soccer by analysing the 1990 World Cup. By defining four groups, British Isles, European, South American and Emerging Nations, and by analysing the respective patterns of play in matches with respect to pitch position, they were able to differentiate between the playing styles of these international groups. They also presented data in a case study of Cameroon, which had a successful World Cup. Cameroon was compared to the other groups to examine the way in which the former had developed as a footballing nation.

The development and growth of integrated software packages incorporating programming languages, graphics packages, databases and word processors have enabled the development of whole systems that immediately access graphics in both the data input, as well as the data processing stages. These systems have great potential to provide quantitative and qualitative feedback.

17.5.4   *Applications of computerized notation systems*

At the World Congresses of Science and Football, a considerable amount of work on computerized analysis of soccer has been presented; this is all collated in the proceedings (Reilly *et al.*, 1988, 1993, 1997; Spinks *et al.*, 2002). Similarly, the proceedings of the World Conferences in Notational Analysis of Sport (Hughes, 1996, 2000; Hughes and Franks, 2001; Hughes and Tavares, 2001) offer a large amount of work on analysis of soccer. A number of these have already been considered because of their contribution to the development

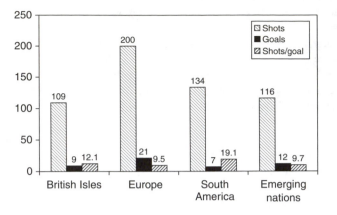

Figure 17.4 Shots per goal in the 1990 World Cup: an example of data output in graphical form, demonstrating the facilitation of assimilation of the data. (Reproduced with permission from Yamanaka *et al.*, 1993.)

of hardware and software as well as their contributions to research. Some significant applications will be reviewed.

Using computer-assisted video feedback and a specific algorithm for the statistics, Dufour (1993) presented an analysis of an evaluation of players' and team performance in three aspects – physical, technical and tactical. The ability of the computerized systems to provide accurate analysis and feedback for coaches on their players and teams was clearly demonstrated.

Gerisch and Reichelt (1993) used graphical representation of their data to enable easier understanding by the coach and players. Their analyses concentrated on the one-on-one confrontations in a match, representing them in a graph with a time-base, so that the development of the match could be traced. Their system can also present a similar time-based analysis of other variables, interlinking them with video so that the need for providing simple and accurate feedback to the players is attractively achieved. Despite the limited amount of data presented, the results and their interpretation were very exciting in terms of their potential for analysis of the sport.

Winkler (1993) presented a comprehensive, objective and precise diagnosis of a player's performance in training and match-play using a computer-controlled dual video system. This was employed to assess physical fitness factors employed in training contexts. In addition, he used two video cameras, interlinked by a computer, to obtain a total view of the playing surface area. This, in turn, permitted analysis of all the players in a team throughout the whole match, on and off the ball – something that not many systems have been able to produce.

In international tournaments, teams are judged on their ability to win matches. To achieve these victories, the teams must have effective ways to win the ball, create successful attacks first by reaching the attacking third of the field, create scoring chances and to complete them by scoring goals with a high efficiency. Luhtanen *et al.* (2001) selected offensive and defensive variables of field players and goalkeepers in the EURO 2000 competition and attempted to relate the results to the final team ranking in the tournament. The final ranking order in the World Cup 1998 tournament was explained by calculating the rank correlation coefficients between team ranking in the tournament and ranking in the following variables: ranking of ball possession in distance, passes, receptions, runs with the ball, shots, interceptions, tackles and duels. Selected quantitative and qualitative sum variables were calculated using ranking order of all obtained variables, only defensive variables and only offensive variables. The means and standard deviations of the game performance variables were calculated and ranking order for each variable was constructed. Spearman's correlation coefficients were calculated between all ranking game performance variables. Only the variable of successful passes at team level explained success in the EURO 2000.

France was the best team in the performance of passes, receptions, runs with ball and tackles. In percentage of the successful passes, France was the top team. The strengths of Italy were in defence. The Italians were best in interceptions and third best in tackles. In the passing activity their position was 15th, but in the percentage of successful passes 2nd. In the overall ranking taking into account all analysed variables, Italy was 13th. This analysis would give Netherlands a better place than third. Netherlands was first in ball possession (8.9 km) and second in the amount of passes and shots and also close to the top place in the corresponding successful executions. Because Netherlands controlled the ball a lot, its players did not have many chances of interceptions or duels. Germany was traditionally strong in having the ball in possession (2nd), in passing play (2nd) and in the number of goal scoring trials (4th). However, the weaknesses were found in the defensive activities of interceptions (16th) and tackles and duels (15th).

17.5.5   *Analysing the game structure*

Data from the 1990 World Cup were analysed with a view to the interpretation of goal scoring and the importance of possession. Previous data may have been misinterpreted because they had not been 'normalized' (i.e. the number of goals scored should be divided by the frequency of number of passes per possession). There were 88% of possessions with four passes or less, and so there should be more goals scored, proportionately, from these possessions, if they were truly the best 'chance'. Observations showed that the passing tactics of a team are dependent upon the skill levels of the players in each team (Figure 17.5), and the tactics best suited to a particular team can be gauged by match analysis (Hughes, 1993).

Usually, notational analysis uses numerical data to study and assess the quality of a match. As far as the analysis of the tactical aspects of the game is concerned, there is a dearth of published research with regards to their theoretical bases. Grehaigne *et al.* (1996) tried to construct a knowledge base about soccer using some qualitative observational tools. In a soccer match, structures and configurations of play should be considered as a whole rather than examined a piece at a time. Systems with many dynamically interacting elements can produce a rich and varied patterns of behaviours that are clearly different from the behaviour of each component considered separately. To that effect, effective space, game action zone and configurations of play were examined to show that this type of analysis is complementary more than opposed to the numerical data analysis systems. This work demonstrates the growing awareness of the need for qualitative factors, investing the quantitative nature of notation data with a far greater wealth of relevance to an overall performance.

Soccer has received a major share of the research by notational analysts over the last five decades and substitutes have been positively or negatively affecting the results of soccer matches since the change in the rules in the sixties. Pearce and Hughes (2001) analysed substitutions during the European Football Championships 2000. First, general characteristics

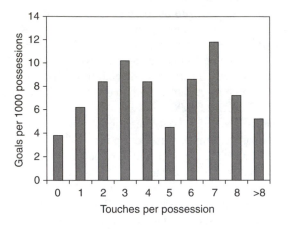

Figure 17.5 The normalized frequency of goals scored (goals per 1000 possessions) with respect to the number of touches per possession.

Table 17.1 A summary of the impact on performance by substitutes

	Negative impact	No impact	Positive impact
Effect on team performance	29.1%	12.5%	58.4%
	(7 substitutes)	(3 substitutes)	(14 substitutes)
Substitute's performance	25%	37.5%	37.5%
compared to player replaced	(6 substitutes)	(9 substitutes)	(9 substitutes)

were recorded from all 176 substitutions that were made during the tournament. Second, 24 substitutes were analysed to assess their impact on their team's performance and whether the substitute's performance was better than the player replaced.

The research was carried out using a computerized notation system in conjunction with video recording. Data collection involved gathering data from a match 15 min prior to the substitution and 15 min following substitution. The data were analysed by comparing the number of actions that occurred by a team/individual during the first 15 min, to the number of actions that occurred in the second period of 15 min. A total of 24 matches were selected from EURO 2000, and substitutions chosen so that the respective teams were equally balanced in winning, losing and drawing situations. Performances between the two periods were evaluated by the use of a matrix that enabled variables to be ranked in order of importance. These matrices were then applied to both the team performance and the respective individual performances for 15 min before and after substitution (see Table 17.1).

The majority of substitutions took place during the second half of the match and primarily for tactical reasons. The most frequent position for substitutions was in midfield, possible reasons for this could include the large work-rates associated with these positions. Results also suggested that substitutions might influence teams' performances with 60% of substitutions analysed having a positive influential effect on performance. The findings suggested that substitutes play an important role in their contribution towards team performance and can provide coaches with increased flexibility.

Hughes (2001) described the recent developments in the technology of analysis and feed-back. The potential of the new integrated computer and video systems to the coaching process was highlighted. The general advantages of using this computer system are:

- Coaches can 'read' the game. They obtain quantity information, which can reveal hidden aspects of the game.
- The program can sort the players of one or both teams by their effectiveness in defence, in attack or generally, as well as get the information on what share of the team effectiveness (in percentages) was due to each of the players individually.
- The program is aimed at giving an evaluation of what the two teams had done, as well as to show the reasons for the victory (failure) of one of the two teams. The comparison is affected on the basis of the points assigned according to the various criteria (developed by the authors): play in attacking (passes, shots, technique proficiency and prowess), defence (defending players, goalkeeper) and active participation (dominance and power in the game).

When these sophisticated analysis systems are truly integrated with the video analysis and editing suites such as Noldus Observer or Sportscode, then the task of the analyst and coach will be more immediate but never simpler.

17.6   Computers and video

The ability of computers to control the video image has introduced exciting possibilities for enhancing feedback. An inexpensive IBM-based system was developed for the team sport of field hockey; this system has been modified to analyse and provide feedback for ice hockey and soccer (Franks and Nagelkerke, 1988). Following the game, a menu-driven analysis program allowed the analyst to query the sequentially stored time–data pairs. Because of the historical nature of these game-related sequences of action, it was possible to perform both post- and pre-event analysis on the data. That is to say, questions relating to what led up to a particular event or what followed a particular event could now be asked. In addition to presenting the sports analyst with digital and graphical data of team performance, the computer can also be programmed to control and edit the videotape record of the game action.

The interactive video computer program could access, from the stored database, the times of all specific events such as goals, shots, set plays and so on. Then, from a menu of these events, the analyst could choose to view any or all of these events within one specific category. The computer may be programmed to control the video such that it finds the time of the event on the video and then plays back that excerpt of game action. It was also possible to review the same excerpt with an extended 'lead in' or 'trail' time around that chosen event.

Recently, more and more feedback is given in the form of edited videotapes. It is obviously more relevant to the feedback between the coach and the player that the players see the elements of their performances that the coach is discussing. This form of feedback is far more immediate to the task of communicating the message from the analyses and obviously easier to understand than tables of data, or even clearly presented colour graphics. Consequently, the modus operandi of a notational analyst working with a coach, has changed radically from that 5 or 6 years ago. The analyses are still completed, either in-event or post-event, but the data are then reinterpreted via edited video tapes – some coaches still want the actual data on their players' performances. As everybody is now 'visually' literate, the expectation of the quality of the videos is that it will be broadcast quality. If it is not, players tend not to watch or give the messages in the video any credence. Consequently, sophisticated computerized video editing suites have become a necessary part of the notational analysts' battery of equipment.

The ability of computers to control the video image has now made it possible to enhance existing sport specific analytical procedures. The pioneering work of Franks and his colleagues was completed without the use of digital video or the much larger memory capacities that are now available for computers. The advent of these has enabled commercial systems for objective observation to be developed for the first time. Most of these involve a generic data input system that can then be tailored to any sport desired. The data entry can be completed in-event or post-event, with the program linking the events to the time code on the computer, the implications of the subsequent analyses can then be rendered by the computer editing the relevant parts of the performance and downloading them to tape for the coach and player. This factor considerably simplifies the searching and editing processes described above. At present the editing facilities are crude and do not offer the broadcast-quality presentations of specific editing suites.

17.7   Computerized modelling of sports

With the increased capacity and speed of the more recent PCs, it is now possible to use some of the large databases that have been gathered at various centres around the world, to provide comparative information for iterative techniques of modelling sports performance. Analysts who have gathered a sufficiency of data have often used these data to suggest normative models of performance at different levels (squash – Hughes, 1986; soccer – Hughes *et al.*, 1988; rugby union – Potter and Carter, 1995). It has not been easy to make the transition from these empirical models to a theoretical or conceptual model of performance.

Grehaigne (2001) analysed the configurations of the game according to positions of the players, their speed, and their directions and proposed a model to analyse the transition between two configurations of play that enables one to take time into consideration as the match evolves. McGarry and Franks (1995) cited Reep's work in soccer since the 1950s, and how the statistical analysis of these data reveals mathematical functions and consistencies of certain behaviours. The conclusions drawn from their work suggested that it would be of benefit to a side to maximize the probabilities of certain actions at the expense of others. Reep and Benjamin (1968) found that the goal : shot ratio was 1 : 10 and thus it would seem fair to suggest that an increase in the number of shots would lead to an increase in the number of goals. Since they also found that most shots came from passing movements with very few passes then the 'long ball' or 'direct style' of play becomes important. Franks and Nagelkerke (1988) found that passing movements leading to goals were even shorter than passing movements leading to shots, hence suggesting that there lies a sub-group within the shots on goal group.

McGarry and Franks (1995) described how sports analysis could move on from being a descriptive process to becoming a predictive one. If there is some level of consistency within the performance, then future performance can be predicted from past matches through stochastic modelling. They sub-divided sports into two sections, those determined by score (squash, tennis, etc. where the result is win or lose) and those by time (soccer, rugby, etc. where the result is win, lose or draw). This is an important distinction when modelling aspects are to be discussed.

The characteristics of score-dependent sports are based largely on a structured sequence of discrete events where the relationship between each event is related to the opponent.

Time-dependent sports are invasive and interactive and can be considered as relatively contingent in a temporary state. The structure of the sport is very important when it comes to deciding what method of modelling one should use to predict performance.

Score-dependent sports can be modelled by simply using discrete event models but the time-based sports need time models since the next event is always dependent on both event and time. McGarry and Franks (1997) tested the so-far untried Poisson model for discrete events in time-dependent sports. They also discussed the importance of the number of competitors involved on the development of a model – the greater the number of competitors then the larger the scope for variability. This simply emphasises the problems coaches face when they view a game. It would appear to be the case that the previous event only becomes important to the coach when a critical event has just occurred. The amount of data is one reason why little modelling has been directed into team games in a conceptual manner. Perl (2001) considered the potential use of neural networks to optimize the combinations of performance indicators for rugby and soccer that predict the result from known data. These combinations are then used to predict the result of a match after 20 min of play, the data being entered in the computer in-event. The initial results were very accurate, except for games with extreme scores.

17.8   Future developments

There are certain developments that will extend notational analysis over the next few years. The first will be the development of 'all-purpose', generic software. Work in some centres has almost reached this point now. Another technological advance that will make computerized notation more easily handled by the non-specialist will be the introduction of 'voice-over' methods of data entry. Taylor and Hughes (1988) demonstrated that this is possible, but relatively expensive at contemporary prices. They expected that voice-interaction should be a natural extension of any computing hardware system.

The integration of both these technological developments with computerized-editing suites, that will enhance video feedback, will enable both detailed objective analysis of competition and the immediate presentation of the most important elements of play. Computerized systems on sale now enable the analysis, selection, compilation and re-presentation of any game on video to be processed in a matter of seconds. The coach can then use this facility as a visual aid to support the detailed analysis. Franks (2000) devised a more detailed model of the feedback process that could be possible with this type of technology.

As these systems are used more and more, and larger databases are created, a clearer understanding of each sport will follow. The mathematical approach, typified initially by Eom (1988) and McGarry and Franks (1994 and 1995), will make these systems more and more accurate in their predictions. At the moment the main functions of the systems are analysis, diagnosis and feedback – few sports have gathered enough data to allow prediction of optimum tactics in set situations. Where large databases have been collected (e.g. soccer and squash), models of the games have been created and this has enabled predictive assertions of winning tactics. This has led to some controversy, particularly in soccer, due to the lack of understanding of the statistics involved and their range of application. Nevertheless, the function of the systems could well change, particularly as the financial rewards in certain sports are providing such large incentives for success.

Technological advances aside, the real future of notational analysis lies in the growing awareness by coaches, players and sports scientists of its potential applications in soccer.

Whether the most sophisticated and expensive of systems is being used, or a simple pen and paper analysis, as long as either system produces accurate results that are easy to understand, then coaches, players and sports scientists will increase their insights into soccer performance.

Summary

Various forms of notation analysis have been used for centuries to assist in the coding of human movements. In contemporary research the technique of motion analysis has been modified to study the movements of soccer players during a game. The advent and development of computers have stimulated sophisticated analysis of patterns of play and sequences of events leading to the creation of goal-scoring opportunities. This type of investigation is complemented by multi-media methods of observing and recording the behaviour of players. Future technological developments are likely to enhance the capture and retrieval of information. These facilities have great potential for assisting coaching staff in making decisions about game tactics. They can also provide performers with insights into many aspects of their own actions.

References

Bate, R. (1988). Football chance: tactics and strategy, in *Science and Football* (eds T. Reilly, A. Lees, K. Davids and W. Murphy), E & FN Spon, London, pp. 293–301.

Benesh, J. and Benesh, R. (1956). *Reading Dance – The Birth of Choreology*. Souvenir Press, London.

Church, S. and Hughes, M.D. (1987) *Patterns of play in association football – a computerised analysis*. Communication to First World Congress of Science and Football, Liverpool, 13–17 April.

Downey, J.C. (1973). *The Singles Game*. EP Publications, London.

Dufour, W. (1993). Computer-assisted scouting in soccer, in *Science and Football III* (eds T. Reilly, J. Clarys and A. Stibbe), E & FN Spon, London, pp. 160–6.

Eom, H.J. (1988). A mathematical analysis of team performance in volleyball. *Canadian Journal of Sports Sciences*, **13**, 139–41.

Franks, I.M. (2000). The structure of sport and the collection of relevant data, in *Computer Science in Sport* (ed. A. Baca), OBV & HPT publishers, Vienna, Austria, pp. 226–40.

Franks, I.M. and Nagelkerke, P. (1988). The use of computer interactive video technology in sport analysis. *Ergonomics*, **31**, 1593–603.

Gerisch, G. and Reichelt, M. (1993). Computer- and video-aided analysis of football games, in *Science and Football II* (eds T. Reilly, J. Clarys and A. Stibbe), E & FN Spon, London, pp. 167–73.

Grehaigne, J.-F. (2001). Computerised analysis of soccer. *International Journal of Performance Analysis Sport (Electronic)*, **1**, 48–57.

Grehaigne, J.F., Bouthier, D. and David, B. (1996). Soccer: the players' action zone in a team, in *Notational Analysis of Sport – I & II* (ed. M.D. Hughes), CPA, UWIC, Cardiff, pp. 27–38.

Harris, S. and Reilly, T. (1988). Space, team work and attacking success in soccer, in *Science and Football* (eds T. Reilly, A. Lees, K. Davids and W. Murphy), E & FN Spon, London, pp. 322–9.

Hughes, M.D. (1993). Notation analysis in football, in *Science and Football II* (eds T. Reilly, A. Stibbe and J. Clarys), E & FN Spon, London, pp. 151–9.

Hughes, M.D. (ed.) (1996). *Notational Analysis of Sport – I & II*, UWIC, Cardiff.

Hughes, M.D. (ed.) (2000). *Notational Analysis of Sport – III*, UWIC, Cardiff.

Hughes, M. (2001). *From Analysis to Coaching – The Need for Objective Feedback*. UKSI Website: www.uksi.com, November.

Hughes, M. and Bartlett, R. (2002). The use of performance indicators in performance analysis. *Journal of Sports Sciences*, Special Edition, **20**, 739–54.

Hughes, M.D. and Franks, I.M. (2001). *Pass.com*. UWIC, Cardiff.

Hughes, M. and Tavares, F. (eds) (2001). *Notational Analysis of Sport – IV*. Faculty of Sports Sciences and Education, Porto.

Hughes, M. and Wells, J. (2002). Analysis of penalties taken in shoot-outs. *International Journal of Performance Analysis Sport (Electronic)*, **2**, 55–72.

Hughes, M.D., Robertson, K. and Nicholson, A. (1988). An analysis of 1984 World Cup of association football, in *Science and Football* (eds T. Reilly, A. Lees, K. Davids and W. Murphy), E & FN Spon, London, pp. 363–7.

Hutchinson, A. (1970). *Labanotation – The System of Analysing and Recording Movement*. Oxford University Press, London.

Jordanov, N. (2001). Computer program for estimation of football games, in *Pass.com* (ed. M.D. Hughes), CPA, UWIC, Cardiff, pp. 253–62.

Luhtanen, P., Belinskij, A., Häyrinen, M. and Vänttinen, T. (2001). A computer aided team analysis of the Euro 2000 in soccer. *International Journal of Performance Analysis Sport (Electronic)*, **1**, 69–77.

Mayhew, S.R. and Wenger, H.A. (1985). Time–motion analysis of professional soccer. *Journal of Human Movement Studies*, **11**, 49–52.

McGarry, T. and Franks, I.M. (1994). A stochastic approach to predicting competition squash match-play. *Journal of Sports Sciences*, **12**, 573–84.

McGarry, T. and Franks, I.M. (1995). Winning squash: predicting championship performance from a priori observation, in *Science and Racket Sports* (eds T. Reilly, M. Hughes and A. Lees), E & FN Spon, London.

McGarry, T. and Franks, I.M. (1997). In search of invariance in championship squash, in *Notational Analysis of Sport – I & II* (ed. M. D. Hughes), CPA, UWIC, Cardiff.

Partridge, D. and Franks, I.M. (1989a). A detailed analysis of crossing opportunities from the 1986 World Cup (Part I). *Soccer Journal*, May–June, 47–50.

Partridge, D. and Franks, I.M. (1989b). A detailed analysis of crossing opportunities from the 1986 World Cup (Part II). *Soccer Journal*, June–July, 45–8.

Pearce, M. and Hughes, M. (2001). Substitutions in Euro 2000, in *Pass.com* (ed. M.D. Hughes), CPA, UWIC, Cardiff, pp. 303–16.

Perl, J. (2001) Artificial neural networks in sports: concepts and approaches. *International Journal of Performance Analysis Sport (Electronic)*, **1**, 100–7.

Pettit, A. and Hughes, M. (2001) Crossing and shooting patterns in the 1986 and 1998 World Cups for soccer, in *Pass.com* (eds M. Hughes and I. M. Franks), CPA, UWIC, Cardiff, pp. 267–76.

Potter, G. and Carter, A. (1995). Performance at the 1995 Rugby World Cup, in *Rugby World Cup 1995* (ed. C. Tau), International Rugby Football Board, Bristol.

Purdy, J.G. (1977). Computers and sports: from football play analysis to the Olympic games, in *Optimal Strategies in Sports* (eds S.P. Ladany and R.E. Machol), North Holland, Amsterdam, pp. 196–205.

Reep, C. and Benjamin, B. (1968). Skill and chance in association football. *Journal of Royal Statistical Society, Series A*, **131**, 581–5.

Reilly, T. (1990). Football, in *Physiology of Sports* (eds T. Reilly, N. Secher, P. Snell and C. Williams), E & FN Spon, London, pp. 371–425.

Reilly, T. and Thomas, V. (1976). A motion analysis of work-rate in different positional roles in professional football match-play. *Journal of Human Movement Studies*, **2**, 87–97.

Reilly, T., Stibbe, A. and Clarys, J. (eds) (1993). *Science and Football II*, E & FN Spon, London.

Reilly, T., Bangsbo, J., Hughes, M. and Lees, A. (eds) (1997). *Science and Football III*. E & FN Spon, London.

Reilly, T., Lees, A., Davids, K. and Murphy, W. (eds) (1988). *Science and Football*. E & FN Spon, London.

Sanderson, F.H. and Way, K.I.M. (1977). The development of an objective method of game analysis in squash rackets. *British Journal of Sports Medicine*, **11**, 188.

Spinks, W., Reilly, T. and Murphy, A. (2002). *Science and Football IV*. Routledge, London.

Taylor, S. and Hughes, M.D. (1988). Computerised notational analysis: a voice interactive system. *Journal of Sports Sciences*, **6**, 255.

Thornton, S. (1971). *A Movement Perspective of Rudolph Laban*. McDonald & Evans, London.

Winkler, W. (1988). A new approach to the video analysis of tactical aspects of soccer, in *Science and Football* (eds T. Reilly, A. Lees, K. Davids and W. Muphy), E & FN Spon, London, pp. 368–72.

Winkler, W. (1993). Computer-controlled assessment and video-technology for the diagnosis of a player's performance in soccer training, in *Science and Football II* (eds T. Reilly, J. Clarys and A. Stibbe), E & FN Spon, London, pp. 73–80.

Withers R.T., Maricic, Z., Wasilewski, S. and Kelly, L. (1982). Match analyses of Australian professional soccer players. *Journal of Human Movement Studies*, **8**, 158–76.

Yamanaka, K., Hughes, M. and Lott, M. (1993). An analysis of playing patterns in the 1990 World Cup for association football, in *Science and Football II* (eds T. Reilly, J. Clarys and A. Stibbe), E & FN Spon, London, pp. 206–14.

18 The science of match analysis

Tim McGarry and Ian M. Franks

Introduction

The analysis of sports performance can be undertaken from several different standpoints. The biochemist may focus on the depletion of substrate used as fuel for the provision of energy to the working muscles during sports competition, while the physiologist may attend to the exercise responses in the respiratory, cardiovascular and neuromuscular systems. The biomechanist – as well as the student of motor control – may be interested in how the kinetic and kinematic profiles that describe skilful acts change as a result of learning or fatigue, while the sports psychologist may investigate various mental aspects that are associated with skilled actions. Even though each discipline (or sub-discipline) differs in its theoretical constructs, dependent variables and tools of measurement, one important feature remains invariant. Each discipline converges upon a performance outcome that has meaning within the context of the sports contest being examined. Match analysis describes sports behaviours at this level of performance outcome, coding the actions of individuals or groups in technical terms that have relevance to players and coaches.

This chapter contains an outline for match analysis using research from various scientific disciplines. The aims are to detail a rationale for match analysis, report on recent developments in the descriptions of sports contests, and, lastly, to demonstrate how some of the information generated using sports analysis might be used in sports practice.

18.1 Information feedback

Information about skilled performance, presented in the form of feedback, is one of the most important variables that affects the learning and hence the proficiency of a motor skill. In some situations a failure to provide such knowledge (or information), or, alternatively, the provision of incorrect (or irrelevant) feedback may prevent learning from taking place. Furthermore, the quality of the information that is provided, as well as the scheduling of that same feedback, has been shown to have varying effects on the learning of motor skills. The presentation of precise information at the correct time will maximize the learning process and hence promote skill acquisition. For a review of the effects of augmented feedback on skill learning see Magill (2001, pp. 235–81).

The information that is gathered from a sports contest is used in sports practice to prepare for the next contest. This process of information renewal occurs as a sports contest unfolds. It also occurs in post-mortem fashion once sport competition has taken place, that is in the interval between sports contests. In each instance the coaching staff analyse past (or present) sports behaviours and make apposite changes in an attempt to improve future sports performance. The widespread technique of 'scouting', where information on an opposing individual or team is gathered in advance of a forthcoming contest, provides a ready example from sports practice of how information gathered from one contest is used in preparation for the next contest. Yet the ability of humans, including well-practised sports practitioners, to recall recent observed occurrences in sports contests is fraught with error (see Franks and Miller, 1986). The systematic recording of sports behaviours was thus introduced as a memory aid in support of the coaching process. For further details on the need for systematic observation aids in match analysis, with particular reference to soccer, see Franks and Miller (1986, 1991). The benefits of systematic observation tools notwithstanding, their introduction to sports practice was resisted by those who held to the traditional view that experienced coaches could observe freely and report accurately on the key aspects of sports performance. Today, however, many sports organizations receive support from a variety of services within the sports sciences, including those of match analysis.

18.2   Sports performance (match) analysis

Soccer has a (relatively) long history of match analysis. Reep and Benjamin (1968) reported data from an extensive set of English League and World Cup matches taken from 1953 through 1967. These data, recorded using systematic observation methods, included the frequency of passing sequences of varying lengths, shots at goal and goals scored. The frequency of passing sequences, reported in tabular form by Reep and Benjamin (1968, table 3), are presented in Figure 18.1. The frequency proportions of passing sequences of more than seven passes reported by Reep and Benjamin are not presented in Figure 18.1. The x-axis represents the length of unbroken passing sequences, that is the number of consecutive passes within a team before the chain was broken. The y-axis represents the frequencies of their occurrence expressed as a relative proportion of the total number of frequencies from that data set. The number of games from which these data were drawn, as reported by Reep and Benjamin (1968, table 3), is presented in parentheses (see legend, Figure 18.1).

It can be seen from Figure 18.1 that the longer the passing sequences (beginning at zero passes) the lower the frequency of their occurrence. That longer passing sequences yield lower frequencies is only to be expected, given that the likelihood of the chain being broken would be predicted to increase as the length of the chain increases. These data on the frequency of passing sequences can be described using a negative binomial function of the form p_0, p_1, p_2 through p_n. The terms p_0, p_1, p_2 through p_n represents the probabilities of passing sequences of various lengths, from zero, one, two through n passes, respectively. The stability of the negative binomial distribution (see Figure 18.1, Reep and Benjamin, 1968, table 3), together with the longevity of the 1/9–10 goal/shot ratio (Reep and Benjamin, 1968, table 4) led Reep and Benjamin to conclude that chance exerts a significant influence on the outcome of a soccer match. These data were to have a profound impact on English soccer.

The finding that some events in a soccer match appear to follow a probability structure led to the implementation of 'direct play' by some coaches in English football. This term is used to describe a style of play that tends to assault the defence of the opposing team with direct attacks once possession of the ball has been won. This method of direct play stands in contrast to the alternative stylistic approach of traditional continental soccer, which aims to retain ball possession whilst probing for an opening on which to base an attack. The merits and demerits of each approach were contested in debates within the soccer community for some years.

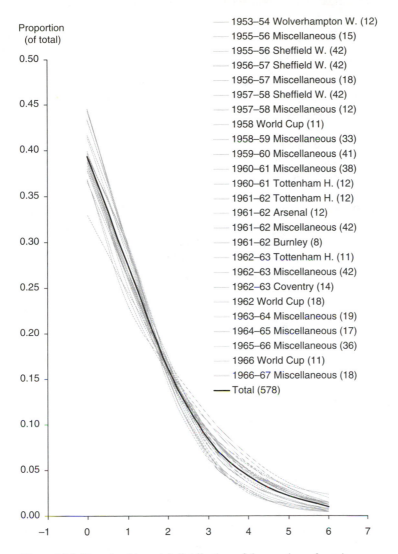

Proportion
(of total)

1953–54 Wolverhampton W. (12)
1955–56 Miscellaneous (15)
1955–56 Sheffield W. (42)
1956–57 Sheffield W. (42)
1956–57 Miscellaneous (18)
1957–58 Sheffield W. (42)
1957–58 Miscellaneous (12)
1958 World Cup (11)
1958–59 Miscellaneous (33)
1959–60 Miscellaneous (41)
1960–61 Miscellaneous (38)
1960–61 Tottenham H. (12)
1961–62 Tottenham H. (12)
1961–62 Arsenal (12)
1961–62 Miscellaneous (42)
1961–62 Burnley (8)
1962–63 Tottenham H. (11)
1962–63 Miscellaneous (42)
1962–63 Coventry (14)
1962 World Cup (18)
1963–64 Miscellaneous (19)
1964–65 Miscellaneous (17)
1965–66 Miscellaneous (36)
1966 World Cup (11)
1966–67 Miscellaneous (18)
Total (578)

Figure 18.1 Negative binomial distribution of the number of passing sequences in soccer. (The data were first reported in Reep and Benjamin, 1968, table 3.)

The use of direct play by some teams is based on the negative binomial distribution of passing sequences, as well as the remarkably stable goal/shot ratio. Thus, on the basis of probability it would seem that, in the long run, an increase in the number of shots at goal would be expected to boost the tally of goals. In addition, since most shots at goal arise routinely from sequences containing few passes (e.g. ~80% of shots at goal stem from three passes or less), the playing of direct passes into and within the scoring zone may be predicted to maximize the population of shots from which goals are scored. The strategy of direct play would furthermore be expected to decrease the likelihood of a team losing possession in the defending half of the field. This reduction in 'turn-over' in the defending half of the field is important since it is often the case that the opposing team will win possession in the attacking half of the field before going on to shoot at goal. For example, approximately half of the shooting opportunities from the 1953–67 data arose from regained possession in the attacking half of the field (see Reep and Benjamin, 1968, table 4). It is an impressive fact that the soccer statistics first reported by Reep and Benjamin (1968) hold many years on from their initial report (see for example Franks, 1988; Reep, 1989). However, despite the good association that exists between successful performance in soccer and short passing sequences (cf. the direct style of play), the nature of that association is still not well understood. In particular, the existence of a cause (few passes) and effect (goals scored) relationship for these data has yet to be identified.

The recording of quantitative data provides for an objective accounting of sports performance. While this process is necessarily descriptive in its form, the data are sometimes used in a prescriptive fashion to influence future behaviours. The direct style of soccer play described above provides a classic example of how future match strategies were designed on the basis of previous data. For the development of strategies to be valid, however, requires that the data on which they are predicated remain invariant. The wealth of soccer data first reported by Reep and Benjamin (1968) has been demonstrated to meet this important criterion. Invariant data, however, only admit the possibility of identifying valid match strategies, there is no assurance that the aforesaid strategies are valid necessarily on the basis of invariant data.

The presence of invariant structures in sports contests is hinted at in the widespread practice of scouting. On this reasoning the data gathered from a past contest are held to yield useful information for deliberations to a future contest. One example is in identifying tactical strategies for use in a future setting based on the availability of existing information. This process makes use of predictions, or expectations, based on the following lines of inductive reasoning. If a given behavioural tendency is observed under a certain set of conditions, then the same behavioural tendency is expected in the future under the same (or similar) set of conditions. Thus, future sports performances can be (and are) predicted from past data on the basis of probabilities (cf. Reep and Benjamin, 1968). We noted earlier that the sport behaviours must be invariant if this practice is to be valid.

While the existence of invariant behaviours in sport is key if the information from a past contest is to be used reliably to prepare for a future contest, a demonstration of these invariant features in the scientific literature is often found to be wanting. Invariant features in sports contests may be sought at a variety of levels of analysis of behaviour, from the individual player through to the collective interactions of many players. Since the presence of more players rapidly increases the number of possible interactions, it follows that dyadic (i.e. one versus one) sports may exhibit invariant tendencies more readily than team (i.e. many versus many) sports. In addition, the nature of dyadic sports is often more structured than that of team sports, an observation that may also tend to less variability in dyadic sports. For example, possession in dyadic sports is often exchanged in intermittent fashion, whereas, in team sports possession is more likely won and lost in transient fashion as a result of competitive efforts. There are of course some exceptions to this general observation.

On the above reasoning, McGarry and Franks (1996) examined international squash match-play behaviours in an initial search for invariant patterns in sports contests. Specifically, we tried to identify behavioural signatures for the individual using probability matrices as an index of a player's shot selection patterns. The data yielded mixed results. In short, the probability of a player's shot type was more reliable (or stable) when the context of the antecedent shot was more detailed. These findings were taken to indicate that the level of analysis is an important consideration if the information from one contest is to transfer reliably to the next contest. Further data on the search for stable playing patterns in squash were reported in Hughes *et al.* (2000). Notwithstanding, the mixed results from our research prompted us to reconsider the behaviours that occur in sports contests using an alternative system description for squash match-play, as described below.

18.3 Identifying key behaviours (perturbations) in sports contests

Sports analysts have typically sought to record the entirety of behaviours observed in a sports contest and then to examine these data for trends in the patterns of play. This approach to sports analysis tends to treat each datum value as being of equal importance. However, as sports contests are dynamic processes that are constantly changing, it seems likely that not all sports behaviours will be of equal importance to the performance outcome. If key behaviours exist then these features would be considered as critical determinants of sporting success. In squash match-play a well-placed shot that extends the opponent and forces a weak shot response on which to capitalize might be considered key if it perturbs the system and causes instability. Likewise, a penetrating pass in soccer that exposes a vulnerable defence and suddenly creates overload from a previously contained situation might too be considered as critical if the system suddenly becomes unsettled. This type of thinking requires a markedly different method of recording the spatial–temporal relations that take place in a sports contest from the approach that has been undertaken hitherto. We describe one method that we have used in our efforts to analyse squash match-play as a dynamical system. An alternative method of describing a dynamic system is provided by Grehaigne *et al.* (1997). These authors reported a different type of dynamical analysis in their descriptions of the transitions that occur between play configurations in soccer.

McGarry *et al.* (1999) sought to examine our hypothesis on the existence of perturbations in squash contests using human raters. Sixty rallies, selected at random from an existing data set, were analysed using video for possible system transitions between stability and instability. The analysis required an independent observer (six experts and six non-experts) to examine every shot in each of the rally sequences for evidence of what might be considered a system transition, or perturbation. Each independent observer received the following worded instructions at the start of the experiment. 'We consider squash match-play as a behavioural exchange between two players whose behaviours transit between bouts of stable and unstable behaviour as a result of a perturbation. Stable behaviours arise when both players seem to be in control of the rally and unstable behaviours arise when either or both players seem to be out of control in the rally and scrambling. The experimental task is to identify those shots that perturb the system from stability to instability and from instability to stability'. We identified the shot that marked the onset of instability (from stability) as a perturbation *onset*. The shot that marked the return to stability (from instability) was identified as a perturbation *offset*. Thus, an onset identifies the start of an unstable exchange and an offset the (re)start of a stable exchange. Hence each shot within a rally could be identified as being stable or unstable from the perturbation flags that indicated the onsets and offsets, respectively. From these data good inter-rater agreement as to the state of the system at any instant was reported.

Hughes *et al.* (1998) extended the notion of perturbations in sports contests from the dyadic sport of squash to the team sport of soccer. These authors defined a perturbation in soccer as some behavioural event that changes the rhythmic flow between the attacking team and the defending team. In each instance a perturbation in soccer was judged to lead to a shooting opportunity. Thus, on some occasions a perturbation might result in a shot at goal (termed a critical incident), whereas, on other times the perturbation might be assuaged and the shooting opportunity forfeited. Reasons for the damping of identified perturbations were held to be poor attacking skills and/or good defending skills. In an analysis of the 1996 European Soccer Championships, Hughes *et al.* reported significant differences between winning and losing teams in goal/perturbation ratios.

18.4   Sports contests and complex systems

To date, the soccer analyst has tended to focus only on those behaviours that occur 'on-the-ball', for example, a tackle, a dribble, a pass or a shot. The exceptions to this rule are found in the analysis of individual players, as seen in the use of time–motion analysis to assess the physiological work-rate demands of individual positions in soccer (e.g. Reilly and Thomas, 1976). 'On-the-ball' behaviours might be expected to provide a good system description for soccer given that ownership of the ball is key for wresting control of the game. Even so, there are some underlying problems with this method of data recording, not least because the game context can change depending on various factors. By way of illustration, we have highlighted some problems of context in Figure 18.2(a) and (b). In each case team 'X' makes the same forward pass against team 'Y', but the context in which the pass is made is changed. The two cases presented in Figure 18.2(a) demonstrate cooperating and competing (i.e. opposing) players in similar positions with respect to the ball as well as each other; however, the cases vary with respect to the position on the field from where the pass is made. The pass in Case 1 is different qualitatively from that of Case 2. In the first case the pass is less of a threat for the defending team than for the second case, although in quantitative terms the pass might be recorded as being the same in each instance. Likewise, the significance of an intercepted pass will change as a result of where on the field of play the pass is intercepted. The scenario in Figure 18.2(b) demonstrates the problem of context with respect to the position of the opposing players. In Figure 18.2(b), the position of only one opposing player (the last Y defender) has changed, but a very different description is presented of what might be recorded as the same qualitative event. Specifically, Case 2 presents a much less contained (or unstable) situation than that of Case 1. The problem of finding a suitable description for a soccer match, as indicated in the above examples, is further complicated given that the activities that occur 'off-the-ball' (not demonstrated in Figure 18.2(a) and (b)) can have a profound impact on the activities that occur on-the-ball. The solution to this problem is not an easy one.

In soccer, each opposing player (and team) competes for control of the match using the variables of space and time. Thus, each team-mate will cooperate with each other in the shared pursuit of control of these same variables. The cooperative and competitive tendencies that exist among players within and between teams, respectively, would furthermore be expected to occur within local constraints. For instance, a common aim of each player (and team) is to free-up space when possession is won and to tie-up space when possession is lost. This duality of purpose, which alternates as possession changes, is carried out at the various levels of playing units, from individual units, through sub-units to the whole team unit. Importantly, the interactions between players are flexible and transient, they are forged and broken at various times as the players cooperate and compete for possession of the ball (or space) in the context of the game demand.

It is possible to consider the spatial–temporal relations as key components in sports contests like soccer yet they have received little formal attention from the scientific community to date. However, we (McGarry *et al.*, 2002) have begun recently to consider the

(a)   Context: field position

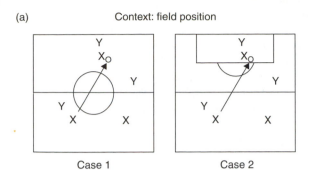

Case 1   Case 2

(b)   Context: player configuration

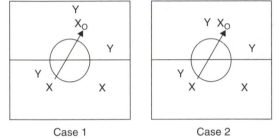

Case 1   Case 2

Figure 18.2   The audit of a soccer match often ignores the context in which the actions occur. (a) The context of the pass changes as a result of the field location in which the pass is made. The pass in the first case is judged as being less important than the pass in the second case. (b) The context of the pass changes as a result of the position of the players. The pass in the first case is judged as being less important than the pass in the second case.

spatial–temporal relations that describe sports contests in terms of a complex system, whose form and function materialize from the non-prescribed interactions that occur between the individual players who comprise the system. The search for a lawful system description of the spatial–temporal relations that occur within a sports contest is of course a challenging undertaking. The prize, however, would be a formal understanding of the patterned relations that are believed to give dynamic structure to sports contests. The type of system description advanced above is consistent with one facet of professional data capture systems for sports analysis now provided commercially. In these systems all of the behavioural events that occur in a soccer game, including the x–y coordinates of every player, as well as the referee and the ball, can be recorded in continuous fashion. The problem for the sports analyst is to locate the dynamic patterns within these data that give rise to the invariant features that exist in sports contests. It is only when invariant (or reasonably invariant) patterns are detected in the data that the information from a past setting can be used reliably by the coach to prescribe behaviours (or strategies) for future competition.

18.5   Information, decision-making and sports analysis: some practical applications for soccer

The objective of the coaching process is to produce sports performances that lead to sporting success. This aim is realized by the coach providing efficient instructions to the athletes in practical settings. The focus of such instructions centres on the development of an array of technical skills that are to be applied in the collective pursuit of a common objective. Often the skills will be developed in the context of a game plan, to which each player is expected to work within. Thus will the coach seek through instruction to maximize *a priori* some of the decision-making choices of his/her charges. To guide his/her instructions the coach will rely on his/her own analysis of sports performance and past experiences. Hopefully, the coach will also make good use of the findings from scientific studies that are based on empirical data. In this section, we describe some research from sports analysis that has yielded some specific training recommendations for soccer practice. The focus of this research is on various aspects of the penalty kick in soccer.

18.5.1   *The penalty kick*

The penalty kick is often a critical determinant of success in soccer. While the odds for success favour the penalty taker instead of the goalkeeper, the outcome of a penalty kick remains uncertain. That the goalkeeper is disadvantaged might be expected given the time constraints under which the goalkeeper must operate. After analysing penalty kicks from the FIFA World Cup Finals between 1982 and 1994, Franks *et al.* (1999) reported that it takes about 600 ms from the instant of foot contact for the ball to cross the goal line, whereas it takes in the region of 500–700 ms for the goalkeeper to complete the movement (i.e. dive) in an attempt to stop the shot. For successful goalkeeping performance the decision to initiate movement (including the direction of the dive) must therefore be made some time before ball contact. Laboratory-based studies have reported 100 ms to be the lower limit for reacting correctly to anticipated visual cues. Thus, the goalkeeper must detect and react to visual cues provided by the penalty taker approximately 100 ms before ball contact if he/she is to stop the shot. The visual cues that are provided outside of this window occur too early or too late on which to make a decision. The goalkeeper should not react too late of course given the constraints of reaction time and movement time noted earlier. Nor, however, should the goalkeeper react too early for two reasons. The first reason is that the goalkeeper will not benefit from subsequent (more reliable) visual cues provided by the penalty taker. For example, Franks *et al.* (1999) reported goalkeepers to be about 46% correct in their direction of movement when they reacted too early. (In the 1982–94 World Cup Finals, the movement of the goalkeeper before ball contact was against the rules but allowed in practice nonetheless.) Excluding any decision to not move, the reported 46% accuracy in prediction of shot direction is consistent with that of 50% that would be expected from chance (i.e. guessing). The second reason for the goalkeeper not to react to early cues is that the goalkeeper will provide the penalty taker with late visual cues before the instant of ball contact.

The decision-making abilities of the goalkeeper within the time constraints mandated by the penalty kick will be improved if reliable visual cues provided by the penalty taker can be detected in sufficient time. Franks and Hanvey (1997) examined the following visual cues provided by the penalty taker for their predictive utility – the starting position, the angle of approach to the ball, the forward or backward lean of the trunk, the placement of the non-kicking foot just prior to contact, and the point of contact on the ball of the kicking foot. The only visual cue found to be both reliable and time efficient was the placement of the non-kicking foot, which pointed in the direction that the ball would travel on more than 80% of the penalty kicks. From the penalty kicks taken in the 1996 European Championships, the placement of the non-kicking foot predicted the direction of the penalty kick in excess of 85%. Furthermore, detection of the placement of the non-kicking foot was estimated to allow a reaction time of 150–200 ms up to the instant of ball contact. These results indicated that a training programme aimed at improving a goalkeeper's decision-making capability within realistic time constraints might yield good dividends for the goalkeeper. In addition to the placement of the non-kicking foot, the angle of the hips has also been reported to provide a valid predictive cue for the goalkeeper (see Williams and Burtwitz, 1993). Placement of the non-kicking foot and the consequent initial angle of the hips would be expected to yield complementary information for the goalkeeper, given the biomechanical constraints that are imposed as a result of anatomical linkages.

18.5.2   Penalty shoot-outs

The nature of the penalty kick necessitates a decision by the goalkeeper in real-time based on the detection of visual cues. However, some decisions in sports contests can be made ahead of time. In soccer the line-up order for a penalty shoot-out provides a good case for *a priori* decision-making on the part of the coaching staff. If the probability of a given player scoring on a penalty kick varies with context – say the scoring probability varies depending on the importance attached to a given penalty kick – then the line-up order that best maximizes the chances of winning the penalty shoot-out poses an interesting question. The problem presented is one of optimisation as illustrated in the following dilemma. To which penalty kick should the best penalty taker be assigned – the first penalty kick in order (hopefully) to get the team off to a good start, the fifth penalty kick, which might be considered as the most important should the shoot-out reach that far, or the third penalty kick, which is more important than the first penalty kick while, at the same time, guarantees that the best penalty taker will take a kick in the shoot-out? McGarry and Franks (2000) analysed all line-up combinations using computer simulations to try to answer the above questions. The analyses indicated that the order of 5–4–3–2–1 represents the best line-up with which to contest a penalty shoot-out –, that is, the fifth best penalty taker should take the first penalty kick, the fourth best penalty taker the second penalty kick and so on. These results hold for a certain set of initial conditions. The best strategy following a tie after five kicks each (i.e. 'sudden death') is to assign the next best penalty taker to the next penalty kick. This is to say that the sixth best penalty taker should be assigned to the sixth penalty kick, the seventh best penalty taker to the seventh penalty kick and so on. This sequence applies for each team regardless of which team wins the coin toss and hence takes the first penalty kick.

The example from the penalty shoot-out exemplifies the point that some of the decisions that must be made in a particular sport context can be prepared for ahead of time. Wider implications for match strategies in soccer emerge if the best line-up for a penalty shoot-out is to be carried out successfully in practice. For instance, following a tied game at the end of extra time it becomes necessary for the coaching staff to identify the best five penalty takers from the available players on the field of play. In practice, this would require an *a priori* ranking of the penalty-taking abilities of each squad player. Importantly, the goalkeepers (i.e. penalty stoppers) should be subjected to the same ranking process too. The intelligent use of substitutions, including goalkeepers, with an eye on a pending penalty shoot-out was therefore advised, circumstances permitting (McGarry and Franks, 2000).

The penalty shoot-out is an important component in the knock-out rounds of soccer competition, as indicated in the following 'statistic'. In each of the 1982–98 World Cups, as well as the 1996–2000 European Championships, at least one team in the final match won a penalty shoot-out in a knock-out round of that same tournament. The only exception to this statistic was the 1994 World Cup in which the final match itself was decided on a penalty shoot-out. If a team is to win an international soccer tournament then there is a fair chance that it will have to win a penalty shoot-out in that same competition. The long-term systematic preparation of the entire squad for a penalty shoot-out would therefore seem appropriate.

Summary

The behaviour of an athlete in sport competition is the product of many complex processes. The aim of various disciplines within the sports sciences is to understand these processes at a fundamental level. Match analysis might help to integrate the separate contributions from the various disciplines. To help illustrate this point the following outlook into future integrated match analysis is offered.

The analysis from the match has been tabulated and after reviewing a brief summary of game statistics the coaching staff is concerned that late in the game 'crosses' from the right side of the team's attack were being delivered behind the defenders and too close to the opposing team's goalkeeper. The result was that the front strikers were not able to contact the ball despite making the correct approach runs (information also gained from the match summary). The coaching staff call for videodisc (immediate recovery) excerpts of each crossing opportunity from the right side of the field in the last 15 min of play. Along with this visual information, the computer retrieves other on-line information that is presented in the inset of the large projected video image. This information relates to the physiological condition of the player(s) under review leading up to the crossing opportunity. In addition, a biomechanical three-dimensional analysis of the crossing technique is presented as each cross is viewed. One player has been responsible for these crosses. Upon advice from the consulting exercise physiologist, coaching staff members have concerns about the telemetred respiration and heart rate of the player. A time–motion analysis of the player's movements in the second half of the game is called for, as well as a profile of the player's fitness level and physiotherapy report prior to the game. These are also retrieved from the same videodisc. After considering the information the coaching staff members record their recommendations for team and individual improvement and move on to the next problem identified from computer analysis. A computer program running in the 'background' is busy compiling instances of good performance (successful crosses) and poor performance that will make up an educational modelling programme for the individual player to view. Also, the expert system of coaching practice is being queried about the most appropriate practice for remedial treatment of 'crossing' in this specific setting. An individual fitness programme is prescribed when another expert system is queried. Lastly, the likelihood of the number of crosses in the final 15 min from the right side of the field being more successful against the next opponent, as well as their expected effect on match outcome, would be predicted for assessment purposes.

All aspects of the above scenario are in place or are under investigation in notational analysis laboratories throughout the world.

References

Franks, I.M. (1988). Analysis of Association Football. *Soccer Journal*, September–October, 35–43.
Franks, I.M. and Hanvey, T. (1997). Cues for goalkeepers. *Soccer Journal*, May–June, 30–8.

Franks, I.M., McGarry, T. and Hanvey, T. (1999). From notation to training: analysis of the penalty kick. *Insight*, **3**, 24–6.

Franks, I.M. and Miller, G. (1986). Eyewitness testimony in sport. *Journal of Sport Behavior*, **9**, 38–45.

Franks, I.M. and Miller, G. (1991). Training coaches to observe and remember. *Journal of Sports Sciences*, **9**, 285–97.

Grehaigne, J.-F., Bouthier, D. and David, B. (1997). Dynamic-system analysis of opponent relationships in collective actions in soccer. *Journal of Sports Sciences*, **15**, 137–49.

Hughes, M., Dawkins, N., David, R. and Mills, J. (1998). The perturbation effect and goal opportunities in soccer. *Journal of Sports Sciences*, **16**, 20.

Hughes, M., Wells, J. and Williams, C. (2000). Performance profiles at recreational, county and elite levels of women's squash. *Journal of Human Movement Studies*, **39**, 85–104.

Magill, R.A. (2001). *Motor Learning: Concepts and Applications*, 6th edn. McGraw-Hill Higher Education, New York.

McGarry, T. and Franks, I.M. (1996). In search of invariant athletic behaviour in sport: an example from championship squash match-play. *Journal of Sports Sciences*, **14**, 445–56.

McGarry, T. and Franks, I.M. (2000). On winning the penalty shoot-out in soccer. *Journal of Sports Sciences*, **18**, 401–9.

McGarry, T., Khan, M.A. and Franks, I.M. (1999). On the presence and absence of behavioural traits in sport: an example from championship squash match-play. *Journal of Sports Sciences*, **17**, 297–311.

McGarry, T., Anderson, D.I., Wallace, S.A., Hughes, M. and Franks, I.M. (2002). Sport competition as a dynamical self-organizing system. *Journal of Sports Sciences*, **20**, 771–81.

Reep, C. (1989). Charles Reep. *The Punter* (The Scottish Football Association). September, pp. 31–7.

Reep, C. and Benjamin, B. (1968). Skill and chance in association football. *Journal of the Royal Statistical Society, Series A*, **131**, 581–5.

Reilly, T. and Thomas, V. (1976). Motion analysis of work rate in different positional roles in professional football match-play. *Journal of Human Movement Studies*, **2**, 87–97.

Williams, A.M. and Burwitz, L. (1993). Advance cue utilisation in soccer, in *Science and Football II* (eds T. Reilly, J. Clarys and A. Stibbe), E & FN Spon, London, pp. 239–44.

19 Information technology

Tony Shelton

Introduction

During our lifetimes, information and computer technologies have significantly transformed the world. Information technology (IT) systems have made our world smaller, safer and ever more efficient. IT has improved communications, education, and health care and given access to multiple information sources for individuals and many different private and public sector organizations. IT systems have helped make life easier and more pleasant for millions of ordinary people by eliminating or automating mundane tasks and everyday chores and by freeing time for the more innovative and creative side of human nature. Over recent decades, there has been a continuing technological evolution in most nations from the earliest pre-industrial, agricultural communities to the more recent industrial societies and today's 'dot com' culture based on the provision, sharing, and management of information. This major technological evolution has had during the last 20 years very profound effects on contemporary sport and leisure (see, for example, the classic review of computers in sport by Lees, 1985). To illustrate these developments for soccer, this chapter focuses on the application of contemporary information technologies to soccer (especially World Cup soccer) and the increasing visibility of the game on the Internet.

19.1 Background and history

Hull (1992) in his book on the history of IT described how, in 1943, development work started on the Electronic Numerical Integrator and Computer (ENIAC) for the US Army. In the same year, Alan Turing and his colleagues at the British Code and Cipher School developed the Colossus machines. In 1948, Turing and his research team working at Manchester University built the world's first electronic, digital, programmable computer.

The earliest computers in the 1940s and 1950s required enormous rooms and air-conditioned, dust-free environments. These early machines were used for 'number crunching' of large amounts of arithmetical data. They were slow, highly expensive, and often unreliable. By the late 1950s, computers, despite their problems, were in widespread use in many commercial organisations for everyday applications such as accounts, payroll and stock control as well as for more sophisticated applications such as statistical analyses.

Early time-sharing techniques in which a single computer could be accessed from many remote but integrated terminals (later computer networks) were introduced during the 1960s. This fusion of communication and computing technologies led to the development of modern 'information technologies'. The eventual development of computer games such as the early versions of Space Invaders can be traced to these developments (Cornwall, 1986).

By 1971, the Intel 4004 chip had taken the integrated circuit one step further by locating all the components of a computer on a minuscule chip. Previously the integrated circuit had to be fashioned for a special purpose, now one microprocessor could be programmed to meet any number of complex demands. Whereas, by the early 1980s, arcade video games such as Pac Man and home video game systems such as the Atari 2600 had created massive consumer interest for programmable home computers.

Since the 1980s, satellite and cable television have made networks possible in which access to many forms of home information and entertainment are available on subscription. In addition, in many countries viewdata systems, in which information is stored on computer databases, can offer the public a wide range of TV services (including soccer results and general news).

More recently during the 1990s, while the data processing speed, storage capacity, and the accuracy of small computers have been constantly increasing, their cost and size has been constantly decreasing. Stern and Stern (1993) noted that by 1993 personal computers were in 40% of homes in America and Europe. In today's modern world, digital mobile phones are in widespread use and portable computers with built-in cellular telephones are becoming available. Notebook computers have been available since the early 1990s and powerful computers which people can wear on their wrists or hang around their necks will be available in the near future.

The Internet has linked computers worldwide into a single massive information cluster (see Leiner *et al.*, 2000). The development of the Internet has revolutionized computing and communications and web addresses such as http://www.fifa.com are in normal everyday use. Initially, Licklider and Clark (1962) described the interactions that could be enabled through global networking. In discussing the 'Galactic Network' concept, Licklider and Clark envisioned a globally interconnected set of computers through which users could access data and programs from any site. The estimated size of the World Wide Web has now surpassed 1.5 billion pages. This rapid and dramatic growth of the Internet over the last decade has been accelerated by the rapid adoption of Web browsers allowing users very easy access to Web-based information. Today, the Internet is changing and evolving to provide new services such as audio and video streams. The widespread availability of powerful affordable computing and communications in portable form is making possible new and novel applications (e.g. Internet, telephone and television).

At present, the most popular current use for computer networks such as the Internet is for accessing electronic mail, or e-mail, which allows users to type in a computer address and send messages through networked terminals across the world. Increasingly, today's computers are able to accept spoken word instructions (voice recognition) and enact intelligent reasoning. For example, intelligent computer-based expert systems assist coaches by providing access to expert coaching diagnoses. The ability to translate a foreign language is also increasingly possible with contemporary computers.

Electronic mail, computer conferencing, and the World Wide Web have allowed communications to expand rapidly. Electronic communication has become common and for many users, these interactions often seem more intimate, secure and convenient than normal face-to-face communication. However, recently, there has been a minor reversal in this expansion. For example in 2001, the UK Consumers' Association found that the number of people who named e-mail as their favourite form of communication had reduced, in the space of the previous year, from 14% to 5%. Meanwhile, the numbers that preferred face-to-face meetings jumped from 39% to 67%. Old-fashioned 'snail mail', in contrast was growing in volume by 2% a year.

Since the spring of 2000, many major technology firms have been bankrupt and shares in high-technology, telecommunications and Internet companies suffered sharp falls on the world stock markets that were accelerated by the attacks on America on Tuesday, 11 September 2001. The collapse of technology share values and particularly the so-called dotcom Internet share valuations suggest that Europe, Japan, and North America may be already in the post-Internet era for IT and digital media.

19.2   The Federation Internationale de Football Association (FIFA) World Cup

The FIFA World Cup has always brought memorable and exciting action to radio and TV audiences all over the world ever since the first event in 1930 in Uruguay. The 2002 tournament attracted the attention of soccer fans, as an unprecedented event in the FIFA World Cup history. The tournament was the first World Cup of the new millennium, the first to be held in Asia, and the first to be co-hosted by two nations, The Republic of Korea and Japan.

Held every 4 years, soccer organizations in each host country manage each World Cup. This organization presents multiple computing and communications problems. For 2002, these problems had to be overcome to ensure a successful and exciting competition for both players and billions of TV spectators. A predicted cumulative TV audience in the region of 40 billion should have tuned in to this event.

In the final spectacular game of the 1998 World Cup, billions of TV viewers had seen the host nation France dominate the defending champion Brazil in a remarkable 3–0 win. The French midfield star Zidane, who earlier in the tournament had been banned for two matches, played superbly and scored twice against Brazil. The Brazilian star forward, the 21-year-old Ronaldo was treated in hospital hours before playing in the game.

In 1998, for the 32 national teams, the millions of spectators at the 10 stadia for the matches and a total 37 billion TV viewers, the FIFA World Cup was the greatest single sports event in the world. This World Cup also served as a major focus and showcase for French expertise in the field of IT and telecommunications (more details are available at the website: http://www.worldcuparchive.com/CUPS/1998/wc98index.html).

Information technology applications were vital for the sale of tickets, the organization of the staff, transport and accommodation logistics, the accreditation and access control and the management of the venues and equipment. Journalists and television viewers were provided with details, statistics and results of the matches via IT. Multimedia workstations enabled both the international media and the public to have rapid access to information.

The 12 000 journalists who worked in the 10 press centres in each of the stadium locations across France and the International Media centre in Paris could access a computer-powered Intranet system. This provided them with full e-mail facilities, game and team statistics, and essential background information that helped them report on the tournament. For the first time in the history of the World Cup, video clips of highlights of each of the 64 matches were available on specially configured PCs in the press centres. The 1998 operations system scheduled the greatest number of teams and games since the World Cup began. It contained the on-line service for purchasing tickets and the www.france98.com official website.

The ticketing system, developed by Electronic Data Systems (see Sullivan, 1998) and implemented on Hewlett-Packard servers, helped to sell advance tickets one year earlier than first planned. The ticketing system was an important application because the revenue from ticket sales represented more than half of the entire operating budget. The system held reservations, assigned seats, and printed the tickets. Although the tickets for the games sold out by early May 1998, additional tickets became available as teams were eliminated and match seats were returned to the organizing committee.

In its role as the official IT hardware and maintenance supplier to the 1998 World Cup soccer tournament, Hewlett-Packard provided more than 75 different kinds of products to help run the event. Hewlett-Packard computers and printers produced the 2.5 million tickets for the 64 games, created more than 80 000 accreditation badges, and delivered real-time game information to the attending journalists and more than 13 million Internet surfers around the world.

To ensure the success of the 2002 FIFA World Cup, Japan and the Republic of Korea developed an IT system to supply multimedia information through various media, such as cable, radio and the Internet for the host cities. The 2002 FIFA World Cup generated several hundreds of millions of dollars from such corporate sponsors as Coca-Cola and McDonald's. Each major sponsor paid about $40 million to sponsor the games in exchange for major TV exposure for their brands. Coca-Cola was one of several long-term sponsors of the World Cup, having in 1998, signed an 8-year deal to sponsor the FIFA World Cup until 2006.

19.3   TV and soccer

Soccer and other football games are without doubt amongst the world's most popular sports and modern IT has had a considerable impact upon them. Professional soccer teams initially feared the introduction of radio and television as it was considered that it would reduce attendance. In contrast, as we all know, TV broadcasting has generally tended to increase interest in soccer. By 2002, TV has become the major economic factor in professional soccer, which is now obviously dependent upon TV deals. With this emphasis on the importance of TV coverage, let us consider a few examples of the impact of television on modern top-level soccer.

In the 1980s, the auction of UK TV rights to professional soccer was rather sedate and could be started and finished in one day. However, in 1992, the commercial satellite TV organization Sky Sports paid the major share of a groundbreaking £304 million deal to secure the first five years of live Premiership coverage in the UK. In contrast, the BBC paid just £4.5 million per season towards this total amount for delayed 'Match of the Day' highlights.

When the deal was renegotiated in 1996 for four more seasons, Sky had to pay £670 million and the highlights cost the BBC £18.25 million per season. In June 2000, at the height of media popularity for soccer, a new 3-year deal for live FA Premier League football was awarded to Sky Sports from 2001–02 for £1.7 billion. Sky Sports was also awarded a 3-year deal for FA Cup and England internationals from 2001–02. Other deals for the highlights and overseas and Internet rights could easily push the total amount that the League could receive during the 3-year period close to £2.0 billion.

For the 2001–02 season, live Scottish football on Sky switched from Sunday to Saturday evenings at 17:35 h. Both the Scottish Football Association and Sky have claimed that the move was to meet the demands of supporters who disliked the Sunday evening slots. However, the move was also possibly a way of reducing the impact of the showing of the highlights of Premier League soccer on ITV. In 2001, The Premier League also announced the first games to be transmitted on pay-per-view (PPV) television. Therefore, the number of Premiership games scheduled for UK soccer's traditional Saturday afternoon could be further reduced as the PPV games are often transmitted on a Sunday at 14:00 h followed by Sky's live game at 16:00 h.

The *Financial Times* reported in 2002, that UEFA (Union des Assiociations Européennes de Football), the governing body for European soccer, was close to a deal with European regulators to shake up the '€530 million-a-year' market for television rights to the Champions' League. One of the options being considered is to split the rights for each country into at least two main packages, one for pay-television matches and one for free-to-air games, to be sold by UEFA. Some major clubs wanted to retain the right to broadcast their games on the Internet and mobile phones, while leaving UEFA to sell the traditional TV rights.

19.4   The Internet and soccer

19.4.1   *Impact of the Internet*

The impact of the Internet on soccer has been very extensive and rapid. On the Internet, neither advertising space or available airtime determines the dissemination of soccer information, and news can be recorded and left on-line for future reference. Over 11 million soccer-related Web pages exist currently as many supporters have started their own sites as they gain access to the World Wide Web. Most professional soccer clubs have websites that they use to generate revenue or solidify their support and in which player information, game schedules, and ticket information is available on-line. In their on-line sports stores, clubs are displaying and selling miscellaneous soccer merchandise.

On the international scene, soccer is heavily featured on the Internet. For example in China, there are several million mostly young, male Web surfers and Chinese soccer websites played a major role in generating public awareness of the 2002 World Cup. According to SoccerNet (http://www.espn.go.com), an estimated 100 million Chinese TV sets tuned into a recent mid-week match when China played regional rival Japan. The European Information Technology Observatory (http://www.eito.com/) forecasts that China will have 20 million users on-line by 2003, just 1.6% of the population. By comparison, the US, Canada and Sweden already (2002) have over 40% of their populations on-line. China's 20 million estimated users in 2003 will be the same size as the total on-line population of Western Europe in 1997, and larger than the on-line populations of any EU country other than Germany in 2003.

Chinadotcom provided Internet services for Team China in the run-up to the 2002 World Cup. Fans across China and around the world were able to access up-to-the-minute bilingual information on national teams, players, match schedules and results on the official website, (www.team.china.com) built in conjunction with the sponsorship of the Chinese Football Association.

As the cost of broadband Internet access reduces, streaming popular broadcast content such as soccer over the Internet becomes more readily available. A live Internet streaming of the match between Denmark and the Netherlands on 10 November 2001 allowed viewers to watch on-line. Often soccer matches can be heard on live Internet audio broadcasts. In the US, Videosport.com (PPV) offers live video and audio Internet broadcasts of major international soccer events in Europe.

The future of soccer on the Internet may be with a system that lets you watch games in a three-dimensional (3D), graphical rendering. For example, ToPlay Soccer (Intel) lets users choose the camera angle, speed and view of a game. In the near future, soccer teams and TV companies are expected to offer digitized games from their websites. It takes about 24 h to convert a game into the 3D format, while a 20-s clip can be converted in less than an hour, according to Intel representatives. The system uses Internet 3D technology, which reduces the amount of bandwidth needed to view graphics over the Web and can be used with a standard Internet connection. A typical soccer game will consume about 4 MB of drive space, according to Intel.

19.4.2   Some useful soccer websites

www.thefa.com – the FA website is the official homepage of English football on the Internet; www. sportinglife.co.uk/ – good source of daily soccer and sports news; www. scottishfa.co.uk/ – the official Scottish Football Association website; www.fifa.com/ – is the FIFA homepage; www.adidas.com/ – is the Addidas homepage; www.wsc.co.uk/ – when Saturday Comes is an independent UK soccer magazine written from a fan culture perspective; www.uefa.com – is the European governing body website; www.nike.com/ – is the Nike homepage.

A large and growing proportion of all Internet newsgroups is devoted to soccer-related topics. Each new soccer information service, for example, usually has several newsgroup discussion areas. The most popular soccer websites contain fantasy leagues. Participants select teams of real players, and match by match; they see how their soccer teams are doing by checking player statistics. Often participants can bargain for player transfers during the season. Team selection and management are achieved through e-mail and websites.

There is organized soccer betting on several websites where registered on-line users can get a variety of match odds and can place their bets. It is anticipated that soon sophisticated real-time gambling will be linked to games being broadcast on television. The ultimate fate of interactive betting may depend upon how gambling is regulated over the Internet and if on-line gamblers have access to real-time accurate analysis of soccer performances.

19.5   The role of information technology in enhancing performance

Artificial Intelligence (AI) systems (Ignizio, 1991) can easily disseminate expert advice. These systems could aid and accelerate team selection by allowing the very best soccer selectors' knowledge and reasoning abilities to be available on a soccer club's computer systems. Expert systems shells (initially 'empty' AI systems which can be 'filled' with all types of expert knowledge), when coupled with the latest multi-media technologies, also offer the prospect of developing interactive, intelligent systems for skills training in soccer.

Increasingly, IT is used for the gathering, analysis, and display of large sets of data on soccer play and performance and for the simulation of human movement. In the future, AI expert systems could aid human soccer selection and training by analysing databases of player demographics. Databases and other commercial software could also relieve many of the problems of coaching and organizing soccer.

Modern computer-based information and decision support systems allow players' personal, medical and playing details to be made directly available to managers and coaches and in case of injury or illness to be immediately accessible to doctors or to other medical staff. Database systems could also contain an individual player's personal, attendance, fitness and medical records for health monitoring purposes. Such systems would aid the administration of the team by improving both the efficiency of the organization's record-keeping and by improving the organization's internal communications. Modern spreadsheet packages facilitate the analysis of all forms of data and will allow the results to be presented and reported in various tabular or graphic formats for improved managerial decision-making.

Clearly, database systems could also help many clubs in accelerating this selection process by allowing the very best soccer selectors' knowledge and reasoning abilities to be readily available in a club's computer systems. Expert systems shells when coupled with the latest multi-media technologies offer the real prospect of developing interactive, self-paced, intelligent systems for motor skills training in soccer.

In the past, many soccer coaches have been very dependent on their own subjective (and undoubtedly biased) observations of the team's and of individual players' previous performances for evaluation and coaching purposes (Lyons, 1988). A team's coaches can now replay the whole match on videotape in slow motion with full freeze frame facilities and even with computer control (Franks and Nagelkerke, 1988). These useful playback facilities allow accurate and objective coaching advice to be more widely available. In addition, modern video technologies obviously allow coaching information to be presented in a more complete and dynamic format rather than by just offering the players verbal comments.

Video recordings, when edited, can also be made available either for resale or for publicity, advertising/sponsorship or for coaching purposes. The use of video also allows the detailed investigation of players' individual techniques (and any variations in their performance) and abilities, and permits scientific, biomechanical, or kinematic analysis. The use of video is ideal in coaching, it is versatile and inexpensive in use, and tapes can be edited and used.

19.6 Multi-media systems

In addition, interactive multi-media systems including video or CD-ROM technologies, are now being developed for the delivery of entertainment, education and training, information dissemination, or for the marketing or promotion of products or services (Bhatnagar and Mehta, 2001). Smart Cards are used increasingly for gaining access to hotels, leisure centres, and many other areas in which personal or organizational security is required. Systems for the administration of various soccer membership and attendance schemes should improve ground security. It is apparent that the use of credit cards and bar code readers at point-of-sale terminals has dramatically improved the efficiency of supermarkets and other sales organizations and could have similar applications for the marketing of soccer.

Modern software applications such as word processing, spreadsheets, databases, and analytical or statistical software, desktop publishing are in standard, everyday use by major organizations. In the near future, AI systems, multi-media, and the development of virtual reality systems with real-time, full motion soccer action offer outstanding possibilities for highly interactive coaching, training and home entertainment. The use of information technology in soccer is not just science fiction hype but is quickly becoming contemporary reality.

Summary

Information and computer technologies have made a huge impact on many aspects of our lives. Their influences are evident in the organization of major soccer tournaments. These technologies have also affected commercial aspects of the game. Further developments in information technology should have benefits for soccer coaching and coach education as well as home entertainment.

References

Bhatnagar, G. and Mehta, S. (2001). *Introduction to Multimedia Systems*. Academic Press, New York.
Cornwall, H. (1986). *The New Hacker's Handbook*. Century Hutchinson, London.
Franks, I.M. and Nagelkerke, P. (1988). The use of computer interactive video in sport analysis. *Ergonomics*, **31**, 1593–60.
Hull, R. (1992). *In Praise of Wimps: A Social History of Computer Programming*. Alice Publications, Pecket Well, West Yorkshire.
Ignizio, J.P. (1991). *Introduction to Expert Systems: The Development and Implementation of Rule-Based Expert Systems*. McGraw-Hill, London.
Lees, A. (1985). Computers in sport. *Applied Ergonomics*, **16**, 3–10.
Leiner, B.M., Cerf, V., Clark, D., Kahn, R., Kleinrock, L., Lynch, D.C., Postel, J., Roberts, L.G. and Wolff, S. (2000). *A Brief History of the Internet*. http://www.isoc.org/internet/history/brief.html#Introduction.
Licklider, J.C.R. and Clark, W. (1962). *On-Line Man Computer Communication*. AFIPS SJCC, August.
Lyons, K. (1988). *Using Video in Sport*, National Coaching Foundation, Leeds.
Stern, N. and Stern, R.A. (1993). *Computing in the Information Age*. John Wiley and Sons, New York.
Sullivan, K.B. (1998). Net Takes World Cup Global, *PC Week*, June 8.

Part 5

Growth and adolescence

20 Growth and maturity status of young soccer players

Robert M. Malina

Introduction

Given the popularity of soccer throughout the world, there is, surprisingly, relatively little information on the growth and maturation of young players. In this chapter, the growth and maturity status of male soccer players between 9 and 18 years of age are considered, and then the results of two studies of elite Portuguese youth soccer players are treated in more detail. Implications of the growth and maturity data are discussed in the context of selection and retention of soccer players during childhood and adolescence, and risk of injury.

20.1   The young soccer player

There are many participants in youth soccer and many levels of competition. Comparisons of the growth and maturity status of young players at the local level may provide a different picture than comparisons of more elite players of the same age at national or international levels. Many youths participate in soccer programmes for a year or several years, and then move on to other activities as interests change, as skill demands become greater with higher levels of competition, and as competitive sport becomes more selective and exclusive. Hence, the meaning of the term 'young athlete' needs to be specified.

Young soccer players are usually defined in terms of success on school teams, in selected age group (e.g. travel team) and club competitions, and in national and international selections and competitions (e.g. under-12 regional selection or under-16 national selection). Many studies considered in this chapter include young male soccer players who can be classified as select, elite, junior national or national calibre. It was not always possible to estimate the talent levels of the samples. In some cases late adolescent players are labelled as semi-professionals.

20.2   Growth status

Growth status refers to size attained at a given chronological age (CA), most often as height and weight. Mean ages, heights and weights from studies of young male soccer players from about 9 to 18 years of age in Europe and the Americas were drawn from the literature. European data include samples from Austria (Kosova *et al.*, 1991), Belgium (Vrijens and van Cauter, 1985; De Proft *et al.*, 1988), Croatia (Jankovic *et al.*, 1993), Denmark (Lindquist and Bangsbo, 1993; Hansen *et al.*, 1999), Finland (Luhtanen, 1988), Germany (Herm, 1993), Hungary (Farmosi and Nadori, 1981), Italy (Dal Monte *et al.*, 1980; Mazzanti *et al.*, 1989; Viviani *et al.*, 1993), Portugal (Garganta and Maia, 1993; Magalhaes *et al.*, 1997; Malina *et al.*, 2000), the former Soviet Union (Lukyanova and Novocelova, 1964), the

United Kingdom (Baxter-Jones *et al.*, 1995; see also Jones and Helms, 1993), Wales (Bell, 1988), and combined samples from the Czech Republic, France and Germany (Junge *et al.*, 2000). Data for the Americas include samples from Argentina (Barbieri and Rodriguez Papini, 1996), Brazil (Matsudo, 1978; Soares and Matsudo, 1982; Soares *et al.*, 1994), Chile (Donoso *et al.*, 1980), Cuba (Alonso, 1986), Mexico (Peña Reyes *et al.*, 1994), and the United States (Kirkendall, 1985; Cumming, 2002). One report dates to 1964, five date to 1978–82, and the remaining date between 1985 and 2000. The data are cross-sectional, with the exception of a mixed-longitudinal study of young British players (Baxter-Jones *et al.*, 1995), a longitudinal sample of players in the former East Germany (Herm, 1993), and a 1-year longitudinal study of young Danish players (Hansen *et al.*, 1999). Data for the British and German samples were treated cross-sectionally. Since the data for the Danish sample spanned one year, only the initial observations were plotted. Most studies report data for a single age group (e.g. 10 years), or for several ages combined, usually two years (e.g. 11–12 years (although some studies include players of a broader age range in a single group). In some instances, only 'whole year' age categories were reported (e.g. 11 years). It was assumed that this represented the age at the last birthday and the mean age of the sample was assumed to be the mid-point of the interval (i.e. 11.5 years).

Mean heights and weights of the soccer players from Europe and the Americas are plotted relative to the new United States growth charts (Centers for Disease Control and Prevention, 2000) in Figures 20.1 and 20.2, respectively. Selected percentiles of the reference data are shown – 25th (P 25), the median (P 50) and 75th (P 75). Mean heights for the majority of

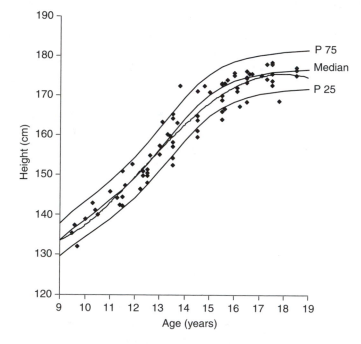

Figure 20.1 Mean heights of young soccer players from Europe and the Americas plotted relative to United States reference data: 25th, 50th (median) and 75th percentiles (Centers for Diseases Control and Prevention, 2000). References for the specific studies of soccer players are cited in the text. The dashed line is the fit of a fourth degree polynomial to the data points.

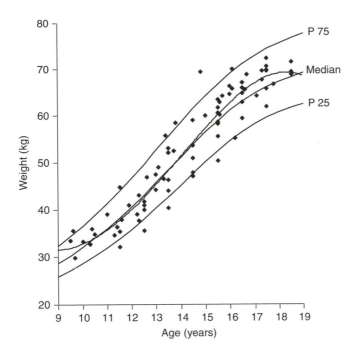

Figure 20.2 Mean weights of young soccer players from Europe and the Americas plotted relative to United States reference data: 25th, 50th (median) and 75th percentiles
(Centers for Diseases Control and Prevention, 2000). References for the specific
studies of soccer players are cited in the text. The dashed line is the fit of a fourth
degree polynomial to the data points.

samples of young soccer players from Europe and the Americas fall within the bounds of
P 25 and P 75. Only one mean height exceeds P 75, while 11 mean heights are below P 25.
A fourth degree polynomial was fitted to the data points for height. The resulting line fits the
reference median quite closely, and a possible age trend is suggested. The means tend to
fluctuate above and below the reference median during childhood and early adolescence
(about 8–14 years). In later adolescence (14+ years), however, many of the mean heights
approximate or are below the reference median.

Mean body weights of young soccer players are plotted relative to the United States reference in Figure 20.2. Mean weights for the majority of samples fall within the bounds of P 25
and P 75. As for height, mean weight for only one sample exceeds P 75, but five mean
weights are at P 75. In contrast to height, only four mean weights are below P 25 of the reference data, although several are on P 25. A fourth degree polynomial was fitted to the data
points. The resulting line approximates the reference median from late childhood into adolescence, about 9–14 years. Subsequently, the line is above the refence median and remains
there into late adolescence. Indeed, in late adolescent players (15+ years), all but three of the
mean weights approximate or are above the reference median.

The overall trend for the body size of young soccer players suggests generally similar
weight-for-height during childhood and early adolescence, but more weight-for-height in
later adolescence. The body mass index (BMI, weight/height^2) is commonly used to describe

weight-for-height relationships in the context of overweight and obesity. The BMI is of limited utility, however, for adolescent males in general and athletes in particular given the considerable growth in fat-free mass during normal male adolescence and the generally larger size and fat-free mass of adolescent athletes. It is more appropriately an index of heaviness (Malina *et al.*, 2002). Most studies of soccer players do not report the BMI, and it is misleading to estimate the BMI for a sample from mean heights and weights.

The weight-for-height relationship in adolescent soccer players is consistent with data for somatotype. Soccer players tend to be muscular or mesomorphic as assessed by the Heath–Carter somatotype protocol (Carter and Heath, 1990). Although there is some variation that is associated with sampling and expected changes with growth and maturation, particularly the growth spurt, mean somatotypes of early and late adolescent soccer players are generally similar to and fall within the ranges of somatotypes of national and international calibre players (Carter, 1988; Carter and Heath, 1990). This tendency emphasizes a potentially important role for physique in the selection or exclusion process. Soccer players also tend to be leaner than the general population. The proportionally high weight-for-height probably reflects a greater amount of lean tissue, specifically muscle mass.

Corresponding data for young soccer players from other areas of the world are apparently limited. Available height and weight data for young soccer players from Japan (Atomi *et al.*, 1986; Satake *et al.*, 1986; Togari *et al.*, 1988) and China (Chen, 1991) were compared with reference values for Japanese boys (Malina, 1994). Mean heights of young soccer players fall close to the reference for the general Japanese population and mean body weights are generally below the reference with one exception. A sample of candidates for the national youth team in Japan with a mean age of 17.5 ± 0.7 years is taller and especially heavier than the reference. Otherwise, the limited data for Japanese youth players indicate similar weight-for-height as the reference. Mean heights of the Chinese players are close to the Japanese reference from 10 through 15 years, but late adolescent Chinese players (16–17 years) tend to be slightly above the Japanese reference for height. In contrast, mean weights of the Chinese soccer players are generally below the Japanese reference. Hence, the sample of Chinese players has less weight-for-height, which is also true in the Chinese population in general.

There is a need for more data from young soccer players in different areas of the world. Such data are probably available in the medical and training records of soccer clubs, but do not ordinarily make their way into the scientific literature.

20.3   Maturity status

Maturation refers to progress towards the biologically mature state. It is an operational concept because the mature state varies with the body system considered. All tissues, organs, and systems of the body mature. Maturation, or the process of maturing, can be viewed in two contexts, timing and tempo. Timing refers to when specific maturational events occur, such as the age at which pubic hair appears in boys, or the age at maximum growth during the adolescent growth spurt. Tempo refers to the rate at which maturation progresses, for example, how quickly or slowly the youngster passes through the adolescent growth spurt. Timing and tempo vary considerably among individuals.

20.3.1   *Assessment of maturity status*

Biological maturity during childhood and adolescence is ordinarily viewed in the context of skeletal, sexual and somatic indicators. Of the three types of indicators, only skeletal maturity can be used during childhood and adolescence. The assessment of skeletal maturity involves matching a hand-wrist radiograph of a child to a set of criteria. Three methods are commonly used, the Greulich–Pyle, Tanner–Whitehouse and Fels methods (see Roche *et al.*, 1988; Malina *et al.*, 2003). The methods vary in criteria for making assessments and procedures used to construct a scale of skeletal maturity from which a skeletal age (SA) (sometimes called bone age in the clinical literature) is assigned. A SA corresponds to the level of skeletal maturity attained by the child relative to the reference sample for each method. Skeletal ages assigned by the three presently used methods are not equivalent.

Skeletal age is expressed relative to the youngster's CA, or the difference between SA and CA can be calculated. Skeletal age may also simply be compared to CA. For example, a boy may have a CA of 10.5 years and an SA of 12.3 years. In this instance, the boy has attained the skeletal maturity equivalent to that of a child of 12.3 years of age and is advanced in skeletal maturity status for his CA. In contrast, a boy may have a CA of 10.5 years but an SA of 9.0 years. Although the boy is chronologically 10.5 years of age, he has attained the skeletal maturity of a child 9.0 years of age; this boy is delayed in skeletal maturity for his CA. The terms advanced and delayed should be used carefully. In these examples, the terms simply mean that the child's SA is advanced or early, and delayed or late, respectively, relative to her/his CA. The terms imply nothing about the factor or factors that underlie the advancement or delay in skeletal maturity.

Indicators of sexual maturity are useful only during puberty when they are overtly manifest. These include genital and pubic hair development in boys, which are ordinarily rated relative to the criteria of Tanner (1962), and testicular volume. They are assessed during clinical examination, although some studies utilize self-assessment. Boys are described as being in a stage of genital and/or pubic hair development. As presented in Table 20.1, stages range from 1 (pre-pubertal) to 5 (post-pubertal or mature). Stages 2–4 indicate intermediate stages of the pubertal state. Stage 2 represents the initial, overt manifestation of the characteristic, while stages 3 and 4 represent more advanced stages. Note that the stages of genital and pubic hair development are not equivalent, for instance, genital stage 3 is not equivalent to pubic hair stage 3, and ratings for each characteristic should not be combined to yield an overall maturity score.

During puberty, boys of the same CA can be grouped by stage of genital or pubic hair development to provide an estimate of variation in body size associated with maturity status within an age group. It is important to emphasize that such comparisons are valid only among youth of the same CA or within a relatively narrow age range. Age groups are usually

Table 20.1 Range of CAs within each stage of genital maturation in a sample of boys[a]

Genital stage	Chronological age (years)
G1	10.00–12.08
G2	10.75–13.17
G3	11.75–14.58
G4	12.83–14.83
G5	13.75–14.92

Note
a Adapted from Haschke (1983).

defined in terms of a single year (e.g. 12.0–12.99 defines 12 years of age). It is probable that older children within a single age group differ in body size and performance from younger children within the same age group. This issue is often overlooked.

It is also of limited utility to group youth by stage of a secondary sex characteristic to the exclusion of CA unless the age range of the group is relatively narrow. This procedure may reduce the variation within the sample to some extent, but variation independently associated with CA is overlooked. Variation in CA within a sample of boys 10.0–14.9 years of age classified by stages of genital maturity is shown in Table 20.1. The range of CAs within each stage is minimally 2 years, and is greatest in genital stage 3 where the variation approximates 3 years, 11.7–14.6 years.

Studies of young athletes often characterize a sample as pre-pubertal, assuming that the youth do not vary in maturity status. Pre-pubertal simply means that the boys show no overt manifestations of secondary sex characteristics (i.e. genital or pubic hair stage 1). However, pre-pubertal boys do in fact vary in biological maturity, and the only indicator that is applicable is skeletal maturity. Among a sample of boys 8.2 ± 0.1 years, for example, SAs varied between 6.3 and 9.4 years (Malina *et al.*, 2003).

Somatic maturity refers to the timing of maximum growth in height during the adolescent spurt, age at peak height velocity. Longitudinal data beginning at 10 or 11 years of age and that span at least 5 years and preferably a broader span during adolescence are required to derive valid estimates of age at peak height velocity and peak velocity of growth at this time. Estimates of the age at peak height velocity are generally mathematically derived. Estimates for samples of European and American adolescents were recently summarized (Malina *et al.*, 2002).

20.3.2   *Maturity status of young soccer players*

Available data dealing with the biological maturity status of youth soccer players are based on SA and secondary sex charactersitics. Data for age at peak height velocity are more limited.

Several studies of skeletal maturity in Belgium (Vrijens and van Cauter, 1985), Italy (Mazzanti *et al.*, 1989; Cacciari *et al.*, 1990), Mexico (Peña Reyes *et al.*, 1994), and Japan (Atomi *et al.*, 1986; Satake *et al.*, 1986); a study of sexual and skeletal maturity in Italy (Cacciari *et al.*, 1990); and a short-term longitudinal study of testicular volume in Denmark (Hansen *et al.*, 1999) suggest that among soccer players 10–13 years of age, boys of all maturity levels, that is, advanced (early), average ('on time') and delayed (late) are represented. The SA approximates, on average, CA at these ages. The short-term longitudinal (1 year) study of Danish players indicates a difference between elite and non-elite players. Elite players (11.9 ± 0.5 at initial observation) were advanced in genital maturation (testicular volume) compared to non-elite players (11.6 ± 0.7 at initial observation), and this difference persisted during the interval of the study (Hansen *et al.*, 1999). The mean testicular volumes of the non-elite players approximated those of the early pubertal players 10.0–11.9 years of age reported by Cacciari *et al.* (1990).

With increasing age and experience, boys advanced in skeletal and sexual maturity are more common in samples of youth soccer players, especially 14+ years of age (Cacciari *et al.*, 1990; Peña Reyes *et al.*, 1995). The observations of young Danish players indicate that the difference in testicular volume between elite and non-elite soccer players increased over the course of the 1-year longitudinal study (Hansen *et al.*, 1999), and the mean testicular volume of the elite players (12.9 ± 0.5 years) was slightly less than that of pubertal Italian players 12.0–13.9 years of age (Cacciari *et al.*, 1990). This would suggest that the sample of elite players was well into puberty and the adolescent growth spurt.

Skeletal age tends to be, on average, in advance of CA in soccer players 14+ years of age, and some adolescent players attain skeletal maturity at a younger CA than the reference samples upon which the methods for assessing skeletal maturity were developed (see below). Conversely, later maturing boys are lacking among mid- and late-adolescent soccer players.

The trend for skeletal maturity is shown in Figure 20.3 and Table 20.2 for youth soccer players from two urban centres in Mexico (Peña Reyes *et al.*, 1995). The figure shows the plot of SA and CA for individual boys. During childhood and early adolescence, there is considerable variability in the sample as SAs fall above and below the line of identity. After about 13.5 years of age, most of the SAs are above the line of identity, indicating more advanced skeletal maturity in the older adolescent soccer players. The SA data for Mexican players were also partitioned into categories of average, advanced and late maturity status (Table 20.2). Average or an SA 'on time' for CA, was defined as an SA within ±1.0 year of CA. Advanced or early was defined as an SA ahead of or older than CA by more than 1.0 year, and delayed or late was defined as an SA behind or younger than CA by more than 1.0 year. Using a band of plus and minus 1.0 year allows for the error associated with the assessment of SA and accommodates the broad range of youth who are average or 'on time' in maturity status. The young soccer players as a group were advanced in SA, and the difference between SA and CA, on average, increased with age. Although numbers were small, there was only one late maturing boy (SA lagged behind CA by more than 1 year) among youth soccer players >13.0 years of age.

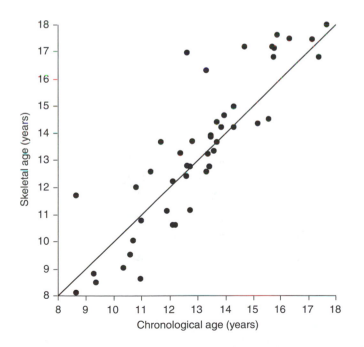

Figure 20.3 Skeletal ages of individual Mexican youth soccer players plotted relative to their CA. The diagonal line is the line of identity, SA = CA. (Redrawn after data reported in Peña Reyes *et al.*, 1994.)

Table 20.2 The difference between SA and CAs (SA − CA, mean ±
standard error) and distribution of Mexican soccer players
by CA group and maturity status within each age group[a]

Age group	n	SA − CA difference (years)[b]	Maturity category[c]			
			L	A	E	M
11–12	12	0.27 ± 0.42	3	8	3	0
13–14	14	0.49 ± 0.29	0	12	2	0
15–16	10	0.62 ± 0.34[d]	1	3	5	1

Notes
a Adapted from Peña Reyes *et al.* (1994).
b Skeletal age minus CA. SA is based on the Fels method.
c Maturity categories are as follows: L = late ('delayed') = SA younger than
CA by >1.0 year; A = average ('on time') = SA within =1.0 year of CA;
E = early ('advanced') = SA older than CA by >1.0 year; M = mature =
skeletally mature or adult.
d n = 9 because the one skeletally mature player is not included in the calculation of SA−CA. He is simply labelled as mature or adult and no SA is
assigned.

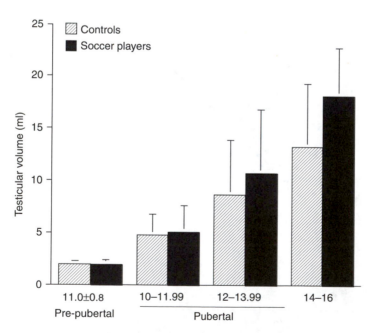

Figure 20.4 Mean testicular volume (plus one standard deviation) of Italian youth soccer players
and controls (non-athletes). The pubertal players were grouped into 2-year age
categories. (Drawn from data reported in Cacciari *et al.*, 1990.)

The testicular volume of Italian youth soccer players are compared to corresponding data for control subjects (non-athletes) in Figure 20.4 (Cacciari *et al.*, 1990). Early pubertal players and non-athletes 10–11 and 12–13 years of age did not differ, on average, in testicular volume (Figure 20.3), and also in stage of pubic hair development, SA and body size. However, pubertal players '14–16 years' of age (note, the decimal age range was not given, although it can be assumed from the text that 14 refers to 14.0 years) were advanced in testicular maturation. The mean testicular volume for the pubertal players, 18.2 ± 4.5 ml, approximated the reference value for Swiss boys >17 years of age, 18.6 ± 4.0 ml (Zachman *et al.*, 1974). The 14–16-year-old players were also advanced in skeletal maturity and stage of pubic hair development, and were taller and heavier than the control (Cacciari *et al.*, 1990).

Information on the age at peak height velocity of young soccer players are limited. Longitudinal data for 32 Welsh (Bell, 1993) and 8 Danish (Froberg *et al.*, 1991) soccer players give estimates of 14.2 years in each sample. This age does not differ from estimated ages at peak height velocity for European adolescents (Malina *et al.*, 2002). The results for these two small samples appear to differ somewhat from those for cross-sectional studies of skeletal and sexual maturation, which suggest a trend for boys advanced in maturity status to be more successful in soccer in later adolescence. The results, however, are not that inconsistent. Advanced skeletal and sexual maturity of soccer players is more apparent among those 14.0 years of age and older, ages when most boys have already passed peak height velocity.

Data for 33 Belgian youth soccer players indicate an estimated age at peak height velocity of 13.8 ± 0.8 years (Janssens *et al.*, in preparation). Players in this longitudinal study were followed for 5 years, but ages at the start of the study varied from 10.4 to 13.7 years. As a result, ages at peak height velocity could not be estimated for some boys. Among boys for whom complete longitudinal records were available, peak height velocity was already experienced before the start of the study in 25 players (i.e. it probably occurred between 10.4 and 13.7 years, and was not yet reached by 18 boys). Allowing for the boys among whom peak height velocity already occurred and did not occur, the overall mean for this sample of soccer players is probably slightly earlier than mean ages at peak height velocity in two samples of non-athlete Belgian males, 14.0 and 14.2 years (Malina *et al.*, 2003).

The estimated age at peak height velocity for a sample 83 Japanese junior high school soccer players is 13.7 ± 1.1 years (Nariyama *et al.*, 2001). This age is later than corresponding ages at peak height velocity in Japanese adolescent boys (Malina *et al.*, 2003).

20.3.3   *Elite Portuguese soccer players 11–16 years*

Height, weight and skeletal maturity were assessed in 135 elite youth soccer players 10.7–16.5 years of age (Malina *et al.*, 2000). All subjects were members of one of the eight best teams in their respective age categories. Mean heights and weights of 11–12-year-old players approximated the reference medians for the United States (Centers for Disease Control and Prevention, 2000); in contrast, mean heights and weights of 13–14 and 15–16-year-old players were slightly above the reference values. On average, SA approximated CA in the 11–12-year-old players, 12.4 ± 1.3 and 12.3 ± 0.5 years, respectively. On the other hand, mean SA was in advance of mean CA in the two older groups: 14.3 ± 1.2 and 13.6 ± 0.7 years, respectively, in 13–14-year-old players, and 16.7 ± 1.0 and 15.8 ± 0.4 years, respectively, in the 15–16-year-old players. Seven boys (16%) in the oldest age group were already skeletally mature and were not included in calculating SA–CA differences. They are simply classified as adult and no SA is assigned.

Table 20.3 The difference between SA and CAs (SA − CA, mean ± standard error) and distribution of Mexican soccer players by CA group and maturity status within each age group[a]

Age group	n	SA − CA difference (years)[b]	Maturity category[c]			
			L	A	E	M
11–12	12	0.27 ± 0.42	3	8	3	0
11–12	63	0.11 ± 0.14	13	37	13	0
13–14	29	0.67 ± 0.18	2	16	11	0
15–16	43	0.98 ± 0.15[d]	1	14	21	7
National team	19		0	8	8	3
Others	24		1	6	13	4

Notes
a Adapted from Malina *et al.* (2000).
b Skeletal age minus CA. SA is based on the Fels method.
c Maturity categories are as follows: L = late ('delayed') = SA younger than CA by >1.0 year; A > average ('on time') = SA within ±1.0 year of CA; E = early ('advanced') = SA older than CA by >1.0 year; M = mature = skeletally mature or adult.
d $n = 36$ because the seven skeletally mature players are not included in the calculation of SA − CA. They are simply labelled as mature or adult and no SA is assigned.

The players were also classified as late, average ('on time'), and advanced in maturation based on the difference between SA and CA (Table 20.3). The same criteria as used for the sample of Mexican soccer players described earlier were used (see Table 20.2). Among the 11–12-year-old Portuguese players, the percentages of late and early maturing boys were equal (21%). Among 13–14-year-old players, the percentages of late and early maturing boys were, respectively, 7 and 38%. Corresponding percentages of late and early maturing 15–16-year-old players were 2% and 49%, respectively. As noted above, seven (16%) of the 15–16-year-old players were already skeletally mature (i.e. biologically adult). Thus, the contrast between late and early maturing 15–16-year-old players was 2% and 65% respectively. Clearly, with increasing CA, later maturing boys are noticeably lacking in this sample of elite youth soccer players (Malina *et al.*, 2000).

When the SA and CAs of the young Portuguese players are plotted relative to each other, SAs are distributed above and below the line of identity (Figure 20.5). This finding is especially true in late childhood and early adolescence up to about 14.0 years of age. Subsequently, SAs tend to fall more frequently above the line of identity which indicates that SA is in advance of CA or that biological maturity of the adolescent soccer players is advanced.

The skeletal maturity data for the samples of Mexican and Portuguese youth soccer players are generally consistent. Both data sets suggest that among players 10–13 years of age, boys of all maturity levels are represented. However, with advancing age and experience, boys advanced in skeletal maturation are more common among adolescent players 14+ years of age.

The sample of 15–16-year-old Portuguese players included the national selection ($n = 19$) for this age group. Members of the national selection were, on average, slightly taller and

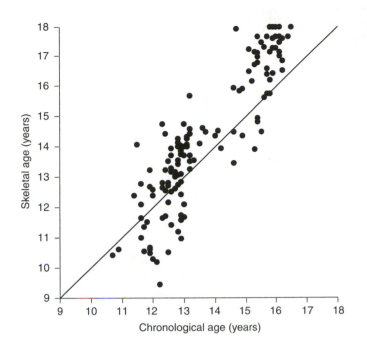

Figure 20.5 Skeletal ages of individual elite Portuguese youth soccer players plotted relative to their CA. The diagonal line is the line of identity, SA = CA. (Redrawn after data reported in Malina *et al.*, 2000.)

Table 20.4 Growth and maturity characteristics (mean ± standard deviation) of 15–16-year-old Portuguese national team members by position[a]

Variable	Position			
	Forward (n = 4)	Midfield (n = 7)	Defence (n = 6)	Goal (n = 2)
CA (years)	16.0 ± 0.2	15.8 ± 0.2	16.1 ± 0.1	16.0
SA (years)	17.2 ± 0.3	16.3 ± 0.7	17.1 ± 0.4	
SA − CA (years)	1.2 ± 0.4	0.5 ± 0.5	1.0 ± 0.4	
Height (cm)	169.9 ± 6.6	174.7 ± 2.7	176.9 ± 4.6	175.7
Mass (kg)	66.9 ± 4.9	64.9 ± 2.8	68.0 ± 8.3	76.4

Note
a Adapted from Malina *et al.* (2000). One skeletally mature player in each position is not included for SA.

heavier than non-select players ($n = 24$), 174.5 ± 4.8 versus 172.2 ± 6.5 cm and 67.5 ± 6.3 versus 63.2 ± 5.6 kg, respectively. The two groups, however, did not differ in skeletal maturity. Comparisons of the national team members by position are shown in Table 20.4. Forwards ($n = 4$) and defenders ($n = 6$) were more advanced in skeletal maturity than midfielders ($n = 7$). There was a gradient in mean stature from forwards (shortest) to defenders (tallest), whereas defenders were heaviest, followed by forwards and then midfielders. Forwards appeared to have proportionally more weight-for-height compared to players at the other two positions (Malina *et al.*, 2000).

20.3.4   Elite Portuguese soccer players 13–15 years

The age, height and mass of 69 elite Portuguese soccer players 13.2–15.1 years of age by position are summarized in Table 20.5. The general trend in height and mass by position (largest to smallest) in this sample of soccer players is, on average, from forwards to defenders to midfielders. The number of players was too small to classify defenders and midfielders into central and lateral positions. Height does not differ significantly by position, but forwards are significantly heavier than midfielders but not defenders. The trend in size by position among the 13–15-year-old players (Table 20.5) contrasts that in the 15–16-year-old national team members (Table 20.4). The contrasting trends need to be viewed in terms of age and maturity-associated variation. In samples of adolescent non-athletes, maturity-associated variation in body size is most marked between 13 and 15 years of age (Malina *et al.*, 2003) so that more mature 13–15-year-old boys are generally taller and heavier than less mature boys. Among the 13–15-year-old players, those in the attack position are slightly older and tend to be more mature than those at the other positions. This trend is shown in Table 20.6, which summarizes the distribution of stages of pubic hair development in this

Table 20.5 Age, height and mass (mean ± standard deviation) of elite 13–15-year-old Portuguese soccer players by position[a]

Variable	Position		
	Defence (n = 29)	Midfield (n = 30)	Forward (n = 10)
Age (years)	14.2 ± 0.6	14.2 ± 0.6	14.7 ± 0.5
Height (cm)	169.2 ± 7.5	165.4 ± 9.0	170.8 ± 9.9
Mass (kg)	57.3 ± 7.8	54.5 ± 9.8	61.4 ± 9.2

Note
a Adapted from Malina *et al.* (in preparation).

Table 20.6 Distribution of stages of pubic hair development in elite 13–15-year-old Portuguese soccer players by position[a]

Stage of pubic hair	Defence (n = 29)	Midfield (n = 30)	Forward (n = 10)
1 (pre-pubertal)	1	5	0
2	3	5	2
3	5	7	1
4	11	7	3
5 (mature)	9	6	4

Note
a Adapted from Malina *et al.* (in preparation).

sample of 13–15-year-old soccer players. Overall, 58% of the players are in pubic hair stages 4 and 5. All five stages are represented among midfielders and defenders, and four of the five stages (except stage 1) are represented among the small sample of attackers. However, 70% of forwards and defenders are in pubic hair stages 4 and 5, which suggests that as a group they are maturationally in advance of midfielders in whom all five pubic hair stages are about equally represented.

20.4   Implications

Many factors are involved in successful athletic performance during childhood and adolescence. Physical characteristics as reflected in growth and maturity status are important, but they are not the only determinants of successful performance. They are part of a complex matrix of biocultural characteristics related to the demands of the sport of soccer. Three issues related to the growth and maturity characteristics of child and adolescent soccer players are subsequently discussed. The first two deal with issues related to selection and injury, respectively. The third deals with the implications of maturity-associated variation for performance.

20.4.1   Selection and exclusion

The presently available data suggest that the sport of soccer systematically excludes late maturing boys and favours biologically average and early maturing boys as CA and sport specialization increase. This fact may reflect selection (self, coach or some combination), differential success of boys advanced in maturity status, the changing nature of the game (more physical contact is permitted in older age groups), or some combination of these factors. It is also possible that late maturing boys selectively drop-out as age and sport specialization increase.

The growth and maturity characteristics of young soccer players may influence the processes of exclusion and inclusion, that is, some boys will be excluded from participation in soccer and others will be included on the basis of their growth and maturity status. These processes probably vary among countries as well as among clubs within a country. Nevertheless, an important question about talent identification and selection among youth soccer players needs more detailed study: what influence do the processes of inclusion and exclusion have on maximizing the possibility that boys with the greatest potential for success at the highest levels of competition will still be participating in soccer and be available for selection to high level teams in their late adolescent or perhaps young adult years? The question is related to the identification of a select few participants who might bring local, regional and especially national recognition through the success of the respective teams at each level. A related question deals with the talented, late maturing boy who is generally smaller and not as strong as his age peers: How can the small, late maturing boy who is exceptionally skilled be nurtured and in some cases protected so that he will persist in the sport? This youngster needs reassurance that he will eventually go through his growth spurt and sexual maturation, and catch up in height to the other boys.

Another question deals with the potential health benefits of regular physical activity during childhood and adolescence: what influence do the processes of inclusion and exclusion have on the likelihood that most boys will continue to play soccer and receive the health-related benefits of this form of physical activity throughout adolescence and into adulthood? This question is related to a philosophy of well-being associated with health benefits that may be received by the general populace through participation in sports.

At first, it might seem that solutions to these vastly different questions must be different. The desire to win often drives the participatory process in sport. This is especially true in soccer where national and international tournaments are available for children 10 years of age and younger. Many coaches, even at local levels of competition, experience pressures to win. These pressures result in a tendency for coaches, parents and programme administrators to directly or indirectly select early maturing players, who can help youth teams win now, and exclude late maturing players, who might have the potential to become national caliber performers when they reach full physical maturity.

Stature is not as important in soccer as it is in some other sports. This is evident by the success that has been achieved by some relatively short players at the international level. However, it is often the case at the youth level, that the taller and more physically mature individuals are often chosen for select teams and are provided with better coaching and more exposure to the sport because they are perceived to be the better players. At the time of selection, they are likely to be the best players because of the size, strength and power advantage associated with early maturation in boys. It is virtually impossible, however, to select players on the potential that they might develop into outstanding athletes. Furthermore, parents of the physically mature players would likely have a strong argument against selecting others on the basis of potential.

The talent identification process can be exacerbated by the use of 2-year age categories (e.g. 8–10, 10–12, 12–14, 14–16 and 16–18 years) that are defined for local and international competitions. Under this arrangement, for example, a late maturing 8-year-old, who was born in the early part of the 'soccer year' and has an SA of 7 years, would have a difficult time competing against an early maturing 10-year-old, who was born in the late part of the 'soccer year' and has an SA of 12 years. Even for a potentially talented player, the 5 years of maturational difference would be difficult to overcome in competing for a position on a select team at this age level. Note that among 11–12-year-old elite players discussed earlier, SA–CA differences ranged from −2.77 years to +2.57 years, a range of variation of 5.34 years.

Youth players, who are not chosen for an advanced team and the associated special coaching, quickly get the message that they will 'not make it' or 'do not have what it takes'. They probably drop-out of soccer and pursue other activities. In many instances, decisions as to who will be selected to play, and the converse – who will be excluded, are often placed in the hands of individuals with little or no training in understanding the growth and maturation of children and their potential influences on the selection process and performance. This is generally the case in youth programmes operated by volunteer administrators and coaches.

The processes used to select players should perhaps be changed. They should be directed at providing the best possible training throughout childhood and adolescence to all individuals interested in participating in soccer and to encourage them to continue to play irrespective of how successful they might be during their childhood years. This approach should lead to a broader base of individuals participating in soccer through adolescence into adulthood, thus fulfilling the health-related benefits of soccer. It should also provide a larger pool of individuals which may include a greater number who potentially could be the best players.

20.4.2   *Growth and maturity status, and injury in soccer*

Many discussions of injuries in young athletes often focus on risk factors, or the characteristics and/or conditions that might place a youngster at risk of injury (Micheli, 1983, 1985; Caine and Lindner, 1990). Potential risk factors related to characteristics of the young athlete (internal factors) include: (1) physique; (2) problems in structural alignment; (3) lack of

flexibility; (4) lack of muscular strength or strength imbalance; (5) the adolescent growth spurt; (6) maturity-associated variation; (7) injury history; and (8) behavioural factors.

Potential risk factors associated with the sport environment include: (1) inadequate rehabilitation from prior injury – loss of conditioning, flexibility and strength; (2) training errors – improper technique, lack of adequate instruction, use of inappropriate drills, lack of conditioning; (3) playing conditions – structural hazards: goal posts, fences, sprinklers; surfaces: uneven, wet; environment: lighting, heat/cold, humidity, lightning; proximity to spectators; (5) age groups – size, maturity and experience mismatches in broad age groups; (6) coach behaviours – inappropriate drills and techniques, poor instruction, forced participation after injury or incomplete rehabilitation; (7) parent behaviours – unrealistic expectations, pushing a child too fast, having a child 'play-up' in an older age group; and (8) sport organizations (administrators, coaches and officials) – focus on winning, increased tolerance for aggression and body contact. Needless to say, interactions between internal and external risk factors are important.

Of particular relevance to the present discussion is the adolescent growth spurt and maturity-associated variation in the context of injury. As noted earlier, individual differences in the timing and tempo of the growth spurt and sexual maturation are considerable, and boys differing greatly in size and maturity often compete on a regular basis. Maturity mismatches in size and strength are commonly singled out as potential risk factors for injury.

What is it about the growth spurt that may place an adolescent athletes at risk? The association between increased prevalence of injuries and the adolescent growth spurt has been long recognized (Dameron and Reibel, 1969; Peterson and Peterson, 1972; Burkhart and Perterson, 1979; Bailey *et al.*, 1989). The term association needs to be emphasized. There are no prospective or longitudinal data that relate injuries to parameters of the adolescent growth spurt. Youths who present to a clinic with an injury are ordinarily seen only on this occasion and it is virtually impossible to estimate where a youth is in his growth spurt based on one observation.

Longitudinal data on growth in bone mineral during the adolescent growth spurt indicate that the peak velocity of bone mineral accrual occurs after peak height velocity by more than 1 year, on average (Iuliano-Burns *et al.*, 2001). The lag in bone mineral accrual relative to linear growth may suggest a period of skeletal 'fragility' which might contribute to the increased occurrence of injuries (sport and non-sport) during the adolescent spurt. This suggestion, however, needs to be verified with longitudinal observations on the growth, maturity and body composition characteristics of young athletes.

Other changes during the adolescent growth spurt also need consideration. Loss of flexibility is indicated as a risk factor. Flexibility, however, is joint specific and is a highly individual characteristic. The range of motion of some joints increases during puberty in contrast to the general suggestion that flexibility decreases. Loss of flexibility in athletes during adolescence may be sport specific, such as, for example, the loss of quadriceps flexibility in soccer players (Kibler and Chandler, 1993). Although flexibility and strength (static and explosive) are not related, it has been suggested that an imbalance between strength and flexibility may lead to abnormal movement mechanics, which in turn may be a risk factor for injury. Peak gains in muscular strength and power occur, on average, after peak gains in height and closer in time to peak gains in body weight (Malina *et al.*, 2003). Does this contribute to the strength imbalance described in some adolecent athletes?

As noted earlier, maturity-associated variation in body size, strength, power and other performance characteristics are magnified during adolescence. Hence, an important question is the following: what is the extent of maturity mismatches in size, strength and power as a factor in injuries? Unfortunately, the maturity status of youth soccer participants has not be systematically related to injury.

Three studies on injuries among soccer players have considered biological maturity status, but they are largely descriptive. Vidalin (1988) described 11 youth soccer players, 12.2–15.7 years of age, who were injured during a season. Nine of the 11 injured players had SAs that would categorize them as late maturers; the other two were average or 'on time' in skeletal maturity. In a mixed-longitudinal study of young soccer players in England, on the other hand, there were no differences in the incidence of injury among soccer players in different stages of puberty (Sports Council, 1992; Baxter-Jones *et al.*, 1993; Rowley, 1993).

Using grip strength and height as an indicator of maturity, Backhous *et al.* (1988) reported more injuries in a summer camp for soccer among boys who were classified as 'tall and weak' ('skeletally mature but muscularly weak') than among those who were 'tall and strong' ('mature') or 'short and weak' ('immature'). Note, however, that biological maturity was not assessed. The authors assumed that height was directly related to stage of secondary sex characteristic development. This assumption does not allow for individual differences in the timing and tempo of the adolescent growth spurt and for genotypic differences in height. Further, weight-for-height relationships were not considered in the analysis.

Although the results of the presently available studies are limited, they suggest that maturity status per se, or size and strength variation associated with maturity status during adolescence, may be a risk factor for injury in soccer. There is a need for systematic study of injuries of young soccer players in the context of their growth, maturity and perhaps behavioural, fitness and performance characteristics.

20.4.3   *Maturity and performance*

What is the contribution of variation in body size and sexual maturity to performance characteristics deemed important for soccer: aerobic resistance, speed and explosive power? This question was addressed in the sample of Portuguese players 13–15 years of age (Malina *et al.*, in preparation). Aerobic resistance was measured with an intermittent endurance task in which players alternated between high- and low-intensity exercise. Speed and explosive power were measured by means of a 30-m sprint and vertical jump, respectively. Multiple linear regression was used to estimate the relative contribution of CA, stage of sexual maturity, height, weight and years of formal training in soccer to three physical performance variables. Results suggest that body size, stage of sexual maturation, and years of training accounted for 16–50% of the variance in the three performance tasks. The results highlight the interactions among body size, maturity status and experience in the physical fitness of adolescent soccer players.

Midfielders scored, on average, highest on aerobic resistance, but lowest in speed and power. Forwards, on average, scored slightly higher in speed and power compared to defenders, but scored lowest in aerobic resistance. When viewed by stage of sexual maturation, performances on each of the three tasks increased, on average, with stage of pubic hair development. The trends were more apparent in aerobic resistance and speed, and to a lesser extent in power (Malina *et al.*, in preparation).

Summary

Presently available information on the growth and maturity status of young male soccer players are summarized, and the results of two studies of elite Portuguese youth players are considered in more detail. Mean heights of young soccer players tend to fluctuate above and below the reference median during childhood and early adolescence (about 8–14 years), but in later adolescence (15+ years), many of the mean heights approximate the reference median or are below the reference median. Mean weights approximates the reference median from late childhood into early adolescence, about 9–13 years. Subsequently, mean weights tend to be above the reference median and remain there into late adolescence. Among soccer players 10–13 years of age, boys of all maturity levels, that is, advanced (early), average ('on time') and delayed (late) are represented, but some data for 11–12-year-old players suggest that elite players are maturationally advanced compared to the non-elite. With increasing age and experience, boys advanced in skeletal and sexual maturity are more common in samples of youth soccer players, especially 14+ years of age.

The results suggest that the sport of soccer systematically excludes late maturing boys and favours average and early maturing boys as CA and sport specialization increase; on the other hand, it is possible that later maturing boys selectively drop out of soccer as interests change and sport specialization increases. Implications of the growth and maturity data are discussed in the context of selection and retention of soccer players during childhood and adolescence, and injury.

The preceding discussion is limited to boys. Soccer is growing in popularity among girls and women from the local to the international levels. At present, data on the characteristics of young female soccer players are extremely limited (e.g. Siegel, 1995; Siegel *et al.*, 1996). This sample of reasonably select soccer players 10–18 years of age tended to approximate, on average, the United States reference in height and weight, but had proportionally longer legs. The median age at menarche was similar to the general population. Issues related to growth and maturity characteristics, talent identification and injuries among female players surface on a regular basis and need to be addressed in a systematic manner.

References

Alonso, R.F. (1986). Estudio del somatotipo de los atletas de 12 anos de la EIDE occidentales de Cuba. *Boletin de Trabajos de Antropologia*, April, 3–18.

Atomi, Y., Fukunaga, T., Yamamoto, Y. and Hatta, H. (1986). Lactate threshold and VO_2 max of trained and untrained boys relative to muscle mass and composition, in *Children and Exercise XII* (eds J. Rutenfranz, R. Mocellin and F. Klimt) Human Kinetics, Champaign, IL, pp. 53–8.

Backhous, D.D., Friedl, K.E., Smith, N.J., Parr, T.J. and Carpine W.D. (1988). Soccer injuries and their relation to physical maturity. *American Journal of Diseases of Children*, **142**, 839–42.

Bailey, D.A., Wedge, J.H., McCulloch, R.G., Martin, A.D. and Berhhardson, S.C. (1989). Epidemiology of fractures of the distal end of the radius in children as associated with growth. *Journal of Bone and Joint Surgery*, **71A**, 1225–31.

Barbieri, C. and Rodriguez Papini, H. (1996). *Informe Final Proyecto Antropometrico Torneos Juveniles Bonaerenses* Final Provincial 1996. Direccion de Impresiones del Estado y Boletin Oficial, Provinica de Buenos Aires, Buenos Aires, Argentina.

Baxter-Jones, A.D.G., Helms, P., Maffulli, N., Baines-Preece, J.C. and Preece, M. (1995). Growth and development of male gymnasts, swimmers, soccer and tennis players: a longitudinal study. *Annals of Human Biology*, **22**, 381–94.

Baxter-Jones, A.D.G., Maffulli, N. and Helms, P. (1993). Low injury rates in elite athletes. *Archives of Disease in Childhood*, **68**, 130–2.

Bell, W. (1988). Physiological characteristics of 12-year-old soccer players, in *Science and Football* (eds T. Reilly, A. Lees, K. Davids and W.J. Murphy) E &FN Spon, London, pp. 175–80.

Bell, W. (1993). Body size and shape: a longitudinal investigation of active and sedentary boys during adolescence. *Journal of Sports Sciences*, **11**, 127–38.

Burkhart, S.S., and Peterson, H.A. (1979). Fractures of the prosimal tibial epiphysis. *Journal of Bone and Joint Surgery*, **61A**, 996–1002.

Cacciari, E., Mazzanti, L., Tassinari, D., Bergamaschi, R., Magnani, D., Zappula, F., Nanni, G., Cobianchi, C., Ghini, T., Pini, R. and Tani, G. (1990). Effects of sport (football) on growth: auxological, anthropometric and hormonal aspects. *European Journal of Applied Physiology*, **61**, 149–58.

Caine, D.J. and Lindner, K. (1990). Preventing injury to young athletes. Part 1: Predisposing factors. *Canadian Association of Health, Physical Education and Recreation Journal*, **56**, 30–5.

Carter, J.E.L. (1988). Somatotypes of children in sports, in *Young Athletes: Biological, Psychological, and Educational Perspectives* (ed. R.M. Malina), Human Kinetics, Champaign, IL, pp. 153–65.

Carter, J.E.L. and Heath, B.H. (1990). *Somatotype: Development and Application*. Cambridge University Press, Cambridge.

Centers for Disease Control and Prevention (2000). *National Center for Health Statistics CDC Growth Charts: United States*. http://www.cdc.gov/growthcharts.htm.

Chen, J.D. (1991). Growth, exercise, nutrition and fitness in China, in *Human Growth, Physical Fitness and Nutrition* (eds R.J. Shephard and J. Parizkova), Karger, Basel, pp. 19–32.

Cumming, S.P. (2002). A biopsychosocial investigation of self-determined motivation in recreational and travel youth soccer. Doctoral dissertation, Michigan State University, East Lansing.

Dal Monte, A., Leonardi, L.M., Sardella, F., Faina, M. and Gallippi, P. (1980). Evaluation test of the alternate aerobic-anaerobic potential in subjects at development age, in *1st International Congress on Sports Medicine Applied to Football, Proceedings Vol. II* (ed. L. Vecchiet), D. Guanella, Rome, pp. 788–94.

Dameron, T.B. and Reibel, D.B. (1969). Fractures involving the proximal humeral epiphyseal plate. *Journal of Bone and Joint Surgery*, **51A**, 289–97.

De Proft, E., Cabri, J., Dufour, W. and Clarys, J.P. (1988). Strength training and kick performance in soccer players, in *Science and Football* (eds T. Reilly, A. Lees, K. Davids and W.J. Murphy), E & FN Spon, London, pp. 108–13.

Donoso, H., Quintana, G., Rodriguez, A., Huberman, J., Holz, M. and Godoy, G. (1980). Algunas caracteristicas antropometricas y maximo consumo de oxigeno en 368 deportistas seleccionados Chilenos. *Archivos de la Sociedad Chilena de Medicina del Delporte*, **25**, 7–17.

Farmosi, I. and Nadori, L. (1981). Ifjusagi labdarugok alkati es motorikus vizsgalatanak nehany eredmenye. *Testnevelesi Foiskola Kozlemenyei*, **1**, 173–9.

Froberg, K., Anderson, B. and Lammert, O. (1991). Maximal oxygen uptake and respiratory functions during puberty in boy groups of different physical activity, in *Children and Exercise: Pediatric Work Physiology XV* (eds R. Frenkl and I. Szmodis), National Institute for Health Promotion, Budapest, pp. 265–80.

Garganta, J. and Maia, J. (1993). A comparative study of explosive leg strength in elite and non-elite young soccer players, in *Science and Football II* (eds T. Reilly, J. Clarys and A. Stibbe), E & FN Spon, London, pp. 304–6.

Hansen, L., Klausen, K., Bangsbo, J. and Muller, J. (1999). Short longitudinal study of boys playing soccer: parental height, birth weight and length, anthropometry, and pubertal maturation in elite and non-elite players. *Pediatric Exercise Science*, **11**, 199–207.

Haschke, F. (1983). Body composition of adolescent males. *Acta Paediatrica Scandinavica* (Suppl), **307**.

Herm, K.-P. (1993). The evidence of sportanthropology in training of young soccer player, in *Science and Football II* (eds T. Reilly, J. Clarys and A. Stibbe), E & FN Spon, London, pp. 287–91.

Iuliano-Burns, S., Mirwald, R.L. and Bailey, D.A. (2001). The timing and magnitude of peak height velocity and peak tissue velocities for early, average and late maturing boys and girls. *American Journal of Human Biology*, **13**, 1–8.

Jankovic, S., Heimer, N. and Matkovic B.R. (1993). Physiological profile of prospective soccer players, in *Science and Football II* (eds T. Reilly, J. Clarys and A. Stibbe), E & FN Spon, London, pp. 295–7.

Janssens, M., Philippaerts, R.M., Van Renterghem, B., Stoops, F., Cauwelier, D., Vaeyens, R., Matthys, D., Bourgois, J. and Vrijens, J. Physical fitness in adolescent soccer players related to peak height velocity, (in preparation).

Jones, A.D.G. and Helms, P. (1993). Cardiorespiratory fitness in young British soccer players, in *Science and Football II* (eds T. Reilly, J. Clarys and A. Stibbe), E & FN Spon, London, pp. 298–303.

Junge, A., Dvorak, J., Chomiak, J., Peterson, L. and Graf-Baumann, T. (2000). Medical history and physical findings in football players of different ages and skill levels. *American Journal of Sports Medicine*, **28** (Suppl.), S16–S21.

Kibler, W.B. and Chandler, T.J. (1993). Musculoskeletal adaptations and injuries associated with intense participation in youth sports, in *Intensive Participation in Children's Sports* (eds B.R. Cahill and A.J. Pearl), Human Kinetics, Champaign, IL, pp. 203–16.

Kirkendall, D.T. (1985). The applied sport science of soccer. *Physician and Sportsmedicine*, **13** 53–9.

Kosova, A., Hlatky, S., Lilge, W. and Holdhaus, H. (1991). Physical structure and performance of young soccer players. *Anthropologiai Kozlemenyek*, **33**, 267–72.

Lindquist, F. and Bangsbo, J. (1993). in *Science and Football II* (eds T. Reilly, J. Clarys and A. Stibbe), E & FN Spon, London, pp. 275–80.

Luhtanen, P. (1988). Kinematics and kinetics of maximal instep kicking in junior soccer players, in *Science and Football* (eds T. Reilly, A. Lees, K. Davids and W.J. Murphy), E & FN Spon, London, pp. 441–8.

Lukyanova, R.P. and Novocelova, N.I. (1964). Physical development and physical preparation of young athletes in track and field, soccer and basketball. *Theory and Practice of Physical Culture*, **6**, 38–41, 1964 (reprinted in *Yessis Translation Review*, **2**, 18–22, 1967).

Magalhaes, D., Seco, P. and Ribeiro, B. (1997). Physical fitness of young Portuguese soccer players. *Ninth European Congress on Sports Medicine*, Porto, Program and Abstract Book.

Malina, R.M. (1994). Physical growth and biological maturation of young athletes. *Exercise and Sports Science Reviews*, **22**, 389–433.

Malina, R.M., Bouchard, C. and Bar-Or, O. (2003). *Growth, Maturation, and Physical Activity*, 2nd edn, Human Kinetics, Champaign, IL.

Malina, R.M., Eisenmann, J.C., Cumming, S.P., Ribeiro, B. and Aroso, J. (in preparation). Maturity-associated variation in the growth and functional capacities of youth football (soccer) players 13–15 years.

Malina, R.M., Peña Reyes, M.E., Eisenmann, J.C., Horta, L., Rodrigues, J. and Miller, R. (2000). Height, mass, and skeletal maturity of elite Portuguese soccer players 11–16 years of age. *Journal of Sports Sciences*, **18**, 685–93.

Matsudo, V.K.R. (1978). Efeitos do treinamento nas caracteristicsas de aptidao fisica de futebolistas adolescentes e adultos, in *Anais do VI Simposio de Ciencias do Esporte*, Centro de Estudos do Laboratorio de Aptidao Fisica de Sao Caetano do Sul (CELAFISCS), reprinted in CELAFISCS: Des Anas de Contribuicao as Ciencias do Esporte. Centro de Estudos do Laboratorio de Aptidao Fisica de Sao Caetano do Sul, Sao Caetano, Sao Paulo, Brazil, pp. 298–304.

Mazzanti, L., Tassinari, D., Bergamaschi, R., Nanni, G., Magnani, C., Ghini, T., Pini, R., Amendola, C., Drago, E. and Cacciari, E. (1989). Hormonal, auxological and anthropometric aspects in young football players, in *Growth Abnormalities* (eds J.R. Bierich, E. Cacciari and S. Raiti), Raven Press, New York, pp. 363–9.

Micheli, L.J. (1983). Overuse injuries in children's sports: the growth factor. *Orthopaedic Clinics of North America*, **14**, 337–60.

Micheli, L.J. (1985). Preventing youth sports injuries. *Journal of Physical Education, Recreation and Dance*, **56**, 52–4.

Nariyama, K., Hauspie, R.C. and Mino, T. (2001). Longitudinal growth study of male Japanese junior high school athletes. *American Journal of Human Biology*, **13**, 356–64.

Peña Reyes, M.E., Cardenas-Barahona, E. and Malina, R.M. (1994). Growth, physique, and skeletal maturation of soccer players 7–17 years of age. *Humanbiologia Budapestinensis*, **5**, 453–8.

Peterson, C.A. and Peterson, H.A. (1972). Analysis of the incidence of injuries to the epiphyseal growth plate. *Journal of Trauma*, **12**, 275–81.

Roche, A.F., Chumlea, W.C. and Thissen, D. (1988). *Assessing the Skeletal Maturity of the Hand-Wrist: Fels Method* CC Thomas, Springfield, IL.

Rowley, S. (1993). *Training of Young Athletes (TOYA) Study: TOYA and Intensive Training*. The Sports Council, London.

Satake, T., Okajima, Y., Atomi, Y., Asami, T. and Kuroda, Y. (1986). Effect of physical exercise on physical growth and maturation. *Journal of Physical Fitness of Japan*, **35**, 104–10.

Siegel, S.R. (1995). Growth and maturity status of female soccer players from later childhood through early adulthood. Unpublished Master's thesis, University of Texas at Austin, Austin, TX.

Siegel, S.R., Katzmarzyk, P.T. and Malina, R.M. (1996). Somatotypes of female soccer players 10–24 years of age, in *Studies in Human Biology* (eds E. Bodzsar and C. Susanne), Eotvos Lorand University Press, Budapest, pp. 277–85.

Soares, E.A. and Matsudo, V.K.R. (1982). Efeitos do treinamento de futebol sobre a PWC_{170} em escolares. *Revista Brasileira de Ciencias do Esporte*, **4**, 7–10.

Soares, E.A., Matsudo, V.K.R., Ferreira, M., Figueira, A.J. and Leandro de Araujo, T. (1994). Relationship among agility tests in junior high soccer players. *XIX Simposio Internacional de Ciencias do Esporte*, Centro de Estudos do Laboratorio de Aptidao Fisica de Sao Caetano do Sul, Sao Caetano, Sao Paulo, Brazil, p. 135 (abstract).

Sports Council (1992). *TOYA and Sports Injuries*. The Sports Council, London.

Tanner, J.M. (1962). *Growth at Adolescence*, 2nd edn. Blackwell Scientific Publications, Oxford.

Togari, H., Ohashi, J. and Ohgushi, T. (1988). Isokinetic muscle strength of soccer players, in *Science and Football* (eds T. Reilly, A. Lees, K. Davids and W.J. Murphy), E & FN Spon, London, pp. 181–5.

Vidalin, H. (1988). Football. Traumatismes et age osseux. *Estude prospective de 11 cas. Medecine du Sport*, **62**, 195–97.

Viviani, F., Casagrande, G. and Toniutto, F. (1993). The morphotype in a group of peri-pubertal soccer players. *Journal of Sports Medicine and Physical Fitness*, **33**, 178–83.

Vrijens, J. and van Cauter, C. (1985). Physical performance capacity and specific skills in young soccer players, in *Children and Exercise XI* (eds R.A. Binkhorst, H.C.G. Kemper and W.H.M. Saris), Human Kinetics, Champaign, IL, pp. 285–92.

Zachman, M., Prader, A., Kind, H.P., Hafliger, H. and Budliger, H. (1974). Testicular volume during adolescence: cross-sectional and longitudinal studies. *Helvetica Paediatrica Acta*, **29**, 61–72.

21 Identifying talented players

Thomas Reilly, A. Mark Williams
and Dave Richardson

Introduction

Many thousands of youngsters who participate in soccer aspire to play at an elite level. The dream to become a star player and compete on the international stage may even be the foremost motive for playing from an early age. Motivation is often secured by following a contemporary role model, deemed to be one of the game's best players, whether in the local professional team or in international soccer. The reality is that for the majority of youngsters the dream outshines their capabilities and only a select few will achieve public acclaim and success in the game.

It is self-evident that some individuals have natural gifts, being able to run faster or demonstrate exquisite skills compared to their peers. They seem to acquire ball skills readily and have an apparent natural affinity for field games. Even those acknowledged as exceptionally talented in soccer may fail to realize their true potential as adult players, due to lack of commitment to training, poor tolerance of competitive stress or receding interest in play.

Contemporary professional soccer offers attractive material rewards and popular acclaim for those who are successful. Players can command lucrative salaries when they are good enough to move to the top clubs whilst the clubs who had nurtured young star players benefit financially from their transfer. It is understandable that clubs spend considerable effort in attempting to 'spot' potential stars and draw their first team from the development of their own youngsters. The identification of players with good potential helps to ensure that these young players receive specialized coaching and training to accelerate the process of talent development. Furthermore, the reliable identification of talent allows the club to target its resources on the small number of individuals selected for systematic development, ensuring a more effective investment for the future.

Due to the need for club and country to produce young talented players, sports scientists can play a role in working alongside coaches, scouts and administrators, each of whom must underline key elements of the talent identification and development process. Indeed, as professional clubs throughout Europe have systematized their 'football academies' as the cradle for their young players, they have increasingly utilized sport scientists for help in guiding and evaluating their talent identification and development processes.

In this chapter, the role of sports science in helping to identify soccer talent is considered. Key stages within the talent identification process are highlighted and explained. Some examples of talent identification and development systems are reviewed, including models from other sports. A comprehensive multidisciplinary approach embracing physical (anthropometric), physiological, psychological and sociological research is adopted. The implications of this research for talent identification are addressed prior to completion of an overview for long-term planning.

21.1   Identification and development of talent

Talent has several properties that are genetically transmitted and therefore innate. Nevertheless, talent is not always evident at an early age but trained people may be able to identify its existence by using certain markers. These early indications of talent may provide a basis for predicting those individuals who have a reasonable chance of succeeding at a later stage. Very few individuals are talented in any single domain; indeed if all children were equally gifted, there would be no means of discriminating or explaining differential success. Furthermore, talent is specific to that particular domain.

The complex nature of talent is highlighted by these principles. It is not surprising, therefore, that there is no consensus of opinion, nationally or internationally, regarding the theory and practice of talent identification. Usually professional clubs depend on the subjective assessment of their experienced scouts and coaches, employing a list of key criteria. These criteria are set out as acronyms; for example, the key phrase incorporated in the scouting process of Ajax Amsterdam is TIPS, standing for technique, intelligence, personality and speed. Alternative lists include TABS (technique, attitude, balance, speed) and SUPS (speed, understanding, personality, skill).

Whilst the ability of coaches and scouts to interpret these criteria and recognize such aptitudes for success should not be underestimated, augmenting their judgements with sports science contributions can provide a measure of objectivity to the process. Data collected by sports scientists can, at least, help to confirm or question the practitioners' initial intuition with respect to players' strengths and weaknesses.

From a scientific point of view, the pursuit of excellence can be broken down into four key stages (Russell, 1989; Borms, 1996). These can be distinguished as detection, identification, selection and development (see Figure 21.1).

Talent detection refers to the discovery of potential performers who are currently not involved in the sport in question. Due to the popularity of soccer and the large number of children participating, the detection of players is not a major problem when compared with minority sports. Talent identification refers to the process of recognizing current participants with the potential to become elite players. It entails predicting performance over various periods of time by measuring physical, physiological, psychological and sociological attributes as well as technical abilities either alone or in combination (Régnier *et al.*, 1993).

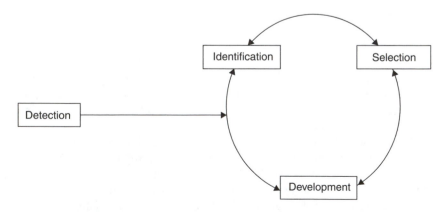

Figure 21.1 Key stages in the talent identification and development process. (Adapted from Williams and Reilly, 2000.)

An attempt is made to match a variety of performance characteristics, which may be innate or amenable to learning or training, to the requirements of the game. A key question is whether the individual has the potential to benefit from a systematic programme of support and training. Talent identification has been viewed as part of talent development in which identification may occur at various stages within the process. Talent development implies that players are provided with a suitable learning environment so that they have the opportunity to realize their potential. The area of talent development has received considerable interest of late, leading several researchers to suggest that there has been a shift in emphasis from talent detection and identification to talent guidance and development (see Durand-Bush and Salmela, 2001). Finally, talent selection involves the ongoing process of identifying players at various stages that demonstrate prerequisite levels of performance for inclusion in a given squad or team. Selection involves choosing the most appropriate individual or group of individuals to carry out the task within a specific context (Borms, 1996). It is particularly pertinent in soccer since only 11 players can be selected to play at any one time.

For many years, scientists have attempted to identify key predictors of talent in various sports (for a review, see Régnier *et al.*, 1993). In this type of research, particularly evident in Australia and the former Eastern bloc countries, there are attempts to identify characteristics that differentiate skilled from less skilled performers and to determine the role of heredity and environment in the development of expertise (see Figure 21.2). This research has embraced various sports science disciplines.

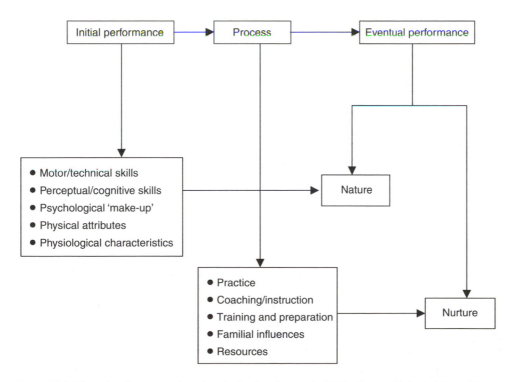

Figure 21.2 The role of nature and nurture in the development of elite players. (Adapted from Bloom, 1976.)

21.2   Talent identification models

Identifying and selecting talented soccer players are not straightforward operations. Detection and identification of talent are more difficult in team games than in individual sports such as running, cycling or rowing, where predictors of performance are more easily scientifically prescribed (see Reilly *et al.*, 1990). Long-term success in soccer is dependent on a host of personal and circumstantial factors, not least of which is the coherence of the team as a whole and the availability of good coaching. These factors make it difficult to predict ultimate performance potential in soccer players at an early age with a high degree of probability.

Nevertheless, there are lessons to be learnt from the talent identification models employed by various nations. Most notable were those used by Eastern European states from the 1960s to the late 1980s. The systems relied on the generation of a comprehensive database of personal and performance variables and formal monitoring of progress and development. The systems were most effective where clear relationships between individual characteristics were established. These were almost exclusively individual rather than team-based sports.

The most systematic trawl for talented sports participants was probably in the former Deutsch Democratic Republic or East Germany. The search for talent was supported by the state because of the political importance attached by the government to success in sport, especially at World Championships and Olympic Games. Yet the national soccer team failed to achieve any success consistently, itself an indication of the difficulty of isolating predictors of future success in this game.

Within the system employed in the Deutsch Democratic Republic, not every individual displaying characteristics of talent was selected for systematic training. Youngsters were selected for specialisation, only on the provision that they:

i   were healthy and free of medical complaints;
ii   could tolerate high training loads;
iii   had a psychological capability for training;
iv   maintained good academic achievement levels.

In more recent years, the Australian Institute of Sport formed the model for some European countries, notably Great Britain, to follow. In preparation for the 2000 Olympic Games in Sydney, the Australian Institute of Sport had paid considerable attention to its own talent identification and development. The major effort was targeted at individual sports such as rowing, swimming, cycling and athletics. A novel approach to talent identification and development was adopted for women's soccer, as reported by Hoare and Warr (2000).

The Australian experience consisted of detecting individuals with athletic ability in field games and selecting them for a fast-track programme of training in soccer skills. Special arrangements were made for their participation in the country's major women's soccer league. Whilst individuals within the team achieved a limited success in the game, the experience did not yield convincing evidence that talent detection and identification was a parsimonious process in soccer.

Matsudo *et al.* (1987) described the pyramid model they used in Brazil that embraced six tiers of performance abilities. The standards of proficiency ranged from physical education classes at the base of the pyramid to international competitors at its apex. Their test battery incorporated anthropometric, physiological and performance profiles and was expressed as 'intended for inferring training and selective factors'. Its use in specific sports was limited, other than directing individuals at an early age to the sports to which they seemed biologically most suited.

The most effective input from sports science into talent identification is likely to be multidisciplinary. Identifying talent for field games at an early age is unlikely to be mono-disciplinary or mechanistic. Successful identification of talent is then followed by selection onto a formal programme for developing playing abilities and nurturing the individual towards realizing the potential already predicted. Eventual success is ultimately dependent on a myriad of circumstantial factors, including opportunities to practice, staying free of major injuries, the type of mentoring and coaching available during the developmental years. Lastly, but not least in importance are personal, social and cultural factors.

Research programmes into talent identification have embraced each of the sports science disciplines. These are considered in turn prior to addressing their implications for multidisciplinary work.

21.3   Research on talent identification in soccer: predictors of talent

21.3.1   *Physical predictors of talent*

Players' anthropometric characteristics (e.g. stature, mass, body composition, bone diameter, limb girth) are related to performance in important and sometimes complex ways (Borms, 1996). The implication is that such measurements may assist in the identification of talent (Carter, 1985). Successful young soccer players, for instance, appear to have similar somatotypes/physiques to older successful performers (see Peña Reyes *et al.*, 1994; Malina *et al.*, 2000). In particular, adult stature, which is commonly used for prediction, is strongly influenced by genetic factors (Lykken, 1992), whilst other physical attributes (e.g. muscle mass, body fat) are seen as being more amenable to training, and dietary influences (see Bouchard *et al.*, 1997; Reilly *et al.*, 2000).

Research evidence indicates that elite youth soccer players have greater biological age (i.e. more physically mature) than their less proficient counterparts and coaches appear to favour players advanced in morphological growth during the selection process (see Panfil *et al.*, 1997; Malina *et al.*, 2000). This trend in favour of children born early in the selection year (e.g. September–December) is apparent in several countries (e.g. Sweden, Belgium, United Kingdom) and persists into adult elite squads (see Helsen *et al.*, 2000). Over 50% of players who attended the English Football Association's National School when it was based at Lilleshall were born between September and December (Brewer *et al.*, 1995). A similar percentage of players selected for the England national team during the 1986, 1990, 1994 and 1998 World Cup qualifying campaigns were born early in the selection year (Richardson, 1998; Richardson and Stratton, 1999). This latter finding suggests a 'residual bias' as a result of selection policies at youth level that favour individuals born in the early part of the academic year (Boucher and Mutimer, 1994). The discrimination bias was particularly evident with goalkeepers and defenders, who tended to be the tallest and heaviest players at adult level (see Franks *et al.*, 1999).

In previous studies, researchers have inferred the issue of enhanced morphological growth based on date of birth only. There is a clear need to assess various anthropometric characteristics (such as height and body mass) and practice profiles along with date of birth before drawing any firm conclusions.

Many of the physical qualities that distinguish elite and sub-elite players may not be apparent until late adolescence, confounding the early selection of performers (Fisher and Borms, 1990). The implication is that the prediction of future elite players from anthropometric measurements may be unrealistic in younger age groups because performance could be affected by the player's rate of physical growth and maturation. Since late maturing children can compensate for any apparent disadvantage in size and strength by working on their technical capabilities or by improvements in other areas (such as agility and muscular power), it is important that the talent identification process is not overly biased toward the early maturing child. Any potential bias can result in late maturing and potentially talented players dropping out of the game at an early age. Furthermore, late maturing players are more likely to miss out on the experience of high quality coaching. The key message is that young players should be selected on skill and ability rather than on physical size. Helsen *et al.* (2000) suggested some potential solutions to the problem of seasonal birth date distribution. A reduction in age-band range and closer matching of players into groups based on maturational rather than chronological age may be fruitful avenues to explore. Furthermore, more flexibility in allowing players to move between age-bands for certain training practices should also be helpful.

21.3.2   *Physiological predictors of talent*

Physiological measures have also been employed in an attempt to identify key predictors of performance (see Jankovic *et al.*, 1997; Panfil *et al.*, 1997; Janssens *et al.*, 1998; Reilly *et al.*, 2000). Jankovic *et al.* (1997) compared successful and less successful 15–17-year-olds using measures of maximal oxygen uptake ($\dot{V}O_2$ max), anaerobic power, grip and trunk strength measures, and heart volume (absolute and relative). They deemed successful players to be those who were later selected in clubs playing in the top league in Croatia, Germany, Italy and England, whilst those considered less successful did not progress beyond regional leagues. The successful players had superior physiological fitness compared to the others. Janssens *et al.* (1998) showed that performance in short (30 m) and prolonged 'shuttle' running discriminated between successful and less successful 11–12-year-old soccer players. Similarly, in a study by Panfil *et al.* (1997) elite 16-year-olds recorded better performance in running and jumping than their less elite counterparts. Such findings led Jankovic *et al.* (1997) to conclude that physiological measures could be useful in predicting later success in soccer (see also Carter, 1985; Panfil *et al.*, 1997). Nevertheless, the possibility remains that in the above studies part of the physiological superiority of the successful players was due to a more systematic approach to training prior to their induction into the specialized under-age squad.

Although physiological measures such as maximal oxygen uptake ($\dot{V}O_2$ max) have been successful in distinguishing between expert and intermediate young players, they may not be sensitive enough to distinguish players already selected and exposed to systematised training for national teams. In a recent study, Franks *et al.* (1999) analysed data from 64 players who attended the English Football Association's National School (14–16 years) between 1989 and 1993. Anthropometric characteristics as well as aerobic and anaerobic measurements were recorded. Players were categorized according to playing 'position' and whether they had signed a full-time 'professional' contract on graduation. No differences were observed between those who were deemed to be more or less successful. In a group of youth players already highly selected, other factors may determine their employability as professionals. It may well be that talent becomes harder to predict in later years since the population of players becomes smaller and more homogeneous, particularly with respect to their physical and physiological profiles. Those who have not developed the requisite characteristics tend to drop out of the sport at an early age.

Physiological measurements may be useful alongside subjective judgements of playing skills for initial detection of talent, but such measures do not appear sensitive performance indicators on a global basis and can not be used reliably on their own for purposes of talent identification and selection. Moreover, while research using twin siblings has indicated that physiological characteristics are highly genetically predisposed, appropriate training can have a pronounced influence (see Bouchard *et al.*, 1997). Although some people may be more favoured genetically than others to adapt and benefit from training, particularly with regard to the relative distribution of muscle fibre types (Simoneau and Bouchard, 1995), physiological responses to exercise are highly dependent on regular training and practice (see Reilly *et al.*, 2000).

Contemporary professional soccer is played at a higher 'tempo' than 10 years ago (Williams *et al.*, 1999). It is likely therefore that physiological correlates of work-rate during games have gained in importance in the context of fitness for soccer. Physiological values indicative of aerobic fitness, such as $\dot{V}O_2$ max, may be more influential in successful performance in the future. Consequently, a relatively high threshold for oxygen uptake capability may be a significant criterion when young players are assessed. However, there is concern regarding the extent to which a high fitness indicator tracks through from childhood to adulthood.

21.3.3   *Psychological predictors of talent*

21.3.3.1   *Psychological profiling*

Intuitively, it is thought that successful players are distinguished from less successful players on the basis of psychological factors. The assumption is that a talented player possesses personality characteristics that facilitate learning/training and competition. Although coaches and scouts may argue that talented and less talented players can be differentiated on the basis of their psychological 'make up', researchers have yet to identify specific personality characteristics, or an overall psychological profile, that are predictably associated with success in sport. No clear or consistent relationship has been demonstrated between personality and expertise (see Vealey, 1992; Auweele *et al.*, 1993, 2000; Morris, 2000). As yet there is no psychological inventory to help select players with more or less potential or talent and it is hardly likely that any single inventory would have complete predictive power.

Initially, researchers examined whether talented performers differed from their less talented counterparts on specific personality dimensions or traits. These traits are believed to be relatively stable over time and reflect a player's predisposition towards certain types of behaviour (e.g. aggression, extroversion, sensation-seeking, neuroticism). No clear relationship has been found between such global personality characteristics and expertise in soccer. Some studies have highlighted differences on one or more variables (e.g. aggression, tough-mindedness), whereas others have not. In general, however, the majority of published research suffers serious deficiencies in important procedural aspects including research design, sampling and testing procedures, analyses and interpretation (Morris, 1995).

Sports psychologists have also examined the importance of more transient or changeable personality characteristics using 'state' (as opposed to 'trait') or interactionist (based on personal and situational factors) approaches including sport-specific measures of anxiety, self-confidence, motivation and attentional style. Typically researchers have reported that talented players are more committed, self-confident and less prone to anxiety, both prior to and during competition, able to employ various psychological coping strategies effectively, more highly motivated, and better at maintaining concentration during performance (Auweele *et al.*, 1993; Durand-Bush and Salmela, 2001). Although these measures have been more successful than earlier 'trait' measures in distinguishing between elite and sub-elite players, there is no consistent evidence to suggest that such personality 'profiling' benefits the talent identification process (see Vealey, 1992; Morris, 2000). It is also questionable whether talent identification should be based on such 'state' or interactional variables, since these can change from day to day. Such measures may not provide a strong indication of typical behaviour (traits). At present, the use of psychological tests for talent identification purposes can not be endorsed scientifically (Fisher and Borms, 1990). It seems unrealistic to expect that expert performance could be explained purely by personality variables. The range of 'personalities' evident at a typical top professional club in any of the major soccer playing nations would confirm this suggestion. It would be a mistake to use for talent identification a psychological test that has not been validated for such purposes. If there is a role for personality profiling in talent identification, it is likely to be in conjunction with other measures, in certain situations only and with the probability of explaining only a small proportion of the variance in performance (Morris, 1995).

Morris (2000) advised that administrators and coaches should allocate resources to the development of psychological skills among the widest range of young players. Although personality traits are to a certain degree inherited, with heritability estimates between 30 and 60% (Plomin *et al.*, 1994; Saudino, 1997), it seems that psychological skills are highly amenable to specialised training. Sports psychologists have made noteworthy progress in this regard and there is an extensive array of resources and suitably qualified sport psychologists to help facilitate this process (see Hardy *et al.*, 1996; Morris, 1997; Williams, 1998). Motivation, anxiety management, concentration, self-confidence and attentional style, amongst others, are psychological skills that can be refined through appropriate training.

21.3.3.2   *Cognitive factors and game intelligence*

The measurement of perceptual–cognitive skills such as anticipation and decision making provides a promising approach to talent identification. Consistent differences emerge when skilled and less skilled players are tested on their anticipation and decision making skills (Williams and Davids, 1995). Such tests, which typically employ film-based simulations of match situations, have been used successfully with both adult and junior players (Williams, 2000). When compared to their less skilled counterparts, skilled players are: (a) faster and more accurate in recognizing and recalling patterns of play; (b) better at anticipating the actions of their opponents based on contextual information; (c) characterized by more effective and appropriate visual search behaviours; and (d) more accurate in their expectation of what is likely to happen given a particular set of circumstances. Skilled soccer players' enhanced performance on such tests is thought to reflect their superior knowledge developed as a result of specific practice and instruction as opposed to any initial differences in visual skills such as acuity, colour vision or depth perception. It is not clear what proportion of perceptual skill is genetically determined compared with that developed through purposeful practice. It may be that talented players are predisposed to acquiring the knowledge structures underlying perceptual and decision making skill in soccer (Williams, 2000). Although genetic influences are likely to determine responsiveness to training, perceptual skill can be improved through specific instruction and practice regardless of one's initial ability. With the advent of less expensive and more 'user-friendly' digital video editing systems, coaches should be able to create simulations for testing and training the skills of anticipation and decision making rather than expect these functions to be ingrained.

Two other cognitive measures proposed as predictors of talent include intelligence and creative thinking (see Morris, 2000). It has been suggested that significant proportions of elite performers are likely to be academic high achievers (English Sports Council, 1995). The concept of intelligence is difficult to define and measure (see Neisser *et al.*, 1996). It seems to have several aspects (e.g. analytic, creative, practical) that may or may not be relevant to soccer. Skilled players often possess a 'game intelligence' that allows them to analyse major features of their opponent's play (see Singer and Janelle, 1999). It is not clear whether this game intelligence is linked to academic intelligence.

21.4   Sociological considerations in talent identification and development

The development of talented players requires appropriate social and environmental support mechanisms. Traditionally, sociological considerations of the 'talent map' have been perceived as secondary to the conventional laboratory based scientific disciplines. However, sociological considerations place greater emphasis on the nurturing of talent, environmental factors and the availability of opportunity.

21.4.1   *Readiness*

The concept of readiness recognizes that learning is more rapid and enjoyable when readiness exists. Readiness to learn occurs following an accumulation of events and/or experiences enabling the learner to acquire additional information, skills or values (Seefeldt, 1996). Yet, many young children are forced into competitive soccer before they are ready to do so. On average a child is first exposed to competitive soccer between the ages of 6 and 8 years, although there are increasing incidences of exposure as young as 3–4 years. These occurrences arise primarily from the assumption that early exposure enhances a child's chances of later success. However, early exposure to competitive soccer may also increase the risk of injury and drop out. Readiness for competitive youth soccer applies if there is a match between a child's stage of growth, maturity and development and the level of demand presented by the sport.

The social and cognitive capacity of most children under the age of 12 ensures that few possess any real concept of their role within competitive sport. Several authors have emphasized that very young children cannot and do not compete because they are incapable of or uninterested in social comparison (Roberts, 1980; Scanlan, 1988). A lack of competitive motivation should not prevent participation. Ideally, game structures and adult expectations of performance should be modified to meet the developmental capabilities of the children.

Ultimately, it is the child who must practice, be receptive to coaching and compete. The motivation of a child to practice and compete is an essential pre-requisite for successful performance (i.e. self-selection). The key to success does not lie in how early a child is exposed to competitive soccer, but that the child concerned is 'optimally' ready to be involved. This optimisation must also account for the readiness to accept, and cope, with failure and/or limited progression within the sport.

21.4.2   *Significant others*

Implicit within a child's sporting, educational and life experiences are the influences of significant others (i.e. key relationships that are peculiar to a child's individual development).

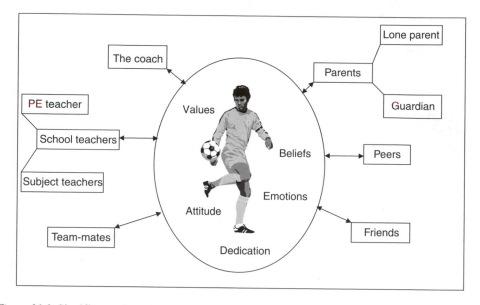

Figure 21.3 Significant others that may influence a young player's values, beliefs, emotions, attitude and dedication toward sport performance at any given time.

Each of these relationships may have an impact and/or influence on a child's values, beliefs, emotions, attitude and dedication toward soccer (i.e. decision-making and thought processes in relation to participation, individual goals and performance expectations). These relationships are highlighted in Figure 21.3.

Children have traditionally been viewed as empty vessels into which families poured their social values, customs and cultures. Children's actions have been depicted as the result of effective or ineffective parenting (Hill and Tisdall, 1997). Today's children are more aware of their participative rights within the family environment.

Significant others may modify their behaviour depending on their individual perception of the child, their relationship with the child, their feelings toward children in general and their knowledge of influences such as age, gender and birth order. The introduction of the coach exposes the child to a new type of authority. It has been argued that the behaviour of coaches and their involvement with a child are more important in the development of talent than are initial ability levels (Carlson, 1988, 1993). Good coaches know when to push players and when to reduce the intensity of training and their expectations. Moderation of effort and potential 'drop out' may occur if harmony does not exist between coach and player. Some effort should be made to prevent this loss of talent by encouraging the player to return to systematic training. Elite players are more likely to think highly of their coaches from early years of involvement (Carlson, 1993). The relationship is based on shared interests in accomplishing a task (e.g. soccer performance) rather than on a personal or emotional issue. Talented youngsters may see the coach as 'the gatekeeper' to future progression, achievement and success within their sport. The potential impact of the coach on the child's socialisation is immense (Richardson, 1999).

The responsibility and impact of these significant others toward the social, academic and sporting development of talented youngsters demands further investigation. A child's social development may be greatly enhanced by the experiences and opportunities afforded by top-level sport performance. If the appropriate support mechanisms are not in place (i.e. determined and provided by significant others) then the child's social development may be impeded. Social restrictions and authoritarian structures embedded within a child's talent programme may restrict peer relationships, development opportunities outside of soccer and ultimately affect participation and performance.

21.5   Environmental and cultural considerations

Children's behaviour patterns, emotions and attributes are affected by their surrounding environment (e.g. school, home, club) and the nature of this environment (e.g. supervised, unsupervised, structured or unstructured). Children's perception of themselves within such environments will dictate their decision-making processes, behaviour and ultimately performance. Young players must be afforded the opportunity of familiarization in order to ensure consistent performance.

Children are impressionable social participants readily influenced, not only by significant others, but by social trends, social status, popular and traditional culture. An important determinant of success in any sport, and particularly in soccer, is socialization into the particular culture (Carlson, 1993). The identification and development of soccer talent in any country may therefore call for a cultural analysis to complement the behavioural and biological approaches. Combining these variables with race and religion provides a mass of structures, rules and opportunity for youngsters to either embrace or disregard (Richardson and Reilly, 2001). The impact of youth culture (e.g. fashion and music) and its increasing impact on the socialisation process has been established (Hill and Tisdall, 1997). Today's society exposes children to a multitude of social choices, with an increasing 'freedom' of choice, resulting in confusion and disorientation. The potential impact of these experiences on a talented youngster's performance at any one time demands further investigation.

Social class and socio-economic status can also affect a young player's development (e.g. Power and Woolger, 1994; Kirk *et al.*, 1997). The financial constraints limiting access to coaching and training sessions, games, facilities, kit, equipment plus various other supporting considerations may offer the middle and upper classes a competitive advantage. Children from single parent families and ethnic minorities have also been reported to be socially disadvantaged in this regard. The implications of this competitive advantage are beginning to emerge. In the UK, soccer has traditionally been seen as the preserve of the 'working classes'. However, today soccer attracts a number of players from 'middle-class' backgrounds or as Revell (1999) remarked:

> ... soccer recruitment has moved out of the backstreets of the industrial north and into the middle-class suburbs ... Some insiders talk of the gentrification of the game, with everything from seat prices to recruitment policies aimed at a middle-class market.
>
> (pp. 2, 3)

It appears that those most suited to the demands of soccer and capable of adapting to them physically, physiologically, psychologically, sociologically and, perhaps, economically are more likely to be successful. The dynamic and transient relationship that exists between a child, his/her peers, parents, coach, teacher in conjunction with natural growth and sexual maturation, and its impact on performance at any given time, is a major consideration in the development of talented youngsters.

Three important phases in the development of expertise have been identified: initiation, development and perfection (Bloom, 1985). Table 21.1 shows the generalized characteristics observed during these career progressions in elite youth performers. The suggestion is that social environments help to shape young talented individuals across the early, middle and late stages of their careers. The implication is that various situational factors and the role of family members and mentors override the natural ability of the performer. Creating an appropriate environment in which to nurture talent may play a more significant role in the development of expertise than does heredity (Salmela, 1996). Inherent talent may be necessary, but it is insufficient for the prediction of eventual playing level. The ideas of Bloom on the development of talent were extended in the sports context by Côté (1999) who suggested four distinct stages of participation in sport, namely sampling, specialising, investment and recreation years. At each of the initial levels, children have the potential to move to another level, drop out of the sport, or enter the recreation stage. A key area for further research therefore is to provide guidelines for nurturing and developing players through each of these stages.

Ericsson and colleagues have placed a strong emphasis on the nurturing of talented players (Ericcson *et al.*, 1993; Ericsson and Charness, 1994; Ericsson, 1996; Ericsson and Lehmann, 1996). They have argued that talent plays only a limited role in the development of expertise. The level of performance is assumed to be directly related to accumulated practice and regardless of natural abilities or genetically predisposed traits, at least 10 years of intensive practice is thought necessary to acquire the skills and experience needed to become an expert within any domain. Ericsson and co-workers suggested that natural ability is not a pre-requisite to the development of expertise, rather expertise depends on the amount of time spent on 'highly structured, effortful' activity with the specific goal of improving performance (see also Howe *et al.*, 1998; Simonton, 1999). Motivation, commitment and hard work are pre-requisites for exceptional performance. Although this 'environmentalist' view is contrary to empirical evidence supporting the contribution of genetics to expert performance (see Bouchard and Lortie, 1985; Bouchard *et al.*, 1997, 1998; Current Directions in Psychological Science, 1997), a supportive learning environment, effective

Table 21.1 Potential characteristics of talented soccer players and their coaches and parents at various stages of their careers (Adapted from Bloom, 1985)

Individual	*Career progression*		
	Initiation	*Development*	*Perfection*
Player	Joyful Playful Excited Special	Hooked Committed	Obsessed Responsible
Coach	Kind Cheerful Caring Process-centred	Strong Respecting Skilled Demanding	Successful Respected/feared, Emotionally bonded
Parents	Shared excitement Supportive Sought mentors Positive	Made sacrifices Restricted activity	

practice and high quality coaching can help overcome perceived shortcomings in initial ability levels (see Helsen *et al.*, 2000).

To this end, future research should focus on the nature of practice conditions and the facilitatory role that coaches play in the development of expert performers (see Helsen *et al.*, 2000). In particular, a clearer understanding of the learning process and the nature, type and frequency of practice leading to expert performance at each stage of development is required. An over-emphasis on performance-outcome measures, particularly in young performers, should be replaced by process measures of performance as more appropriate long-term predictors of potential (see Davids *et al.*, 2000). The development of structured practice regimens and staged performance standards that developing players would aim to achieve, irrespective of chronological age, would also be indicative of progress.

21.5.1 *Injury*

Players' potential to succeed may be determined in part by their susceptibility to injury, both acute and overuse, as a result of practice or competition (Singer and Janelle, 1999). Injuries may be attributed to extrinsic factors such as the behaviour of an opponent, playing surface and so on, or to factors intrinsic to the individual such as personality, biomechanical defects and fitness. The prevention and detection of injuries should therefore be a constant concern in any system of selection and player development.

It is important that players are screened to help detect any factors which predispose them to an increased risk of injury and to identify those players who may have contra-indications to exercise and sport, such as cardiac or respiratory abnormalities. Intrinsic anatomical predisposing factors include a range of skeletal malalignments (e.g. pes planus/cavus, genu valgum/varum), leg length discrepancy, muscle weakness/imbalance, poor flexibility, joint laxity and instability (Renstrom, 1998). There is evidence also of a psychological entity known as the 'injury prone athlete' (e.g. see Sanderson, 1981; Bergandi, 1985), although it is not clear whether this susceptibility is largely innate or transient.

Clear guidelines are required to prevent overuse injuries in children as a result of 'overplaying' and inappropriate training frequency or intensity. The capability to recover, both mentally and physically, from injury also determines the level of achievement over time. This ability may be genetically influenced both physically, in terms of the biological repair of tissue, and psychologically through a capacity to cope (Grove, 1993). To this end, it is essential to have access to appropriate sports medicine and psychological support.

21.6 Overview and implications

It is difficult to draw firm conclusions with regard to the practical utility of a systematic or scientific approach to talent identification, as opposed to relying purely on subjective assessment of performance as evidence of giftedness. Research concerned with the physical and physiological predictors of talent has highlighted a number of potentially important measures such as somatotype, aerobic fitness and anaerobic power. Nevertheless, there is no clear consensus regarding the relative importance of these measures in predicting talent in soccer. Fitness and anthropometric profiling can generate a useful database against which talented players may be compared in order to identify strengths and weaknesses. These measures can also be helpful to scouts and coaches in confirming their initial perceptions of a player's strengths and weaknesses. Whilst such profiling may be used for the purposes of 'selecting in' (i.e. including players within a population pool because they have certain desirable physical and/or physiological attributes), the use of such measures for not selecting players because of perceived shortcomings is questionable and arguably unethical. The individualistic nature of expertise ensures that shortcomings can often be remedied through appropriate interventions (e.g. behavioural modification, diet, training) and that weaknesses in certain areas can often be compensated by strengths in others. At the very least, the systematic collection of such measures, particularly, from childhood through adolescence, would ensure that coaches and others are better informed about how physical and physiological factors affect the development of young soccer players. Longitudinal research of this nature would also help determine the predictive utility of these tests with young players.

Psychological profiling for the purpose of talent identification or selection can not be scientifically endorsed. Although the importance of factors such as self-confidence, concentration and motivation has been highlighted, presently there are no psychological tests that achieve adequate levels of discrimination. It would be more appropriate for professional clubs to devote resources towards psychological support (e.g. advice, counselling) and improving players' mental skills. Contrary to popular opinion mental skills such as motivation, anxiety control and self-confidence can be refined through psychological intervention techniques delivered by accredited sports psychologists. The monitoring and longitudinal profiling of psychological characteristics over time may be helpful in determining the extent to which these variables remain stable over the course of adolescence. Soccer-specific tests of anticipation, decision-making and creative intelligence may offer more predictive utility for talent identification purposes. At elite levels, these cognitive factors may be more important than physical/physiological attributes once players reach hypothesized threshold values in the latter. Since performance on these perceptual tests is at least partly dependent on playing experience (Abernethy, 1988; Williams and Davids, 1995), by definition such measures may not be used for detecting talent. However, preliminary data suggest that players may be differentiated on the basis of such skills after comparatively brief periods of exposure, implicating the potential use of such tests for talent identification purposes (see Williams, 2000). Moreover, there is evidence to suggest that perceptual and cognitive skills are amenable to practice and instruction as part of a holistic approach to talent development (see Williams and Grant, 1999).

Sociologists have questioned the validity of relying solely on physical, physiological or psychological measures to predict potential in sport. In their view, having supportive parents, a stimulating and permissive coach and the dedication and commitment to spend hours and hours practising and refining skills are the real determinants of excellence. According to this approach, there are no early predictors of adult performance. Instead, players should be provided with access to appropriate facilities and opportunities for meaningful practice. Investment in high quality coaches and coach education systems is crucial. Technical support in terms of sports science and sports medicine is essential to ensure that players have the opportunity to fulfil their potential. Clubs have a responsibility to invest in youth, providing today's children with the opportunity to be nurtured into tomorrow's 'superstars'. To this end, a more equitable balance between players' current salaries and investment in talent identification and development is paramount. Also, more opportunities must be provided for young players, at the possible expense of migrant players (see Maguire and Pearton, 2000), by perhaps restricting the number of over-age players at club level.

Avenues for future research have been highlighted throughout this review, but some specific suggestions are outlined. First, a promising direction would be to adopt a multidisciplinary approach to identifying talent. The complexity of talent and the methodological problems associated with its identification precludes the use of a monodisciplinary approach. Future research should employ a comprehensive battery of physical, physiological, psychological and sociological measures (e.g. see Hoare and Warr, 2000; Reilly *et al.*, 2000). This more structured and holistic approach would account for a greater proportion of the variance between talented and less talented players, promoting greater accuracy and improved understanding of the talent identification process (Auweele *et al.*, 1993; Régnier *et al.*, 1993; Prescott, 1996). A comprehensive database is required to develop a criterion-based model or 'talent profile' that may help predict future performance. Moreover, different factors may predict performance at various age levels and consequently, any such model would need to be age-specific (Régnier and Salmela, 1987).

In most previous research, a cross-sectional rather than longitudinal approach has been adopted. Improved understanding may arise from monitoring players over a protracted period of time to determine the key changes that occur as a result of development and practice and how these factors contribute to expert performance. Some possible predictors of talent that could be included in such a battery are highlighted in Figure 21.4. Clubs should permit greater communication between sports scientists and those involved in the talent identification process. Scientists need to determine the nature of the subjective and implicit criteria that coaches/scouts use to identify talented players. Although most coaches and scouts feel that they can 'spot a good player', outlining the criteria upon which such decisions are made is more problematic. Structured interviews with practitioners would enable scientists to quantify and document such criteria, thereby facilitating coherence between the theory and practice of talent identification (e.g. see Côté *et al.*, 1995).

Conclusions about young talented male players cannot necessarily be generalized to females. It is likely that the worldwide growth in female participation in soccer will continue well into the future. It is important therefore that research into talent identification and development is extended to address issues related to young female soccer players.

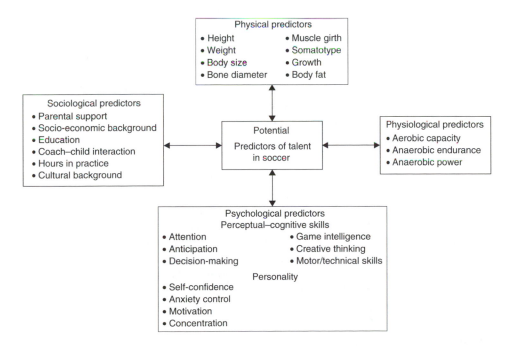

Figure 21.4 Potential predictors of talent in soccer from each sport science discipline. (Adapted from Williams and Reilly, 2000.)

There are also important ethical and educational issues which transcend all aspects of the talent identification and development process (see Fisher and Borms, 1990). Such concerns are commonly neglected in the literature on talent identification (Borms, 1996). A child's overall development and well-being should be of primary concern to those involved in the process of talent development and identification. The pursuit of excellence should not occur at the expense of the child's physical and emotional health, growth and development. Appropriate familial, educational and sociocultural environments are keystones in a balanced approach to child development and especially where talent development is concerned.

21.7   Some key considerations in talent identification and development

- Physiological and anthropometric profiling of players should occur to establish a comprehensive database for talent identification. Profiling should be undertaken at each age group and involve longitudinal assessment. Use of these measures in isolation for talent selection or identification should not be endorsed.
- Whilst psychological profiling would help to establish a comprehensive database on elite players, such tests should not be employed for talent selection or identification. Implementation of mental skills training programmes through accredited sports psychologists should prove more profitable.
- Perceptual and decision making skills can be developed through appropriate practice and instruction. Soccer-specific measures of 'game intelligence' may prove useful as part of a holistic, multidisciplinary, approach to talent identification.
- Systematic observation criteria should be developed to help coaches/scouts identify talented players. Although subjective assessment criteria are already employed by several clubs, these need to be endorsed and refined through collaborative research between sports scientists and those involved in the talent identification process.
- The process of developing talented young players cannot rely solely on technical and physical ability. The potential impact of a young player's life experiences on his/her sporting performance at any given time must be considered. In addition to the physical, physiological, psychological and technical considerations of young talent, the process of talent development must recognize the complexity of childhood, the multitude of social variables, and their potential influence on performance at any given time.

Summary

The pursuit of excellence in soccer may be broken down into the key stages of talent detection, identification, selection and development. Various talent identification models have been promoted and adapted to suit local soccer and cultural contexts. Nature and nurture interact in the process of realizing soccer potential although not in a simple or predictable fashion. Scientific disciplines can be employed to inform personnel concerned with the processes of developing young soccer players; there are many sociological considerations also, especially related to supportive mechanisms. A multidisciplinary scientific approach towards providing a support system for practitioners is likely to be most productive in attempts to help gifted young players realize their aspirations.

References

Abernethy, B. (1988). The effects of age and expertise upon perceptual skill development in a racquet sport. *Research Quarterly for Exercise and Sport*, **59**, 210–21.

Auweele, Y.V., Cuyper, B.D., Mele, V.V. and Rzewnicki, R. (1993). Elite performance and personality: From description and prediction to diagnosis and intervention, in *A Handbook of Research on Sports Psychology* (eds R. Singer, M. Murphy and L.K. Tennant), Macmillan, New York, pp. 257–92.

Auweele, Y.V., Mele, V.V., Nys, K. and Rzewnicki, R. (2000). Personality and (elite) sport: Towards a reconceptualising of the personality concept in sport psychology, in *A Handbook of Research on Sports Psychology*, 2nd edn (eds R. Singer, C. Hausenblas and C.J. Janelle). Macmillan, New York.

Bergandi, T.A. (1985). Psychological variables relating to the incidence of athletic injury. *International Journal of Sport Psychology*, **16**, 141–7.

Bloom, B.S. (1976). *Human Characteristics and School Learning*. McGraw-Hill, New York.

Bloom, B.S. (1985). *Developing Talent in the Young*. Ballantine, New York.

Borms, J. (1996). *Early identification of athletic talent*. Keynote address to the International Pre-Olympic Scientific Congress, Dallas, USA.

Bouchard, C. and Lortie, G. (1985). Heredity and endurance performance. *Sports Medicine*, **1**, 38–64.

Bouchard, C., Malina, R.M. and Perusse, L. (1997). *Genetics of Fitness and Physical Performance*. Human Kinetics, Champaign, IL.

Bouchard, C., Daw, E.W., Rice, T., Pérusse, L., Gagnon, J., Province, M.A., Leon, A.S., Rao, D.C., Skinner, J.S. and Wilmore, J.H. (1998). Familial resembalnce for VO_{2max} in the sedentary state: the heritage family study. *Medicine and Science in Sports and Exercise*, **30**, 252–8.

Boucher, J. and Mutimer, B. (1994). The relative age phenomenon in sport: a replication and extension with ice-hockey. *Research Quarterly for Exercise and Sport*, **65**, 377–81.

Brewer, J., Balsom, P. and Davis, J. (1995). Seasonal birth distribution amongst European soccer players. *Sports Exercise and Injury*, **1**, 154–7.

Carlson, R. (1988). The socialization of elite tennis players in Sweden: an analysis of the players' backgrounds and development. *Society of Sports Journal*, **5**, 241–56.

Carlson, R. (1993). The path to the national level in sports in Sweden. *Scandinavian Journal of Medicine and Science in Sports*, **3**, 170–7.

Carter, J.E.L. (1985). Morphological factors limiting human performance, in *Limits of Human Performance* (eds D. Clarke and H.M. Eckert), pp. 106–17. American Academy of Physical Education Papers No. 18. Human Kinetics, Champaign, IL.

Côté, J. (1999). The influence of the family in the development of talent in sport. *The Sports Psychologist*, **13**, 395–417.

Côté, J., Salmela, J., Trudel, P., Baira, A. and Russell, S. (1995). The coaching model: a grounded assessment of expert gymnastic coaches' knowledge. *Journal of Sport and Exercise Psychology*, **17**, 1–17.

Current Directions in Behavioural Genetics (1997). *Current Directions in Psychological Science*, **6**, (August).

Davids, K., Lees, A. and Burwitz, L. (2000). Understanding and measuring co-ordination and control in kicking skills in soccer: implications for talent identification and skill acquisition. *Journal of Sports Sciences*, **18**, 703–14.

Durand-Bush, N. and Salmela, J.H. (2001). The development of talent in sport, in *A Handbook of Research on Sports Psychology*, 2nd edn (eds R. Singer, C. Hausenblas and C.J. Jannelle), Macmillan, New York, pp. 269–89.

English Sports Council (1995). *Training of Young Athletes Study: Identification of Talent*. The Sports Council, London.

Ericsson, K.A. (1996). *The Road to Excellence: The Acquisition of Expert Performance in the Arts and Sciences, Sports and Games*. Lawrence Erlbaum Associates, Mahwah, NJ.

Ericsson, K.A. and Charness, N. (1994). Expert performance: its structure and acquisition. *American Psychologist*, **49**, 725–47.

Ericsson, K.A., Krampe, R.T. and Tesch-Römer, C. (1993). The role of deliberate practice in the acquisition of expert performance. *Psychological Review*, **100**, 363–406.

Ericsson, K.A. and Lehmann, A. (1996). Expert and exceptional performance: evidence of maximal adaptation to task constraints. *Annual Review of Psychology*, **47**, 273–305.

Fisher, R.J. and Borms, J. (1990). *The Search for Sporting Excellence*. Sport Science Studies 3. International Council of Sport Science and Physical Education, Karl Hoffman, Germany.

Franks, A., Williams, A.M., Reilly, T. and Nevill, A. (1999). Talent identification in elite youth soccer players: physical and physiological characteristics. *Journal of Sports Sciences*, **17**, 812.

Grove, J.R. (1993). Personality and injury rehabilitation among sports performers, in *Psychological Basis of Sports Injuries* (ed. D. Pargman), Fitness Information Technology, Morgantown, WV, pp. 99–120.

Helsen, W.F., Hodges, N.J., Van Winckel, J. and Starkes, J.L. (2000). The roles of talent, physical precocity and practice in the development of soccer expertise. *Journal of Sports Sciences*, **18**, 727–36.

Hill, M. and Tisdall, K. (1997). *Children and Society*. Longman, New York.

Hoare, D.G. and Warr, C.R. (2000). Talent identification and women's soccer: an Australian experience. *Journal of Sports Sciences*, **18**, 751–8.

Howe, M.J.A., Davidson, J.W. and Sloboda, J.A. (1998). Innate talents: reality or myth? *Behavioural and Brain Sciences*, **21**, 399–442.

Jankovic, S., Matkovic, B.R. and Matkovic, B. (1997). Functional abilities and process of selection in soccer. Communication to the 9th European Congress on Sports Medicine, Porto, Portugal, 23–26 September.

Janssens, M., Van Renterghem, B., Bourgois, J. and Vrijens, J. (1998). Physical fitness and specific motor performance of young soccer players aged 11–12 years. Communication to the 2nd Annual Congress of the European College of Sport Science. *Journal of Sports Sciences*, **16**, 434–5.

Kirk, D., Carlson, T., O'Connor, A., Burke, P., Davis, K. and Glover, S. (1997). The economic impact on families of children's participation in junior sport. *Australian Journal of Science and Medicine in Sport*, **29**, 27–33.

Lykken, D.T. (1992). Research with twins: the concept of emergenesis. *Psychophysiology*, **19**, 361–73.

Maguire, J. and Pearton, R. (2000). The impact of elite labour migration on the identification, selection and development of European soccer players. *Journal of Sports Sciences*, **18**, 759–69.

Malina, R.M., Peña Reyes, M.E., Eisenmann, J.C., Horta, L., Rodrigues, J. and Miller, R. (2000). Height, mass and skeletal maturity of elite Portuguese soccer players aged 11–16 years. *Journal of Sports Sciences*, **18**, 685–93.

Matsudo, V.K.R., Rivet, R.E. and Pereira, M.H.N. (1987). Standard score assessment on physique and performance of Brazilian athletes in a six tiered competitive sports model. *Journal of Sports Sciences*, **5**, 49–53.

Morris, T. (1995). Psychological characteristics and sport behaviour, in *Sport Psychology: Theory, Applications and Issues* (eds T. Morris and J. Summers), John Wiley, Chichester, pp. 2–28.

Morris, T. (1997). *Psychological Skills Training: An Overview*, 2nd edn. British Association of Sport and Exercise Sciences, Leeds.

Morris, T. (2000). Psychological characteristics and talent identification in soccer. *Journal of Sports Sciences*, **18**, 715–26.

Neisser, U., Boodoo, G., Bouchard, T.J., Boykin, A.W., Brody, N., Ceci, S.J., Halpern, D.F., Loehlin, J.C., Perloff, R., Sternberg, R.J. and Urbina, S. (1996). Intelligence: knowns and unknowns. *American Psychologist*, **51**, 77–101.

Panfil, R., Naglak, Z., Bober, T. and Zaton, E.W.M. (1997). Searching and developing talents in soccer: a year of experience, in *Proceedings of the 2nd Annual Congress of the European College of Sport Science* (eds J. Bangsbo, B. Saltin, H. Bonde, Y. Hellsten, B. Ibsen, M. Kjaer and G. Sjøgaard), HO + Storm, Copenhagen, pp. 649–50.

Peña Rayes, M.E., Cardenas-Barahona, E. and Malina, R.M. (1994). Growth, physique and skeletal maturation of soccer players 7–17 years of age. *Auxology, Humanbiologia Budapestinensis*, **25**, 453–8.

Plomin, R., Owen, M.J. and McGuffin, P. (1994). The genetic basis of complex human behaviours. *Science*, **264**, 1733–9.

Power, T. and Woolger, C. (1994). Parenting practices and age-group swimming: a correlational study. *Research Quarterly for Exercise and Sport*, **65**, 29–39.

Prescott, J. (1996). Talent identification and development in female gymnasts. *Coaching Focus*, **31**, 12–13.

Régnier, G. and Salmela, J.H. (1987). Predictors of success in Canadian male gymnasts, in *World Identification Systems for Gymnastic Talent* (eds B. Petiot, J.H. Salmela and T.B. Hoshizaki), Sport Psyche Editions, Montreal, QC, pp. 143–50.

Régnier, G., Salmela, J.H. and Russell, S.J. (1993). Talent detection and development in sport, in *A Handbook of Research on Sports Psychology* (eds R. Singer, M. Murphy and L.K. Tennant), Macmillan, New York, pp. 290–313.

Reilly, T., Secher, N., Snell, P. and Williams, C. (1990). *Physiology of Sports*. E & F N Spon, London.

Reilly, T., Bangsbo, J. and Franks, A. (2000). Anthropometric and physiological predispositions for elite soccer. *Journal of Sports Sciences*, **18**, 669–83.

Renstrom, P.A. F.H. (1998). An interaction of chronic overuse injuries, in *Oxford Textbook of Sports Medicine*, 2nd edn (eds M. Harris, C. Williams, W.D. Scanish and L.J. Michele), Oxford Medical Publishers, Oxford, pp. 633–48.

Richardson, D. (1998). A head start? England's World Cup babies (1986–1998). *Insight – the FA Coaches Association Journal*, **1**(4), 28.

Richardson, D. (1999). Responsibility of the coach in the academic development of the young player. *Insight – The F.A. Coaches Association Journal*, **2**(3), 19.

Richardson, D. and Reilly, T. (2001). Talent identification, detection and development of youth football players – sociological considerations. *Human Movement, Polish Scientific Physical Education Association*, **1**, 86–93.

Richardson, D. and Stratton, G. (1999). Preliminary investigation of the seasonal birth distribution of England World Cup campaign players (1982–98). *Journal of Sports Sciences*, **17**, 821–2.

Roberts, G. (1980). Children in competition; a theoretical perspective and recommendations for practice. *Motor Skill: Theory into Practice*, **4**, 37–50.

Russell, K. (1989). Athletic talent: from detection to perfection. *Science Periodical on Research and Technology*, **9**, 1–6.

Salmela, J.H. (1996). Expert coaches' strategies for the development of expert athletes, in *Current Research in Sport Sciences* (eds V.A. Rogozkin and R. Maughan), Plenum Press, New York, pp. 5–19.

Sanderson, F.H. (1981). The psychology of the injury prone athlete, in *Sports Fitness and Sports Injuries* (ed. T. Reilly), Faber & Faber, London, pp. 31–6.

Saudino, K.J. (1997). Moving beyond the heritability question: new directions in behavioural genetic studies of personality. *Current Directives in Psychological Science*, **4**, 86–90.

Scanlan, T.K. (1988). Social evaluation of the competition process: a developmental perspective, in *Children in Sport*, 3rd edn (eds F.L. Smoll, R.A. Magill and M.J. Ash), Human Kinetics, Champaign, IL, pp. 135–48.

Seedfeldt, V. (1996). The concept of readiness applied to the acquisition of motor skills, in *Children and Youth in Sport: A Biopsychosocial Perspective* (eds F.L. Smoll and R.E. Smith), Brown and Benchmark, London, pp. 49–56.

Simoneau, J.A. and Bouchard, C. (1995). Genetic determination of fiber type proportion in human skeletal muscle. *Federation of the American Societies of Experimental Biology*, **9**, 1091–5.

Simonton, D.K. (1999). Talent and its development: an emergenic and epigenic model. *Psychological Review*, 106, 435–57.

Singer, R.N. and Janelle, C.M. (1999). Determining sport expertise: from genes to supremes. *International Journal of Sport Psychology*, **30**, 117–51.

Vealey, R. (1992). Personality and sport: a comprehensive review, in *Advances in Sport Psychology* (ed. T.S. Horn), Human Kinetics, Champaign, IL, pp. 25–59.

Williams, A.M. (2000). Perceptual skill in soccer: implications for talent identification and development. *Journal of Sports Sciences*, **18**, 737–50.

Williams, A.M. and Davids, K. (1995). Declarative knowledge in sport: a byproduct of experience or a characteristic of expertise? *Journal of Sport and Exercise Psychology*, **17**, 259–75.

Williams, A.M. and Grant, A. (1999). Training perceptual skill in sport. *International Journal of Sport Psychology*, **30**, 194–220.

Williams, A.M. and Reilly, T. (2000). Talent identification and development in soccer. *Journal of Sports Sciences*, **18**, 657–67.

Williams, M., Lee, D. and Reilly, T. (1999). *A Quantitative Analysis of Matches Played in the 1991–92 and 1997–98 Season*. The Football Association, London.

Williams, J.M. (1998). *Applied Sport Psychology: Personal Growth to Peak Performance*, 3rd edn. Mayfield, Mountain View, CA.

Index